MAPPING DESIRE

•

This is the first book to explore sexualities from a geographical perspective. The nature of place and notions of space are of increasing centrality to cultural and social theory. *Mapping Desire* presents the rich and diverse world of contemporary sexualities, exploring how the heterosexed body has been appropriated and resisted on the individual, community and city scales.

The geographies presented here range across Europe, America, Australasia, Africa, the Pacific and the imaginary, cutting across city and country and analysing the positions of gay men, lesbians, bisexuals and heterosexuals. The contributors bring different interests and approaches to bear on theoretical and empirical material from a wide range of sources.

The book is divided into four sections: cartographies/identities; sexualised spaces: global/local; sexualised places: local/global; sites of resistance. Each section is separately introduced. Beyond the bibliography, an annotated guide to further reading is also provided to help the reader map their own way through the literature.

David Bell is Research Fellow and **Gill Valentine** is Lecturer in the Department of Geography at the University of Sheffield.

MAPPING DESIRE

geographies of
sexualities

•

Edited by

David Bell & Gill Valentine

London and New York

First published 1995
by Routledge
11 New Fetter Lane, London EC4P 4EE

Simultaneously published in the USA and Canada
by Routledge
29 West 35th Street, New York, NY 10001

Typeset in Sabon by
Solidus (Bristol) Limited
Printed and bound in Great Britain by
Butler & Tanner Ltd, Frome and London

British Library Cataloguing in Publication Data
A catalogue record for this book is available from the British Library

Library of Congress Cataloguing in Publication Data
Mapping desire: geographies of sexualities/edited by David Bell and
Gill Valentine.
p.   cm.
Includes bibliographical references and index.
1. Sexuality. I. Bell, David. II. Valentine, Gill.
HQ23.M358 1994
306.7–dc20   94–34825

ISBN 0–415–11163–3
0–415–11164–1 (pbk)

CONTENTS

•

FIGURES AND PLATES

•

FIGURES

PLATES

ACKNOWLEDGEMENTS

•

We would like to thank: all the contributors for the effort and enthusiasm they put into producing their chapters and providing illustrative material; Ruth Holliday for trouble-shooting technical problems; Beth Martin for retyping Gregory Wood's chapter; Graham Bowden and Nick Scarle from the Cartographic Unit at the Department of Geography, University of Manchester for redrawing Figures 11.2, 13.1, 16.1, 16.2, 18.1, 18.2, 18.3; Michelle Keegan (Plates 17.1–17.4) and Peter Erbland (Plates 18.1–18.4) for providing their work to illustrate Chapters 17 and 18 respectively; John Bell for his diligent proof reading; and Caroline Cautley, the desk editor, for all the work she put in to creating the final product.

We are grateful to Boyz Own Productions for permission to reproduce an advertisement (Plate 1.1) and AIDS Vancouver for permission to use photographs of their safer sex material (Plates 16.1, 16.2, 16.3, 16.4). Every attempt has been made to obtain permission to reproduce copyright material. If any proper acknowledgement has not been made, we would invite copyright holders to inform us of the oversight.

A research grant provided by the Geography Division of Staffordshire University enabled the production of this volume. We are both also grateful to the Economic and Social Research Council for support during various stages of our 'careers'.

Finally, but perhaps most importantly, we are indebted to Tristan Palmer, our commissioning editor, for his enthusiastic commitment to this project and Jon Binnie and Julia Cream for their support and friendship.

INTRODUCTION: ORIENTATIONS

•

David Bell and Gill Valentine

The 'sex lines' listed in this advert (Plate 1.1) from London's free gay paper *Boyz* mark one way in which we can read the space of a city as sexed and sexualised: as Paul Hallam (1993) discusses in *The Book of Sodom*, London's streets are a powerful source of (homo)erotic imagery. In one sense, then, the landscapes of desire which this book seeks to address are the eroticised topographies – both real and imagined – in which sexual acts and identities are performed and consummated. This book might not be the best or only way to 'Discover the truth about sex in the city', but it should at least provide an introduction to ways in which the spaces of sex and the sexes of space are being mapped out across the contemporary social and cultural terrain.

Of course, the London Boys telling and selling their tales over the phone will not have the same meaning for everyone. We need to think about locally sexualised spaces – what Stephen Pfohl (1993: 192) calls '"vernacular" erotic geographies' – if we are going to avoid doing violence to the multitude of experiences and expressions of 'sex' in 'space'. Consider the sharp contrast between the London Boys advert and the two drawings of 'home' also shown here (and, indeed, the sharp contrast between those two homes). These pictures (Plates 1.2, 1.3), from research by Lynda Johnston on New Zealand lesbians' feelings about home (reported later in the book in a collaborative chapter with Gill Valentine), give us a very complex representation of the divided space of the (heterosexual) 'family home' and what might be called (in the rhetoric of the UK's anti-gay Section 28 legislation) the lesbian and gay 'pretended family' home. Through subtle signifiers of heteronormativity ('Dad' washing the car, 'Mum' in the kitchen), the family home (Plate 1.2) is depicted as a place of walls, of separation, but also of surveillance and discipline (see also Colomina 1992 on the architecture of domestic space).

The 'pretended family' home, however, is a very different image (Plate 1.3): instead of the people being lost in the space of the home, their bodies almost constitute the home, with only a sketched roof above to offer shelter. But what is perhaps more remarkable

Plate 1.2 Portrait of a heterosexual 'family home'
Source: Lynda Johnston

about this drawing is the copresence of lesbians, gay men and a baby. Where once the rigours of sexual politics demanded separatism not only from heterosexist culture but also from the opposite sex, the 'pretended family' home now shatters those stereotypes, bringing lesbians and gay men together.

Very different places, then: the fetishised cityscape of London, and the intimate space of the home. It is our intention, in bringing together the authors in this collection, to present a set of equally different places and spaces, from the city to the desert island, from Jakarta to Amsterdam, from the red-light district to the merchant bank, from body to community, from local to global. And far from paying banal lipservice to these landscapes of desire, we hope to bring a range of theoretical and empirical perspectives, drawn not only from geography but also from much further afield, together to inform our thinking about the ways in which the spatial and the sexual constitute one another. To begin this process, we review work firstly by geographers and secondly from scholars beyond our discipline's (porous) boundaries, before moving on to think about the

Plate 1.3 Portrait of a 'pretended family' home
Source: Lynda Johnston

project of researching and teaching sex, sexuality and sexual identity from within the academy – and from within the geographical academy in particular.

PUTTING SEXUALITIES ON THE MAP

The women's, gay and civil rights movements emerged in North America and Europe in the 1960s and 1970s on a wave of social and political upheaval. But despite a growing awareness amongst geographers in the following decade of the need to study the role of class, gender and ethnicity in shaping social, cultural and economic geographies, sexualities were largely left off the geographical map (Bell 1991).

Some of the first geographical works on homosexualities suggested that lesbians and gay men lead distinct lifestyles (defined to a lesser or greater extent by their sexuality and the reactions of others to that sexuality) which have a variety of spatial expressions creating distinct social, political and cultural landscapes. This research has focused almost exclusively on contemporary western societies and has largely followed attempts within urban sociology (e.g. Levine 1979a) to apply ideas from the work of the Chicago School of Human Ecologists (notably Park 1928 and Wirth 1928) to map 'gay ghettos'. For example, Lyod and Rowntree (1978) studied the migration patterns of lesbians and gay men, arguing that they cluster in communities in specific parts of US cities for reasons of avoidance, defence, attack and preservation. Likewise, Harry (1974: 246) argues that 'by migration the relatively isolated gay may be able to replace the impersonality of small town life (for him [sic]) with the interpersonal warmth and cultural affinity of gay life in the big city'. Such explanations mirrored the arguments geographers used in the 1970s and early 1980s to account for concentrations of ethnic groups within cities (Boal 1976). This work has since been heavily criticised and largely rejected out of hand because of its 'racist' (and we might add heterosexist) assumptions (Jackson 1987). Other preliminary attempts to map gay regions and neighbourhoods were made by Barbara Weightman (1981) and by Hilary Winchester and Paul White (1988), who categorised lesbians alongside criminals, ethnic minorities and down-and-outs [sic] as neglected marginalised groups within the inner city.

Most of these studies relied on indirect information, such as directories of lesbian and gay venues and bars, to locate 'gay communities'. In particular, the institutions and leisure services used by lesbians and gay men, especially the gay bar, were an easy target for researchers unable to or uninterested in getting their hands dirty talking to informants. Barbara Weightman (1980: 9), for example, investigated the symbolism of gay bars – claiming that 'gay bars incorporate and reflect certain characteristics of the gay community: secrecy and stigmatisation. They do not accommodate the eyes of outsiders, they have low imageability, and they can be truly known only from within'. This work follows in the footsteps of a number of 'classic' sociological studies of gay bars, for example, by Gagnon and Simon (1967), Achilles (1967) and Harry (1974), that painted a picture of promiscuity and sexual exploitation. These studies have subsequently been heavily criticised for their patronising, moralistic and 'straight' approach to lesbian and gay social and sexual relations. This approach is beginning to be redressed through more sex-positive work that is based on ethnographic research and interviews with lesbians and gay men, such as Jon Binnie's (1992a, 1992b) studies of leather bars in the East End of London and in Amsterdam and Alison Murray's chapter on lesbian sex workers.

The impact that gay communities have on the urban fabric at a neighbourhood level has been at the heart of much of the recent US work on sexualities (Castells 1983; Castells and Murphy 1982; Knopp 1987, 1990a, 1990b; Lauria and Knopp 1985; McNee 1984; Winters 1979; Wolf 1979). This literature has highlighted how gay men have taken over and gentrified areas such as West Hollywood in California, establishing not only gay housing areas and businesses but also, in the face of hostility and oppression, a power base where the 'gay vote' is significant (Knopp 1990a). In Chapter 10 Larry Knopp begins to try and establish a theoretical framework for this early work by examining the relationship between sexualities and aspects of urbanisation in contemporary western societies. This builds on some of his recent attempts to examine the role of sexuality within the spatial dynamics of capitalism (Knopp 1992).

British geographers have been less caught up in this concern with gay commercial and residential bases which perhaps reflects differences between the geographies of US and UK gay (and academic) communities. One exception is Jon Binnie's work on Soho in London and Amsterdam. Part of his chapter examines the emergence of Old Compton Street in the West End of London as a gay commercial district – nicknamed 'Queer Street'; and the development of Amsterdam, one of the gay capitals of Europe, as a location of international lesbian and gay tourism.

These gay commercial/neighbourhood bases in the US and Europe are predominantly populated by gay men (Castells 1983) and dominated by institutions of gay male culture. In this sense much of the 1980s work on sexuality has primarily produced geographies of gay men. Castells has claimed that the absence of similar territorially based lesbian communities reflects the fact that 'women are poorer than gay men and have less choice in terms of work and location' (1983: 140). This is borne out by Maxine Wolfe's (1992) assertion that there are fewer commercial spaces for lesbians than gay men because lesbians, like heterosexual women, rarely own their own businesses because of their lack of economic resources. Wolfe further argues that lesbian bars usually have a short life span, stating that

> Many lesbian bars are not 'places' in the sense of a consistent physical location, which one could design or decorate permanently. Often they are 'women's

nights' at other bars. A few years ago in London, the 'lesbian bars' moved continually on different nights of the week (being held in the private, usually basement, party spaces in heterosexual pubs) to protect women from being beaten up.

Wolfe 1992: 151

But Castells (1983) also goes one step further in his explanation for the lack of lesbian communities, suggesting that there are gender differences in the ways that men and women relate to space. He argues that men try to dominate and therefore achieve spatial superiority, whilst women have less territorial aspirations, attaching more importance to personal relationships and social networks.

Adler and Brenner (1992) have challenged Castells' claims about lesbians. Their work in an anonymous US city suggests that lesbians do create spatially concentrated communities but that 'the neighbourhood has a quasi-underground character; it is enfolded in a broader countercultural milieu and does not have its own public subculture and territory' (Adler and Brenner 1992: 31). Linda Peake's (1993) study of lesbian neighbourhoods in Grand Rapids, Michigan and Gill Valentine's (1995a) work on a town in the UK provide further evidence that lesbian spaces are there if you know what you are looking for. In both research areas there are 'lesbian ghettos' but they are ghettos by name and not by nature. There are no lesbian bars, stores or businesses in these neighbourhoods, neither are there countercultural institutions such as alternative bookstores and co-operative stores. The lesbians in these towns leave no trace of their sexualities on the landscape. Rather there are clusters of lesbian households amongst heterosexual homes, recognised only by those in the know. Both Peake and Valentine suggest that women learn about these areas and make contacts with people in the neighbourhoods through the 'lesbian grapevine' – a process neatly captured by the title of Tamar Rothenberg's chapter about a lesbian community in Park Slope, Brooklyn: 'And she told two friends . . .'.

Rothenberg's chapter also adopts a more critical approach to the use of the term 'community' than some of the earlier work on gay geographies. She points out how geographers have tended to use community and neighbourhood synonymously to refer to a geographically bounded area inhabited by close-knit networks of people, most of whom know each other or share common interests. More recently, geographers have begun to wake up to the problems of eliding these two terms and have begun to draw on Benedict Anderson's (1983) concept of 'community' as an imagined but discursive reality. In his work on nations as 'imagined communities' Anderson suggests that nations are imagined 'because the members of even the smallest nation will never know their fellow members, meet them, or even hear of them, yet in the minds of each they carry the image of communion' (Anderson 1983: 15). This concept is discussed in relation to lesbians and gay men (at local, national and international scales) not only in Rothen-berg's chapter but also in the chapters by Jon Binnie and David Woodhead.

Despite the attention paid to *visible* gay communities like San Francisco, the reality is that most gay men and lesbians live and work not in these gay spaces but in the 'straight' world where they face prejudice, discrimination and queerbashing (Bell 1991). The hegemony of heterosexual social relations in everyday environments, from housing and workplaces to shopping centres and the street, is increasingly the subject of geographical research (Davis 1991, 1992; Valentine 1993a, 1993b) and has led Larry Knopp (forthcoming) to outline the need for lesbians' and gay men's oppressions to be recognised by geographers (e.g. Harvey 1992) who are taking up the cudgels again in the battle for social justice within the city.

Lesbians, like heterosexual women, are economically marginalised and are less likely than gay men to own their own homes (Anlin 1989; Egerton 1990; Wolf 1979). Housing, according to Egerton, is therefore the 'single most chronic practical problem' facing many lesbians (Egerton 1990: 79). Whilst gay men often have more economic resources at their disposal, they also commonly encounter discrimination getting life insurance and endowment mortgages because of commercial paranoia about AIDS. Both groups have also been hit hard by recent housing legislation in the UK and the lack of provision of accommodation for single homeless young people.

Bell (1991) and Johnson (1992) have argued that most housing in contemporary western societies is 'designed, built, financed and intended for nuclear families' (Bell 1991: 325) and that whilst lesbians and gay men can subvert conventional housing layouts to articulate their lifestyles there is no housing that is primarily designed for those who want, for example, to live in 'non-conventional' co-operative or collectively organised households. Notwithstanding this, Egerton (1990) and Ettorre (1978) have both documented lesbian-feminist housing experiments of squatting and communal living. Some of these issues are addressed in Johnston and Valentine's chapter about the performance and surveillance of lesbian identities in the home.

Feminists in geography and elsewhere (e.g. Cockburn 1983; L. Johnson 1994) have been flagging the ritualised performance of gender identities in the workplace and the ways that the sexual division of labour is inscribed on workers' bodies for the last decade. More recently research has drawn attention to the fact that such performances of masculinity and femininity make sense only within a heterosexual matrix. The disciplinary practices that regulate the performance of heterosexuality within 'City' workplace environments are the subject of Linda McDowell's chapter. Sally Munt also examines another aspect of the way different spaces are (hetero)sexed when she compares the experience of being a lesbian in Brighton, the gay capital of the South, with living in Nottingham, a place she characterises as possessing a 'rugged romantic masculinity'.

But as Jerry Lee Kramer points out in his chapter on lesbian and gay lives in rural North Dakota, most of the geographical work on sexual identities and the sexuality of space remains firmly located in the urban (reflecting the discipline's general obsession with the city). Indeed social constructionist arguments about the development of gay identity suggest that this is predicated upon the opportunities offered by city life – anonymity and heterogeneity as well as sheer population size. Even the suburbs and small towns have been passed over in the rush to study the glamorous and sexy city. One of the few exceptions is the now rather outdated work of Fredrick Lynch (1987). He documented some of the problems gay men have developing social and sexual relations with other men in middle-class suburban environments. This problem is felt even more closely by rural lesbians and gay men, whose only openings for expressing their sexuality may come from episodic encounters in public toilets or highway rest areas (Corzine and Kirby 1977; Humphries 1970) or infrequent trips to neighbouring towns' bookstores and porn cinemas (D'Augelli and Hart 1987). This social isolation also goes hand in hand with a lack of access to even the most basic resources such as gay newspapers and book stores and, more importantly, safer sex advice and AIDS support services (Rounds 1988).

Despite these obvious structural limitations, however, lesbians and gay men also choose to live in rural environments. Indeed, Anlin's (1989) research found that the country was an 'ideal' or 'fantasy' place for lesbians to live. In particular, it seems to offer an escape from many of the oppressive aspects of contemporary urban life. For lesbian feminists in the 1970s rural communes offered a chance to establish alternative ways of life away from the man-made city and a whole circuit of women-only farms developed in the US (Cheney 1985; Faderman 1992; Lee 1990; Woolaston 1991). This rural utopian theme is at the centre of many literary and filmic representations of lesbian and gay lifestyles – witness *Desert Hearts* and the explosion of 1990s cowgirl movies. The lifestyles of lesbians and men living in rural areas and the recreational uses of rural sites by gay men and lesbians for leisure activities (and as a setting for sex) are therefore urgently in need of geographical attention. There are many points at which this work will inform and be informed by mainstream work on the 'rural idyll', cultural constructions of rurality, and ecofeminist and green politics.

In the early 1990s geographical work, particularly in the UK, has turned away from an obsession with defining and locating gay residential and institutional communities towards a concern with identity politics. In this respect it mirrors the general cultural and postmodern theoretical turn within human geography as a whole. But this work has also been heavily influenced by the explosion of literature on sexualities and sexual identities in cultural studies and lesbian and gay studies (Plummer 1992).

A whole body of work is emerging in geography that explores the performance of sexual identities and the way that they are inscribed on the body and the landscape. This work has overturned all the old binaries: heterosexual–homosexual; public–private and so on. There is now a greater recognition of the multiplicity of sexualities and the fluid and contextual nature of sexual identities. This book, for example, includes chapters that discuss transsexuality (Julia Cream), bisexuality (Clare Hemmings), sadomasochism (David Bell) and butch-femme lesbian identities (Alison Murray, Sally Munt).

The body was incorporated into the discipline in the 1970s through the work of humanistic geographers, notably David Seamon (1979). Drawing on the phenomenology of Merleau-Ponty, Seamon used repetitive focus groups to explore the ways in which people experience their bodies and move through space both individually and in 'body ballets'. Missing from this research, however, was any recognition of the significance of bodies as being gendered, sexualised, coloured, aged and so on (Longhurst 1994). In the 1990s the body is once again on the geographical agenda. Now, the work of cultural theorists, such as Elizabeth Grosz (1993) and Judith Butler (1990), who view the body as a constantly reworked surface of inscription, is being applied to geography. Julia Cream, for example, argues elsewhere (1993) and in her chapter in this book, that what it means for the body to be sexed is contextual in time and place. This is borne out by Louise Johnson's (1990, 1993) empirical work on an Australian textile mill in which she links class and gender relations to the deployment of sexed bodies within the workplace, and Robyn Longhurst's (1994) pioneering work on the visibly pregnant female body as geography's 'Other'.

Social and cultural geographers are now recognising that the body is politicised (P. Jackson 1993; Pile 1993) and are viewing it as a site of struggle and contestation (Dorn and Laws 1994). Deviant bodies – disabled rights activists, pregnant women, the sick, the elderly – are challenging the medicalisation of the body and are being incorporated in geographical study, for example, through the work of Robyn Longhurst (the pregnant body) and Vera Chauniard (disability). While medical geography has lagged behind in this retheorisation of the body, critical perspectives are beginning to emerge (e.g. Dorn and Laws 1994). Key players in this relationship between bodies and social/political institutions are activist movements such as ACT-UP and Disabled Rights groups. Dorn and Laws (1994: 108–9) argue that these movements 'often emerge from an experience of place' (see also Michael Brown's discussion on AIDS work in Vancouver in Chapter 16). Certainly, they have used transgressive strategies in everyday public places as an intrinsic part of their political project (Cresswell 1994) – the politics of transgression hold many possibilities for future geographical work.

Masculinity is another 'new boy' on the geographical agenda. A late addition to work on geography and gender, early studies on masculinities have tended to shy away from issues of sexuality. Glen Elder and Greg Woods begin to turn the corner in this respect with their very different approaches to masculinity in their chapters on, respectively, male homosexuality in South African mining compounds, and the landscape of the desert island in fiction and film. The possibilities for thinking about the social construction and performance of masculinities in space will certainly develop further, drawing on insights from the emerging field of 'critical studies in masculinity' (as opposed to the politically suspect 'men's studies').

Gregory Woods' and Tracey Skelton's chapters also reveal an engagement with popular cultural productions, something which geographers are showing considerable interest in. Work which looks at the intersection of sexualities and popular culture includes Aitken and Zonn's (1993) paper on Peter Weir's films, work by Jon Binnie (1993b) on 'boy bands' and by Gill Valentine (1995b) on 'the space that music makes' for the lesbian listener, Peter Jackson's readings of masculinity in advertising (1991, 1994), and feminist geographers casting their gaze over painting and film (Massey 1991). Given the vibrancy of work outside geography on sexualities and popular culture (e.g. Dyer 1992; Frank and Smith 1993; Griffin 1993), we can expect a lively proliferation of geographical perspectives on these issues in the future.

The diversity of identities outlined within this volume expose the many ways of 'being' and 'doing' sexuality. Contrast, for example, the managed identities of some of the women in Valentine and Johnston or Rothenberg's chapters with the radical and 'dangerous' performance of lesbianism in Alison Murray's chapter on butch-femme sex workers. These differences highlight some of the tensions between lesbians and gay men who want to assimilate within heterosexual society and the more radical and challenging nature of queer politics. And they also demonstrate the need for geographers not to allow the pendulum of research to swing from its original focus on 'safe' vanilla expressions of lesbian and gay sexualities to exclusively focusing on the radical chic of sadomasochism and leather. Rather it is important to recognise the mutiplicity and contradictory ways different sexualities are lived out. A closer look across other chapters within this book (and at what is missing) also reveals the many other lines of difference that fracture so-called lesbian and gay communities and the sexualities that lie outside the homosexual–heterosexual dualism. Most notably the geographies of sexualities in this book are primarily the experiences of 'whites' within contemporary US and UK cities. This reflects the very limited amount of work on sexuality in other social and cultural contexts; yet cross-cultural variations in sexualities are obviously extremely diverse and complex (Plummer 1992). Elsbeth Robson's (1991) work is one of the rare exceptions to this rule from within geography. In her paper 'Space, place and sexuality in Hausaland, Northern Nigeria' she draws on the work of anthropologists to consider the conflicts and challenges of different forms of sexuality; the importance of state and other social institutions in the support of alternative forms of sexuality to heterosexuality; and the effect of HIV and AIDS on hegemonic and alternative sexualities. It is largely in the terrain of anthropology and Asian studies that some of the most geographically relevant and exciting work on 'nonwestern' sexualities has been taking place. For example, Alison

Murray's (1991) research in Indonesia includes a paper called 'Kampung culture and radical chic in Jakarta'. This assesses the impact of a cultural thaw within the country on the performance of urban styles and subcultures.

Sexuality is – at last – finding a voice as a legitimate and significant area for geographical research. Editorials in several of the major geographical journals and reviews of the state of specific areas within the discipline in the early 1990s have singled out sexuality as a theme that will be an important focus for geographical work in the next decade (e.g. Bondi 1992a, 1993). In particular, Thrift and Johnston (1993) argue in *Environment and Planning A* that sexuality will be to geography in the 1990s what class and gender were to the discipline in the 1980s.

Initially much of the work on sexuality and space was concentrated within social and cultural geography, but it is now spilling out into other areas of the discipline. At the vanguard of this expansion are Jon Binnie (1993c) and David Bell (1995), who have stretched the frontiers of political geography with their challenging work on nationalisms and sexualities; and sexual citizenship respectively. In particular, David Bell's chapter in this book takes up the themes of perversion, citizenship and intimacy, and uses state and legal regulation of sadomasochism and public sex to unpack the complex relationships between the public, the private and the pervert. Economic geography is also beginning to feel the wind of change, not only from the work already mentioned by Larry Knopp and Linda McDowell but also from Jon Binnie's (1993a) attempt to open up the pink economy as an area for research. And following Chris Philo's (1992) recent paper detailing the absence of sexuality amongst a litany of missing 'others' from work on the 'country', rural geography is likely to be the next area of the discipline to take sexuality on board.

Whilst these areas of geography are opening up to the possibilities of sexualities, one area of the discipline – feminist geography – is perhaps in need of a divorce or at least a separation from sexuality. As the study of sexuality has made inroads into the discipline, this growing body of work has been constantly tagged on to feminist geography. From conferences to books and journals, gender and sexuality are being joined at the hip as the Siamese twins of geography (see, for example, Jackson 1989). The underlying problem with this eliding of gender, sexuality and feminism is in the ways in which these terms are used. Gender appears to be commonly used to refer to women (though this is changing with the growth in work on masculinity) whereas sexuality is often used as shorthand for dissident sexualities, usually (male) homosexuality, in the same way that lesbianism and feminism are often used to stand in for one another within popular culture. The danger with this is that many expressions of sexuality are actually an anathema to certain versions of feminism (witness the recent vitriolic debates around sadomasochism and pornography amongst feminists). Similarly, the political correctness of 1970s and early 1980s lesbian feminism, that often policed lesbian and gay spaces, has lost its grip on these 'communities' (we use that term advisedly). And so geographers need to bear in mind that feminism and sexuality is not such a neat equation after all, and should tread more carefully when conflating genders and sexualities in their work as well as lumping the work of and on sexual dissidents under the umbrella of feminist geography.

A big absence from geographies of sexualities is, ironically, the dominant sexuality within contemporary societies – heterosexuality. Feminist geographers have lagged behind other disciplines in examining the role of love, romance and desire in women's oppression. Although there is some work within geography about sexual violence and child abuse (Cream 1993; Valentine 1989) most of this does not sufficiently develop the role of domestic violence and male expressions of heterosexuality in constraining women's lives. Where feminist geographers are making an immensely important contribution is in critiquing the masculinism of the geographical imagination. Doreen Massey's 'Flexible sexism' (1991) exposes David Harvey's malestream worldview, while Gillian Rose (1993b) contemplates the place of women within geography, and Linda McDowell (1990) catalogues the workings of 'Sex and power in academia'.

The only other time heterosexual relations (with the exception of discussions about its hegemony by sexual 'others') are the actual focus of geographers' work is within social and cultural work on prostitution. Symanski (1974, 1981) was one of the earliest geographers to produce a 'geography of prostitution' outlining the different types of clientele and female prostitutes working in different places within California. He argues that prostitution reflects the structure of social relations, for example there is a hierarchical class system, from high-class 'call girls' who meet clients through advertisements and in expensive hotels to street walkers, with black women over-represented in the lower status roles. Prostitution is also another example of patriarchy at work, with male clients commonly escaping arrest. This geography has come under fire, however, for example from Peter Jackson (1989: 145), who criticises it 'for freezing what is in fact a highly dynamic situation'.

But the theme of prostitution and the structure of social relations has persisted in more recent work. Ashworth, White and Winchester (1988) have drawn attention to the role of moral and social norms, as much as statute law, in determining the location of red-light districts. The politics of the policing of prostitution have also been scrutinised by Larsen (1992). His work in four Canadian cities between 1977 and 1986 highlights the importance of the class status of interest groups in their ability to exert political influence over the control of street prostitution through lobbying police and politicians. Not surprisingly, he found that prostitution was more likely to be defined as a 'problem' by the authorities when it occurred in middle-class areas. And in a look at the historical geography of nineteenth-century prostitution, Miles Ogborn (1992) takes a more theoretical approach in considering what the 'other' of prostitution tells us not about the 'margins' but about the centre.

However it is disciplines outside geography that are producing more radical geographies of sex work – embracing not only heterosexual prostitution but the whole gamut of the sex industries: from massage parlours and child pornography to rent boys, sex shows and so on. For example, Wendy Lee's work in South-East Asia points out that 'bodies underpin the balance of payments' of countries in this region (Lee 1993: 79). The commercialisation of sex, sex tourism and all the social, cultural, medical, political and economic implications of this are research topics that are crying out for geographers' attention. In Chapter 14, anthropologist Angie Hart deploys geographical notions of social spatialisation to read a Spanish street prostitution 'barrio' as a landscape constituted through the intricate meanings of sex work for both clients and workers, and in Chapter 5, Alison Murray tackles the relationship between sex work and lesbianism, revealing the complexity of identities at play.

In her role as a sex-worker peer educator, Murray has been tackling the issue of safer sex within the industry. For medical geographers, HIV and AIDS would appear to be a prime site of academic work: Peter Gould (1989: 71) has pointed out that 'geographers are well aware that the "where" of an epidemic is as important as the "when"'. Mapping the transmission of the virus has been at the heart of epidemiologists' attempts to trace its origins and establish global typologies. They have identified three distinct geographical patterns of infection, firstly in the Western industrialised nations where gay men and drug users (and their partners) are 'blamed' for the infection rate; secondly in central Africa, the Caribbean and South America where transmission is primarily attributed to heterosexual activity and unscreened blood products; and thirdly in Asia, the Commonwealth of Independent States and the Middle East, where information and modelling is more sketchy (Gadsby 1988).

Much of this work has had dubious social and political overtones. As Kirby says:

> many resources have been expended on a form of 'reverse diffusion' research in an attempt to track the origins of HIV in the US ... what we see, in this costly work, is an attempt to find the scapegoat, the individual who brought the disease to the US.
>
> Kirby 1990: 10

The furore over 'Patient Zero', purportedly the man to bring HIV into the United States, shows how dangerous (as well as irrelevant) this kind of mapmaking can be (Crimp 1987a), while the clinical production of transmission models does violence to the people affected by HIV and AIDS.

This sort of medical geography research is now coming under a barrage of criticism. Dorn and Laws (1994) argue that there is a need for medical geography to be more self-critical of its intellectual heritage and to move closer to debates in social and cultural geography. The challenge for this area of the discipline is 'to interrogate the *embodied* subject positions which are being forged in contemporary society' (ibid.: 106). As numerous other critics (e.g. Kirby 1990) have argued, the importance of spatial studies is not to recreate the past but to understand the future and to use this 'where' to inform and direct public health policies. Vital to this process is an understanding of sexual identities and sexual relations (King 1994). David Woodhead and Michael Brown are two geographers who are playing a leading role in refocusing research on AIDS in geography. In their chapters in the final section of this book, they address the issue of health promotion in London and Vancouver respectively, within the context of understanding men who have sex with men. Later in this introduction we move beyond geography to consider the place of what Jan Zita Grover (1992) calls 'AIDS work'.

Having skipped (rather partially) in this section through the history of geographical work on sexualities, in the next we look at some of the work that has taken place beyond geography, and pick out some of the concepts and theoretical perspectives that are being (re-)incorporated into geography.

A DEEPER LOVE: CONTESTING MONTREAL PRIDE

In 1992, the Montreal Pride parade became an even more hotly contested and troubled space than it had been the previous year.[1] While in 1991 activists from Queer Nation Rose (QNR) and ACT-UP rejected the ghettoised 'official' parade (which ran only through the city's gay village) and instead marched through downtown Montreal, the following year's furore focused on a set of 'rules and regulations' issued by the organising committee. These stipulated, among other things, that

> there was to be no cross-dressing, no exposure of breasts or buttocks, no displays deemed too 'vulgar' or 'erotic', and no flags . . . As if the outlawing of extravagant fashion weren't enough, it was suggested that the preferred attire of parade participants be blue jeans and a white T-shirt.
>
> Namaste 1992: 8[2]

Against what was called on one subsequent oppositional flyer *fagscism*, groups like QNR and ACT-UP began raising the question of what 'Pride' actually meant in the context of this regulated and disciplined parade – Who was it for? Who felt the pride, and what were they proud of? Across the city, sewing machines ran all night and stores ran out of sequins, fishnet and eye-liner. Reacting vigorously to the 'anti-drag, anti-leather and just plain anti-fabulous sentiment' (Namaste 1992: 8) of the organisers, and mobilising around QN (LA)'s famous slogan 'If you're in clothes, you're in drag!', Montreal's sexual outlaws and perverts dressed to kill:

> irreverent combinations of identities proliferated, including fags posing as dykes, dykes dressed as clone fags, and bisexuals pretending to be fags pretending to be lipstick lesbians.
>
> Namaste 1992: 9[3]

In a final act of absurdity and contradiction, the very parade which had banned homoerotica and cross-dressing so as not to offend (straight) spectators[4] ended with entertainment provided by *drag queens*. As Ki Namaste of QNR eloquently puts it:

> This contradiction – that drag is simultaneously disavowed and permitted – is perhaps best understood in terms of its situation and context. That is, drag queens are permitted in certain spaces, among certain people, at certain times . . . [The parade organisers] want their type of drag, in the spaces they designate . . . Clearly, the activist focus on the idea that drag is everywhere threatens precisely the borders, boundaries and limits [of acceptability].
>
> Namaste 1992: 9

This outrageous story is illustrative of some of the key debates within both theory and activism around sexuality and space, and it is our intention in this section to take up with some of these debates, and to read them into geography by suggesting ways in which insights from events such as Montreal Pride can inform our thinking (and our practice) as geographers – something we might designate an act of what Eve Sedgwick (1994) calls 'queer reading'.

Queer reading, like the rupturing and transformative process of opening up lesbian narrative space within heterosexual writing (Farwell 1990), will hopefully do more than add sex(uality) to geography and stir (Binnie 1994). A queer reading of geography, rather, should function to resist the ways in which geographical knowledge is constituted (for example, as pre-discursively heteronormative) by hybridising and retheorising – through what Susan Bordo (1992: 160) calls 'daring epistemological guerrilla warfare: intervention, contestation, resistance, subversion, interrogation'. The process of queer reading, Sedgwick (1994: 3) says, is concerned to 'make invisible possibilities and desires visible; to make tacit things explicit; to smuggle queer representation in where it must be smuggled and . . . to challenge queer-eradicating impulses frontally where they are to be so challenged'. If we are to begin to think about 'sexuality' and 'space' in any serious way, then queer reading of geography will be an essential guide, for as Mark Wigley (1992: 389) says:

Space is itself closeted. The question must shift to the elusive architecture of the particular closets that are built into each discourse, but can only be addressed with the most oblique of gestures.

Such oblique gestures embody a large part of the project of a queer reading of geography. This hybridising and retheorising must, in part, come from an engagement with bodies of work beyond the horizons of geography, but also from engaging with them *as geographers*: while our epistemological heritage can be (and has been) roundly critiqued as masculinist, heteronormative and disembodied (Binnie 1994; Longhurst 1994; Rose 1993b), we remain optimistic that, while discipline boundaries remain intact (if permeable) within the academy, we can begin (to adopt and adapt West and Zimmerman's (1991) phrase) to *do geography differently* while still definitively *doing geography*. In the sections which follow, by referring back to Montreal Pride 1992 as a vivid acting-out (and up) of many important issues which we would like to deploy in our queer reading of geography, we hope to begin visiblising, explicating, smuggling and challenging: as the *Reservoir Dogs* poster says, 'Let's go to work'.

Perverscapes

Of course, a Pride march – or at least one where anti-fabulous rulings don't apply – is important in that it creates an erotic ludic topography along its route. In thinking about this, we might usefully refer to Sue Golding's writing on the urban as a site for queer play (and politics), and then consider other work focusing on an assortment of fleeting, changing and fixed perverse landscapes. In a trio of papers on related themes, Golding (1993a, 1993b, 1993c) discusses the 'impossible geography' of the city as a site for reconfiguring counterhegemonic sexualities. Through a dense deployment of philosophy, physics, politics and pornography, she envisages the 'creative and wild possibilities' (1993b: 217) presented by 'the "elsewhere" of decadent urban life' (1993c: 88) or the 'elsewhere of sexual mutation curiosity' as she also calls it (1993a: 25).[5] As she writes, with typical intensity, in 'Sexual manners', an essay based around the tale of an unsuccessful SM pick-up:

> Now, before I go on and tell you what happened, let me just say that these words: dom, Master, bottom, whore-fem, butch, Daddy-boy, cruising, play, play-mate, and so on, have their place. Or, rather, they take a place and make a place. They make an impossible place take place. They describe, circumscribe, inscribe a spectacular space, a spectacle of space: an invented, made-up, unreal, larger-than-life-and-certainly-more-interesting space that people like myself sniff out

and crave and live in and want to call 'Home'; a home I want to suggest that is
entirely Urban; an urbanness I want to say that is entirely City and not at all –
or at least not exactly – Community; a queer (kind of) city (or better yet,
cities) ...

Golding 1993c: 80

Through the wild circuits of her writing, Sue Golding makes the same point Larry Knopp
talks about in his chapter (and, by intimation, that Jerry Lee Kramer's discussion of
country lesbians and gay men also shows) – that there is an intimate link between the
urban and the sexually deviant. The task of revisioning sexual politics comes from
dispensing with tired notions of community, and instead to 're-cover "urban-ness" in all
its anomie, and rather chaotic, heterogeneity, if we are indeed serious about creating a
radically pluralistic and democratic society' (Golding 1993b: 216). In this sense, the
Montreal Pride parade *has* to take to the streets, out of the ghetto of the gay community,
if it is retain any radical (rather than merely narrowly celebratory) edge. In the
postmodern, postindustrial city, the play of polymorphous decentred exchange in
polymorphous decentred landscapes makes perfect sense.[6]

   Part of Golding's project is the construction of an imagined geography of the perverse
city – perhaps this is partly why her geography is always already 'impossible'. And of
course, there exist many other real, imagined and fictional landscapes of desire which
have played an important role in shaping sexual identities as well as guiding sex tourists
around the globe (the place of Amsterdam as a gay mecca is discussed by Jon Binnie in
this volume; see also Kellogg 1983 on literary visions of homosexuality; also Bravmann
1994 on the place of Lesbos and Greece in the lesbian and gay imagination). Paul
Hallam's recent collection of literary and journalistic writings on Sodom offers an
excellent 'psychogeographical journey' (Hallam 1993: back cover) around Sodom texts
and around the author's Sodom, the city of London:

   I regularly wander Clerkenwell in search of Sodom. It's the part of London I'm
   most at home in. I like its day-for-night inversions. The hospitals and the meat
   market. The shift workers, the uniforms of the porters and the nurses, the
   doctors' coats and the butchers' aprons. On weekdays there's the added
   attraction of city boys in their suits and ties.

Hallam 1993: 15

More than a decade earlier, Edmund White had wandered around pre-AIDS America to
write *States of Desire* (White 1980), a kind of *Whicker's World* for gay men. In the days
before safer-sex slutting became *de rigeur* as a research method (cf. Binnie 1994), White
cruised the USA, chatting (and more) with local gays, checking out bars and scenes,

observing (and participating in) American Gay Life. Today his book reads like that whole series of urban sociological studies of gay sex lives (Humphries 1970; Levine 1979b; Styles 1979; etc.) – like a fiction, or at best like a fragment from a lost age.

It's important here to make a change of direction. Thus far, the work we have been discussing does not, by and large, problematise 'space'. Rather, it builds upon a commonsense notion that space is unencumbered – naked, if you like – and can thus be dressed in any way: any sexual identity can assume space, and space can assume any sexual identity. But, as anyone who has been queerbashed will tell you, 'space is not an innocent backdrop to position, it is itself filled with politics and ideology' (Keith and Pile 1993: 4). As a number of the chapters here show, we need to make it clear that space is *produced*, and that it has both material and symbolic components (Angie Hart's 'barrio', David Woodhead's 'cottage', Greg Woods' 'fantasy islands', Gill Valentine and Lynda Johnston's 'home' are all suggestive of this notion). Further, an occasion like Montreal Pride does not simply (and uncontestedly) inscribe space as 'queer'; rather, it has to *work* at queering space.

Queer space

While the 'official' 1991 Montreal Pride parade was routed through the city's gay neighbourhoods, thereby 'affirming' and 'empowering' Montreal's sexual dissidents without being challenging or confrontational to the city's heteronormative culture, the 1992 rally, and many more like it around the world which take to the (ambient heterosexual[7]) streets (not to mention the 'unsanctioned' protests and happenings orchestrated by activists and the individual actions of countless queers who have sex, or kiss, or hold hands, or make eye contact, or swap phone numbers, or 'pass' in public space), by coming out into straight space, inevitably *queered* the streets; indeed, queered the whole city. Important in this process is that the presence of queer bodies in particular locations forces people to realise (by the juxtaposition 'queer' and 'street' or 'queer' and 'city') that the space around them, the landscape of Montreal (or wherever), the city streets, the malls and the motels, have been *produced* as (ambiently) heterosexual, heterosexist and heteronormative (Bell *et al.* 1994; Bell and Valentine 1995). And, to take the deconstruction a step further, in an act of kinky Garfinkeling (see Cresswell 1994), taking to the streets in such a perverse parade of genderfucking[8] should begin to reveal that this heterosexing of space is a performative act naturalised through repetition – and destabilised by the mere presence of invisiblised sexualities. As Elspeth Probyn (1995) says: 'space is a pressing matter, and it matters which bodies press against it'.

By drawing on the work of Judith Butler (1990, 1991, 1993), we can begin to understand the role of performativity and theatricality in constructing the self, and space, as prediscursively straight. In particular, Butler's (not unproblematic) discussion of the role of parodic acts like drag in highlighting the performativity of all gendered identities can be used to think about the construction of gendered, sexed and sexualised space. By adapting Butler's (1990: 79) discussion of 'subversive bodily acts' to think about *subversive spatial acts* we can see how even the kiss of two men on the night bus home can fracture and rupture a previously seamless (we might ironically say *homogenous*) space (see also Jon Binnie and Tim Davis in this volume). The straightness of our streets is an artefact, not a natural fact, and Pride marches, zap protests and other non- or anti-heteronormative acts make this clear by making it queer. A continued (if cautious) engagement with Butler's notions of performativity and theatricality will further our understanding of these processes.[9]

In an interesting (and refreshing) take on the performance of space as straight, Helen (charles) (1993) discusses being a black lesbian (and caring for a disabled lesbian) in a space performatively constructed as heterosexual – a CenterParcs holiday camp. Advertised as 'the British holiday the weather can't spoil', CenterParcs offer a range of recreational, sporting and leisure activities for holidaymakers, all protected from British inclemency (of the meteorological variety, at least) under a space-age, atmospherically controlled dome. As (charles) says:

> the CenterParcs ethos was such that it embodied distinct codes of heterosexual conduct, which were stereotypes in themselves: respectability in the production and exposure of non-'pretended', always-happy couples and their families, protected by their able bodies and their whiteness. *But it was all contained in the confined space of a man-made village-camp.*
>
> (charles) 1993: 271, our emphasis

Thus, within the constructed space of the holiday camp, with its overarching air of artifice and artificiality, the 'naturalness' of co-residential (nuclear) heterosexual family life is shown to be equally constructed, equally artificial.[10] Contingent on time and space, we need to note that, like identities, space must be retheorised as 'space-time distribution[s] of hybrid subject-contexts constantly being copied, constantly being revised, sentenced and enunciated' (Thrift 1993: 96). Only through the repetition of hegemonic heterosexual scripts in CenterParcs, or on the streets, does space (become and) remain straight.

MONTREAL PRIDE AND THE POLITICS OF SEX[11]

It is also possible to use this reading of Montreal Pride to illustrate a number of contemporary debates in sexual politics. Exploding out from a particular event, political and cultural divisions which impact far beyond the space of Montreal Pride become apparent (in the same way, Tracey Skelton's discussion of the Buju Banton saga and Tim Davis's chapter on St Patrick's Day in Boston tell us a lot about the interplay of identities and communities beyond the particulars they describe). While we would like to acknowledge right from the start that the issues we discuss here are much more complex than we can do justice to, it is to be hoped that readers will not want to pick fights with what is inevitably a series of partial (in both senses of the word) accounts. All that we can hope to do is engage with some of what we see as the most important theoretical and activist issues, under the headings of queer, identity politics, and AIDS work.

Queer

The actions of the rerouted 'unofficial' Montreal Pride parade in 1991, and the (cross-) dressing-up of parade-goers in 1992 might be read as examples of queer political strategies. Queer has become a notoriously indefinable watchword for certain so-called radical theoretical and activist stances in 1990s sexual politics, the first of which was arguably the act of reclaiming the very word 'queer' from its use as homophobic slang to being a label used *by* a variety of sexual dissidents. But how to define queer? A reasonable place to start might be this somewhat lengthy and supremely queer quotation from a pamphlet circulating New York, *circa* 1990, *I Hate Straights*:

> Being queer means leading a different sort of life. It's not about the mainstream, profit margins, patriotism, patriarchy or being assimilated. It's not about being executive directors, privilege and elitism. It's about being on the margins, defining ourselves; it's about genderfuck and secrets, what's beneath the belt and deep inside the heart; it's about the night.
>
> quoted by Alcorn 1992: 21–2

The particular historical moment which gave birth to queer in its current incarnation is widely recognised as being entwined with a number of inter-related happenings: the pitiful response globally of government care agencies to the AIDS crisis is a pivotal one, as is (in the UK) Clause (later Section) 28, one of a series of anti-'gay' legislations thrown up by the homophobic Thatcher regime. Very real attacks like these have called for very real action, since, in the words of AIDS activists ACT-UP, *silence = death* and *action =*

life. Queer politics, then, arose from the recognition that existing political strategies coming from the lesbian and gay community were impotent. In the face of the New Right, a new adversarial politics was needed: it was time for queers to bash back.

Queer also gladly donned the drag rags of postmodernism; it embraced constructionist arguments about gender and sexuality, playing with and disrupting traditional binary oppositions. As Namaste (1992) and Sedgwick (1994) both describe, pride marches and rallies are now profoundly queered in this sense:

> At the 1992 gay pride parade in New York City, there was a handsome, intensely muscular man in full leather regalia, sporting on his distended chest a T-shirt that read, KEEP YOUR LAWS OFF OF MY UTERUS.
>
> Sedgwick 1994: xi

And, in reaction to sex-negative health authority campaigns and the purges of lesbian feminism which had frowned upon certain sexual practices (sadomasochism, for example), queer is proudly pro-sex: Douglas Crimp's (1987b) famous call for 'principled promiscuity' as a sex-positive antidote to government puritanism in the age of AIDS marks queer as an important site of resistance.

But perhaps the most significant, and for many the most seductive aspect of the appearance of queer, was its much-flaunted inclusiveness: queer embraced literally anyone who refused to play by the rules of heteropatriarchy: 'There are straight queers, bi-queers, tranny queers, lez queers, fag queers, SM queers, fisting queers in every single street in this apathetic country of ours', proclaimed the pamphlet *Queer Power Now* (quoted by Smyth 1992: 17). Perhaps even more importantly, as evidenced by Sedgwick's cross-sloganed queen, queer marked a coming-together of gay men, lesbians, bisexuals, sadomasochists, transgender and transsexual people ... Where the pre-queer world of lesbians and gay men was (is) boundary-ridden, queer welcomed (virtually) everyone, or so it claimed (of course, the histories of sexual minorities and outsiders are filled with alliances over particular issues – as David Bell's discussion of the Countdown on Spanner campaign shows – and unions founded under shared oppressions, as Alison Murray describes in relations between sex workers and lesbians).

Part of this new radical inclusiveness inevitably meant a turn away from assimilationist gay male culture (the kind of culture that would suggest only marching through Montreal's gay village, or a Pride uniform of T-shirt and jeans). A poster for the US magazine *The Advocate* (reproduced in the journal *Discourse* [Kader and Piontek 1992]) neatly encapsulates this aspect of queer politics. On the right, a moustached, thirtysomething man says 'I hate it when you use the word "queer"! Your immature tactics are undermining 20 years of gay rights'. On his left, a younger, bandana-wearing man says 'We're queer! You rich white sell out! Don't you know people are dying and getting bashed!'[12] Queer thus offers a critique of 'gay politics' and particularly of its white, middle-class assimilationist bias. Unfortunately, instead of being able to reject divisive attitudes and labels, queer has *become* a label, an orthodoxy with its own hierarchies and exclusions (Eadie 1993b; Kader and Piontek 1992). However, while we might want to signal some caution in overcelebrating what Sedgwick (1994: xii) calls 'the moment of Queer', the burst of anti-queer sentiment (which, in the UK at least, had the pink presses proclaim the Death of Queer and the dawn of a post-queer age) has settled down considerably, leaving us with a rich new vein of theory and activism to deploy as we need – in fruitful notions like queer reading or Butler's (1993) 'critically queer', for example. Allied to the twists and turns of queer's history has been a series of important shifts in our thinking about the whole idea of identity, whether that identity be 'gay' or 'queer'.

Identity politics

Diana Fuss, in a series of engaging observations in *Essentially Speaking* (Fuss 1989: 97–112), develops a critique of the essentialistic deployment of 'gay identity' in the service of 'community-politics'-building which highlights the contradictions and tensions within 'identity politics' and lets this clash with a 'poststructuralist' problematising of 'identity' which would be quite comfortable in the company of queer.

The essentialism versus constructionism debate in lesbian and gay theory has a long lineage, and continues to be played out not only on the pages of academic texts but also through the lives of individual people as they cast and recast themselves as 'lesbian' or 'gay' or 'bisexual' or 'heterosexual' or whatever. The pains of '"coming out" in the age of constructionism', as Paula Rust (1993) termed it, are particularly acute for those whose sexual scripts have exhibited far more fluidity and flux than conventional identity categories allow (and here, queer might have intervened, if it had lived up to its expectations).

Essentialism relies upon arguing that, in this case, being gay is a 'natural fact', something one is born with, something which cannot be changed (no matter how many 'therapeutic' interventions are deployed). Comparisons between sexuality and ethnic identity have been used to 'explain' and 'defend' homosexuality (Epstein 1987), while arguments about a uniquely 'gay sensibility' serve to substantiate claims that being gay is somehow innate. A powerful argument for essentialism comes from biologists and neurologists like Simon LeVay and Richard Pillard, who have attempted to prove anatomical or genetical foundations for sexual behaviour (see Stein 1993). And for a lot of people struggling to make sense of their sexuality, locating it in an 'accident of birth' has considerable potential. But, as John D'Emilio (1992: 187) has said, it can be difficult to bid for power from the essentialist standpoint of 'I can't help it'.

On the other side, social constructionists argue that there are no sexualities existing outside of culture – our sense of self is a product of the world around us. Reading accounts of non-Western, non-modern, non-urban-industrial sexual behaviours and identities, and hybrid forms like those described in Glen Elder's South African mines, shows us just how (spatially) contingent sexuality is. Following Foucault's plotting of the history of sexuality, and aligned with lesbian-feminist notions of sexual identity as a matter of choice, constructionism offers up considerable potential to destabilise not only sexual identities, but also gender, sex and the body, as Julia Cream's chapter shows. From a political perspective, however, rigid adherence to constructionism can be tricky to deploy outside of ivory-tower theorising: at a recent conference which sought to bring the academy and activism together, the former's insistence that all identities are fictions (albeit 'necessary fictions') provoked acute anxieties and outrage in the latter, who weren't too keen to hear that their struggles, their oppressions, were over nothing more than *fictions*. Out of this opposition arises the position of strategic essentialism (see David Woodhead's and Larry Knopp's chapters). By combining the political clout of both essentialist and constructionist thinking, strategic essentialism offers a potentially radical location for identity politics (see also Rose 1993a). From an essentialist perspective, for example, Section 28's charge of local authorities 'promoting homosexuality' and thereby turning children gay can be demolished as nonsensical. As Woodhead explains, one area where strategic essentialism has been useful is in AIDS activism and safer sex education (see also Chapter 5), and it is to AIDS work that our attention now turns.

AIDS work

Perhaps it is in the domain of 'AIDS work' that the previously criticised notions of queer and identity politics find a truly radical political space. As noted above, the AIDS panic has been fundamental in the creation of queer. As Geltmaker (1992: 609) says:

> The emergence of organized political groups of people with AIDS has forced issues of health and illness into a public visibility which threatens traditional assumptions of privacy and public heterosexual privilege. The struggle against the stigmatization of AIDS has forced many gay men and lesbians to reject the relative pleasures of the closet and its legal girdings in discredited notions of constitutional privacy for a radical insistence on the right to be 'queer' on their own terms in public.

The body becomes, in Geltmaker's words, 'a site of public contestation' (ibid.), and groups like ACT-UP and Queer Nation have successfully deployed a queer political praxis – enacted through the body – in response to AIDS (e.g. Crimp and Rolston 1990) as well as responding to 'local issues' such as the Montreal Pride 'fagscism' débâcle.

In further considering responses to AIDS, Cindy Patton (1990: 124) returns to a 'deconstructive identity politics' as 'a resistance to and at the same time a reinstatement of the underground side of a public politics grounded in the sheer threat of social coercion'. As Biddy Martin (1992: 106) writes, there is a version of queer/identity politics which can work to 'counter the government's rigid deployment of identity categories with practices of resignification and intervention'. Such practices have a political potency, she argues, as 'the work of resignification and redescription avoids the trap of celebrating instability for its own sake'. As a strategy of resistance, these practices may well mark a viable space for a politics of sex to be articulated. AIDS work, which is, as Jan Zita Grover (1992: 234) says, 'fueled by so much anger', marks the site where queer politics and identity politics can, through that anger, effect a radical deployment of the personal as political.

These responses to AIDS, together with the multitude of initiatives like those described in Vancouver by Michael Brown, show the spaces of AIDS activism, from the inner space of the body to the virtual landscape of cyberspace (which Brown suggests AIDS helplines occupy), echo back to our queer reading of geography by showing that at every spatial scale (cf. N. Smith 1993) battles over sexual rights and responsibilities are being fought. The AIDS epidemic, simultaneously global and local, brings into sharp and painful focus our theories of space, place and location.

THE SPACE OF THE ACADEMY

If, as Mark Wigley suggests, all space is closeted, then perhaps the space of the academy is among the most closeted of them all. Gay historian John D'Emilio (1989: 435) writes:

> To contemplate the subject at hand, to think – really to think – about gay history and gay historians in relation to the profession is to tap an interior well of pessimism, discouragement, despair, and exhaustion that shocks as well as frightens me.

As Louise Johnson (1994: 110) suggests, these sentiments might equally be applied to geography. In her discussion of feminist geography's futures, she comments on the continued squeamishness within the space of the university about sexual outsiders:

I have agonised for years about the consequences – professional and otherwise – of 'coming out' in print, declaring my own sexuality and building a feminist geography upon my lesbianism. And basically I've seen the risks as too great, the stakes as too high in a homophobic culture and discipline.

The great risks and high stakes to which Johnson alludes are vividly described in Sally Munt's introduction to *New Lesbian Criticism*:

The pressures of being out and teaching lesbian material include dealing with students and staff to whom you always seem to be representing sexual difference in its entirety; who collapse personal identity with theoretical integrity in a totalising motion which can only work against you, whether you are patronised, idealised or stigmatised.

Munt 1992: xiv

At the same time, acknowledging De Vito's (1981) point that our classes are rarely (if ever) all straight, there is an affirmative and empowering side not only to being 'out' in the classroom, but in teaching about sexuality. Of course, there are ways to teach and ways *not* to teach something like sexuality on a geography curriculum. Treating gay neighbourhoods as some kind of exotic place, for example, and thereby depicting lesbians and gay men themselves as exotic, is obviously very disempowering, as are homophobic comments coming from other students (and staff). Monitoring of students' reception of and reaction to teaching material on sexuality is thus essential, as is careful handling of individual students, in order to avoid unwanted breaches of confidentiality on both sides.

The campus environment offers students the potential for developing a clear sexual identity, but a series of surveys in the United States has highlighted anxieties among lesbian, gay and bisexual staff and students about their safety on campus. A report by the Taskforce on Lesbian and Gay Concerns, University of Oregon (1990) showed that nearly a third of staff respondents and just over half the students had been harassed or threatened on campus, and many others had experienced the emotional intimidation of being 'silenced'. Official policies to deal with harassment are often ineffectual, with many not including sexuality. In the face of subtle harassment and discrimination such as having secretaries refuse to type work on sexuality, exclusion from displays of academic work by staff members, censorship and removal of course material from libraries, such gaps in equal opportunities policy appears all the more worrying. And as Munt (1992) notes, career progression within the academy is limited for those with nonconforming sexual identities.

The problems are not limited to what one 'is', but extend to what one 'does', in terms of research. Trying to secure funding for research, struggling to work in an unsupportive environment, and being misrepresented by sensationalising media (as 'wasting' 'public money' researching 'sex') are very real issues for those trying to do research on sexualities (Plummer 1992). In addition, there is always a danger that making people and places visible can work against the interests of both researcher and researched.

Issues of positionality in research have begun to be talked about by those geographers who have been made aware that no research is 'innocent'. The relationships we have with respondents ooze power (Katz 1992), and must therefore be handled sensitively. The privilege of 'insider status' is open to abuse if groups are misunderstood or mis-represented (England 1994), while the potential for appropriating marginal voices is ever present. As Kim England (1994) notes, only through reflexivity can we make visible the power imbalances within research – but even then, we cannot erase them. And at a different level, our research relationships and the way we report them cannot (indeed must not) be kept impersonal and clinical. We must also be reflexive about how we feel about our respondents – owning up if we feel sexually attracted to them rather than struggling to maintain a false front of objectivity (Binnie 1994; Newton 1993a).

If the points made above sound pessimistic, they should not be read as inferring that the project of a queer reading of geography is an impossible task. The chapters which follow will hopefully begin to open up a space for the creative and wild possibilities that researching and teaching of geographies of sexualities offer. We cannot hope here to offer a definitive text, and we are well aware of the many absences from the collection. It is for the research which follows, and which we would hope to play a part in inspiring, to begin to fill those gaps.

NOTES

1 One wonders in which of these years Elspeth Probyn found herself marching uneasily through Montreal, questioning (like most of the marchers, it seems) the 'we' of 'we're here, we're queer, we're fabuuulous' (Probyn 1993: 10). Evidently there's nothing like a Pride march for bringing a lot of hot issues to the surface.

2 Thanks to Alastair Bonnett for sending Namaste's article.

3 See also Eve Sedgwick's 'T-time', in

Tendencies (1994).

4 It seems that the construction of Pride marches for a straight (tourist?) spectator audience is becoming a very important issue for marches in the US and, judging by some footage of Mardi Gras shown on British TV recently, in Australia too (look at who's *watching* the parade).

5 Sue Golding is also notable for causing a minor moral panic at the Institute of British

Geographers annual conference in Swansea, January 1992, by doing 'Reclaiming the "impossible" urban as site specific for a radical democracy' (a version of 'Quantum philosophy') – frightening the geographers with her leathergear and her strange talk of SM scenes and 'vanilla sex'.

6 The term 'polymorphous decentred exchange' comes from Linda Singer's (1993) discussion of sex in the age of AIDS, but it seems apt to couple that with the urban geographers' claim that postmodern cities are polymorphous decentred landscapes.

7 Phrase from Alison Murray (this volume) – Eve Sedgwick's (1994) term, 'ambient heterosexist' might be more appropriate – or even the more inclusive 'ambient heteronormative'?

8 For work on genderfuck, see Kroker and Kroker (1993).

9 For cautions in (mis)using Butler, see Sedgwick (1993).

10 Of course, one can only speculate about the potential destabilising effect that black and disabled lesbians might have had to this space: queer kiss-in at CenterParcs, anyone?

11 Parts of this section and a lot of the ideas expressed in it were aired in David's paper 'The politics of sex: queer as fuck?' presented at the New Theoretical Directions in Political Geography seminar, Birmingham, September 1993.

12 Note that both are white men, signalling who had the power in pre-queer sexual politics, and who has it still in these queer times.

1

cartographies/identities

SECTION ONE

CARTOGRAPHIES/IDENTITIES

Any geographical thinking about sexualities must begin by exploring how sexual identities are constructed and performed across space. Following Foucault's documenting of sexuality's history, we have to begin to map out the geographies of gendered and sexualised subjects. The task of theorising these maps of identity is addressed by the authors collected in this section. From a variety of theoretical and epistemological positions, these cartographies show how the complexities of assuming any sexual identity are spatially contingent, and often contested. From the intimate spaces of the body to the world of work, these first five chapters reflect both diverse experiences and diverse thinking on identities. Julia Cream focuses on the sexed body, and on the disruptive presence of bodies that are out of place, suggesting that the binary logics of the sex-gender system contort and distort our bodies. As Linda McDowell shows, the female body in male workspace is equally out of place. Drawing on material from research in merchant banks, McDowell suggests that for women in the City, all work is sex work, in that the performance of gendered and sexualised bodies rehearses heteronormative and patriarchal scripts. Viewing 'body work' via feminist theory's engagement with Foucault's work on the corporeal makes clear that the coupling of 'Woman' and 'Work' still carries a great deal of ambivalence.

Alison Murray's chapter, 'Femme on the streets, butch in the sheets' is also about women working – but this time about the play of butch-femme and lesbian sex-work. By suggesting that the performance of identities is spatially contingent (what one is on the streets isn't always what one is in the sheets), Murray explores the complex stories of lesbian sex work, where the sex need not be work, and where lesbian identities are recast as the game changes. Within the binary system of sexuality, which would read us all as either heterosexual or homosexual, locating bisexual identities can be equally problematic, as Clare Hemmings shows in Chapter 3. Reading bisexuality into feminist theory proves a difficult but crucial task in reconstituting our thinking about sexual identities. Bodies (and identities) which don't fit are also central to Glen Elder's work on South Africa under apartheid: state and legal discourses around male same-sex contact reveal a contradictory and essentialist interpretation based on the notions of 'situational homosexuality' as opposed to 'gay identity'. The interplay of 'race' and sexuality, further explored in Section Four, is thus a complex set of encodings and recodings of spaces, acts and identities.

2

RE-SOLVING RIDDLES

the sexed body

•

Julia Cream

RIDDLE 1
We all have one.
Most of us wish we had a different one.
What is it?
The Body

RIDDLE 2
Most of us acquire it at birth.
Some change it, others play with it.
What is it?
Sex

Perhaps it is a bad idea to start with riddles. Perhaps this chapter is already riddled with enough twists and turns. But, then again, maybe there are not any simple answers to the composite of my two riddles: 'what is the sexed body?' It sounds a simple enough question, but my inability to provide an easy answer structures this chapter. Maybe it is enough, at the moment, simply to ask the question, a question that has been largely unasked.

The sexed body is not simply there, ready and waiting, for us to examine. It is not something that can be broken down to its constituent parts. We don't simply add 'sex' to the body and we definitely don't add 'the body' to something called 'sex'. What, then, is this thing that we call the sexed body? The sexed body is a construction that requires explanation. It does not simply exist, it is not a starting point. It is already a constructed and particularised view of nature (Eisenstein 1988: 91). The sexed body is an outcome; an outcome, I argue, of both politics and nature, of mind and of matter.

It has taken little less than a decade for Turner's (1984: 30) assertion of the existence of a 'theoretical prudery with respect to human corporeality' to appear outdated and tenuous. The final decade of the twentieth century is witnessing a spiralling of interest in the body.

Theories of corporeality currently abound, but as Susan Foster (1992: 480) cautions:

> These writings seldom address the body I know; instead they move quickly past arms, legs, torso and head on their way to a theoretical agenda that requires something unknowable or unknown as an initial premise.

I do not want to start with the unknown or the unknowable, but when I begin to question what exactly the body is, I come face to face with a way of thinking that is challenging and unknown. It is difficult to 'understand' something so commonsensically irreducible as the materiality of the lived sexed body – whether it be our hearts, our souls or our genitals. The body has figured as 'an irreducible sign of the natural, the given, the unquestionable' (Kuhn 1988: 16) and whilst I want to embrace the idea that sex and our bodies are cultural constructions of the natural, I hesitate to endorse the claim that 'sex is defined by our ideas, not our bodies' (Solomon 1987: 206). For what are our minds if not our bodies?

The body is, undoubtedly, in vogue. We need to ask what, if anything, is new about the body? Why has it become an important issue now? Perhaps, like identity, 'the body' is 'in crisis'. As Mercer (1990: 43) suggests with reference to identity, it only becomes an issue 'when it is in crisis, when something assumed to be fixed, coherent, and stable is displaced by the experience of doubt and uncertainty'. It would be surprising indeed if the body, that most solid and stable concept that we all are, and all have, escaped the implications of the post/late-modern wave of decentring and destabilisation. With the fragmentation and the undermining of the authentic 'self', and the destabilising of all that was once thought fixed, our bodies can no longer be held to be ultimate arbiters of truth, a bedrock upon which we can base our social and cultural truths. Moreover, the fixity and absolute nature of the body has been so deeply entrenched in Western thought that even 'the juxtaposition of the terms "concept" and "body" seems oxymoronic' (Doane 1990: 163). Shilling (1993) also suggests that feminism, the rise of a consumer culture as well as demographic changes have all sought to make the body a site of academic interest.

There appears to be little, or no, historical consistency to the bodies which gather under the rubric of the 'natural' body. There is no consensus on what the body is and what constitutes it. As Synnott (1992: 80) adds 'some people include hair and nail clippings, spilled blood and faeces', others include the shadow and others would not. I remain attracted to a statement made by Herzlich and Pierret (1987: 69) who have suggested that 'there is always something inexpressible about the body; there are cries and whispers that cannot be put into words'. It captures that *something* of my body, and of yours, that I don't want to do away with. I do not wish to endorse a nihilistic position in which 'the body has been turned inside out and exploded out to the surface where experience has become an outer garment' (Emberley 1988: 49). I don't wish to write out the body altogether. The body, and especially a woman's, is not, as Bordo (1991: 120) reminds us (after Kruger), a postmodern playground, but a battleground.

What is clear, however, is that there is no way that a body can escape its social and cultural setting. There is no body outside of its context, recognisable as human. Mary Douglas' (1973) pioneering work on the body proposes that there is no way of considering the body that does not at the same time involve a social dimension. The social body constrains the way the physical body is conceived and bodily experience, in turn, reinforces and mediates understanding of the social. The body is thus seen as possessing no pure, uncoded state, outside the realm of culture (Fuss 1989).

All bodies are sexed. But, what it means to be a sex, or have a sex, has shifted across time and place. The history of the body is being rapidly documented (see for example Duden 1991; Foucault 1980; Gallagher and Laqueur 1987; Laqueur 1990) and we are now beginning to explore its geographies (see Dorn and Laws 1994; Johnson 1989; Longhurst 1994). Corporeal specificity is being placed on the geographical agenda, both in terms of spatially differentiated bodies as well as calls for embodied producers of geographical knowledge (Rose 1993b).

Male and female are posited as mutually exclusive categories (Grosz 1990) and although their meaning may change over space and time, what stays constant is that women and men have to be distinguishable (Lorber and Farrell 1991) . Our sexed bodies are understood to be either male or female. In our society sexual ambiguity is untenable (unlike ambiguous gender or sexuality). Any possibility of adopting a position that is neither female nor male is simply not permissible, it is almost unthinkable. But I want to think about it. I want to think about the bodies that do not 'fit'; those bodies that are 'out of place' in our social and cultural worlds, disrupting categories of identity and space. Particular bodies in particular sites can disrupt traditionally accepted notions of sex and gender. Three such bodies come easily to mind:

TRANSSEXUAL

The transsexual emerged, in the Euro-American context, in the 1950s. A transsexual is a person who 'identifies his or her gender identity with that of the "opposite" gender' (Stone 1991: 281). Their body does not appear to 'fit' their gender. The transsexual was conventionally understood to be a person trapped in the wrong body. Both facilitated by, and in turn facilitating, the development of new medical technologies, the transsexual was able to achieve an expressed gender identity that 'matched' a body coded as either male or female in the postwar era. As Lindemann (1992) notes, 'the gender identity of transsexuals remains unchanged throughout their lifetime, just as is the case with non-transsexuals'. It is the body that is constructed as mutable and plastic (Hausman 1992).

The transsexual reaffirms the correspondence of *an* appropriate body for a particular gender. The 'right' body has to match the 'right' gender (Bolin 1988).

The transsexual emerged as a new gender category (Billings and Urban 1982; Connell 1987) disrupting beliefs that gender and sex were interdependent and inextricably linked. The work of doctors such as Money and Stoller at the Johns Hopkins University established the parameters of debate and provided a dominant framework within which transsexualism was understood as a gender disturbance for which there was a cure. The 'doctors, the psychiatrists, therapists, endocrinologists, surgeons and urologists' (Vincent 1993: 4) rapidly became the agents who could correct the gender dysphoria through the refinement of medical technologies.

The insistence on acquiring the 'right' body has led to accusations of fetishism and an obsession with the genitals. But, as Shapiro points out (1991: 260), transsexuals are in fact 'simply conforming to their culture's criteria for gender assignment'. In other words, it is our culture that demands that a sexed body corresponds to a gender. Male-to-female transsexualism predominates, with most transsexuals wishing to become women. This helps to explain why techniques of vaginal construction are more sophisticated than those for a penis. Shapiro notes that studies of transsexualism are overwhelmingly focused on men who have become women and adds ironically that this represents 'an interesting twist on androcentric bias in research' (249).

INTERSEX

Intersexed infants are babies born with genitals that are neither clearly male nor female. Kessler (1990) has looked at the case management of such children in the USA in the late twentieth century. She argues that their medical management is ultimately dependent on a cultural understanding of gender. Gender is again equated with genitals. Physicians remain the interpreters of the physical body, using medical technology 'in the service of two-gender culture' (Kessler 1990: 25).

In the face of unequivocal evidence that the binary notion of two – and only two – sexes is a myth, medical and scientific orthodoxy have continued to perpetuate the legal fiction of binary gender. Although, as Epstein (1990: 128–9) notes, 'medicine recognises the flexibility of the continuum along which sexual differentiation occurs', this has not resulted in the 'necessary juridical accommodation of those who occupy minority spaces (which are, ironically, its midpoints) on the continuum'.

Epstein's (1990) assessment of how sexual-ambiguous persons have been treated and understood throughout the centuries reveals the way in which they have been 'progressively' labelled as 'monsters' as well as 'anomalies'. She goes on to note that their existence has been increasingly suppressed by medical science as their anomalous bodies are erased.

XXY

In 1966 sex testing was introduced to the Olympic Games. Moving from a 'crude' visible check to more 'sophisticated' scientific buccal smear tests, sportswomen began to be tested for their sex. First introduced to prevent men masquerading as women this controversial test has thrown up the complexities of trying to test biological maleness and femaleness. This is particularly interesting in light of the fact that women are awarded a 'femininity' certificate (Bradshaw 1992), and not one of femaleness. Implicit in the sex testing of sportswomen is the cultural assumption that we do not expect 'women' to be muscular and strong. A number of women have been 'exposed' through sex testing and disqualified on discovering that they had XXY chromosomes, as opposed to a 'woman' that has XX and a 'man' that carries XY chromosomes. It is, however, debatable and highly contentious to assert that a woman with XXY chromosomes is not a woman. The debate over sex testing has forced the issue of whether it is genes, hormones, or culture that determines sex.

In each of these cases, only one sex is allowed. The transsexual becomes either a male or a female, the intersexed person has a sex, and only one, assigned, and the XXY woman has her sex taken away and her 'real' (male) sex exposed. The biological (or what we have come to know as biology) continues to underlie the 'real'. It continues to be the baseline which culture appropriates and adapts. As Connell and Dowsett (1993: 56) point out, 'a biological warrant is found for pre-existing social ideology'. Increasingly, however, not only are biological facts such as 'sex' exposed as being infused with notions of gender, but also 'almost everything one wants to *say* about sex – however sex is understood – already has in it a claim about gender' (Laqueur 1990: 11, emphasis in original).

Our sexed and gendered boundaries have histories and geographies. The point at which a person's sexual identity comes under scrutiny reveals the times and places in which corporeality is specified, as well as the places where weaknesses, and possible entry points for change, exist. The birth of an intersexed child produces an acute problem for parents answering eager inquiries as to the sex of the newborn child. As Kessler's work illustrates, the intersexed child is usually kept within a hospital environment until a gender is (re)assigned. Ambiguity is contained as well as medicalised. The postoperative transsexual often occupies an ambiguous position within the eyes of the law. In Britain, for example, a male–female transsexual cannot marry a man, or be admitted to women's hospital wards, prisons or lavatories (Vincent 1993). Her sexed body is clearly spatially differentiated. The case of the XXY woman is also a phenomenon of late twentieth-century technology. She tends to be spatially confined to the sports arena. As one athlete stated, 'if I hadn't been an athlete, my femininity would never have been questioned' (Bradshaw 1992: 10).

These people are just some of the pioneers placed, often unwillingly, at the frontiers of sex and gender. Whilst one should never underestimate the commitment, pain and suffering through which people must go, as well as the violence and hostility they may face in order to achieve their desired body/gender/sex, theoretically, and hence politically, what are we doing by relying on our flesh to tell us the secrets and reveal the truths of sex and of gender? Those bending the boundaries of sex, pushing the binaries to their limits, continue to be defined as one sex or the other, the one we are becoming or the one that we leave behind.

All these pioneering bodies do, in some way, confirm and perpetuate the strongly held correlation between social gender and biological sex. The rigidity of the sex/gender system is confirmed, not destabilised. We have to maintain a correspondence between gender and biological body. Julia Epstein (1990) argues that the medical technology and surgical and hormonal interventions not only remove ambiguities but that the 'legal fiction of a binary gender' is maintained as an absolute.

If, then, gender *and* sex are historically and geographically variable categories perhaps we need to think of different ways of understanding and talking about our bodies, our sex and our gender. We need to find new questions: questions that require a reappraisal of what it means to 'be' a woman or a man.

Audre Lorde's now classic essay is surprisingly pertinent to this issue. She states:

> For the master's tools will never dismantle the master's house. They may allow us temporarily to beat him at his own game, but they will never enable us to bring about genuine change. And this fact is only threatening to those women who still define the master's house as their only source of support.
>
> Lorde 1984: 112

Lorde was talking about learning how to take our differences and make them strengths, and on the failure of white academic feminists to recognise difference. Her words have great pertinence to those working to redefine sex and gender. Transsexuals, for example, may indeed beat the master 'at his own game', they may confuse and distort sex/gender rules and conventions but, in the long run, in a society that demands that we be one sex or the other, they confirm them. Lorde's last line about the threat to women is also significant. New ways of defining 'the master's house', new ways of theorising (for example, patriarchy, capitalism, racism and heterosexuality), necessarily involve a 'threat' to categories, concepts and strategies, in which both 'woman' and 'women' are firmly embedded and deeply implicated.

CHEATING OR BEING CHEATED?

The introduction of the sex/gender distinction may be seen as one such attempt to retheorise/play the 'master' at his own game: an attempt to create new rules, not bend old ones. Feminist appropriation of sex and gender was an intervention in a Western world that declared women were 'different' because of their biology, because of their 'sex'. In the 1970s, women began to argue persuasively that gender was a culturally constructed notion that varied across time and space. 'Gender' was a useful intervention. It held out the 'promise of enabling an analysis of male privilege as the product of historically and culturally constituted systems of gender inequality, [and] not as the natural outcome of biological differences between males and females' (Yanagisako and Collier 1990: 131). Women argued that they were no longer defined by their biology, no longer delineated by their hormones or their genes.

The strategic assertion, on the part of feminists, that because something is natural (like female biology) it cannot be changed, or that conversely, because something is social (such as gender) it can easily be adapted, also has its problems. It seems easier, for example, to eliminate the menstrual cycle than alter attitudes towards the (polluting) effects of menstruation. It seems to be more 'natural' to adapt prenatal screening than change our attitudes to disability and incorporate an acceptance of disability rights. The social is now appearing as both as mutable *and* as immanent as the natural. We are increasingly being shown that what we had accepted as the 'truth' of science, whether it was our sexed bodies, or the fusion of sperm and ova (Martin 1987), has already within it a gendered perspective that is inseparable from its cultural context. Our understandings of both the social and the natural are as contingent as each other. The time seems ripe for a new way of trying to think through sex and gender.

'Gender' clearly enabled white feminists to engage in debate, allowing the biological to remain 'fixed', neutral, yet capable of being bypassed. Now, it seems, in a similar move, feminists can take up 'the body' in a way never available to them before. The theoretical expansion in conceptualisations of the body has enabled groups which previously avoided it to enter the debate on a different footing. Birke (1991: 448), for example, acknowledges that thinking about the body and biology 'seemed a dangerous move for feminists'. Adrienne Rich (1979: 40) wrote that 'the body has been made so problematic for women that it has often seemed easier to shrug it off and travel as a disembodied spirit'. The 'equality versus difference' debates (see Bacchi 1990; Snitow 1990) also reflect how entrenched the arguments over the importance of biology have become. Now, some feminists are 'willing and able to speak of what was unspeakable, to explore what was once forbidden, and risk positions that were once sacrosanct or untouchable'. They are theorising the body in innovative, experimental and exploratory ways: now that the body is being conceived in ways that are no longer 'associated with immanence, nature and otherness' (Grosz 1991: 2).

The inclusion of the body into feminist analyses provides another point of entry for highlighting feminist racism. Haraway (1991: 157) has noted how white women 'discovered (that is, were forced kicking and screaming to notice) the non-innocence of the category "woman" '. Gender depended on the category of 'woman', which insists on 'the non-reducibility and antagonistic relation of coherent women and men' (137). Spelman (1990: 127) also argues that white women's somatophobia (fear of and disdain for the body) 'historically has been symptomatic of sexist and racist (as well as classist) attitudes [as] certain kinds, or "races", of people have been held to be more body-like than others, and this has meant that they are perceived as more animal-like and less god-like'.

<h2 style="text-align:center">NEW RULES?</h2>

Judith Butler's (1990, 1991) work has been influential in the rethinking of boundaries of gender, sex and desire. Gender is seen as performative, a repeated performance which highlights the instabilities of gender, sex and desire. It interrupts any neat correspondence of sex, gender, desire and the body; positing, instead, radical discontinuity (see also Reich 1992). Sex no longer needs to 'match' gender. If there is no biological corporeal bedrock upon which to build cultural layers of gender, then gender becomes the means by which sex is produced as pre-discursive, 'natural' and prior to culture. There is, for example, no sexed female body awaiting enculturation, or engendering. Instead, gender is understood to be performative, constituting 'the very subject it is said to express' (Butler 1991: 24). Gender becomes a performance, a performance that requires constant repetition: a daily performance that constitutes identity. If gender is no longer assumed to signify, or be restricted by sex, then as Butler (1990: 6) suggests, '*man* and *masculine* might just as easily signify a female body as a male one, and *woman* and *feminine* a male body as easily as a female one'. Butler argues that gender and sex become regulatory fictions, found and experienced today as binary relations consolidated through the practices of heterosexual desire. Sex and gender do not have to correspond with each other 'correctly' as defined by a binary matrix underpinned by heterosexuality.

To understand gender as a performance, and identity as performative, is not, however, to presume that gender is, or can be, chosen. There is no 'sexed body' that 'decides' what gender it is to be. Gender is not a style, or a game that can be played. Gender is not voluntary, rather 'performativity has to do with repetition, very often with the repetition of oppressive and painful gender norms' (Butler in Kotz 1992: 84). Butler has attempted to find a new way of talking about 'the subject'; a subject that is not always already positioned within phallocratic and patriarchal discourse. She suggests that the current feminist subject 'turns out to be discursively constituted by the very political system that is supposed to facilitate its emancipation' (Butler 1990: 2). In an analogous way, the sexed body that is trying to create new radical, discontinuous ways of thinking about sex and gender requires a new vocabulary, a new way of understanding sexed bodies.

FLESHING OUT THE TRUTH

Radical discontinuity (between sex and gender) has important implications for the body. Where does it leave it? What is the relation between sex, gender and the body? What are the relationships between 'race' and the body? Between sexuality and the body? As previously held biological truths about our bodies are increasingly undermined, we need to start thinking about what kind of body and what kind of world we are living in as well as those that we can envisage.

In a provoking essay, Jacqueline Zita (1992) suggests something which, at the moment, seems not only implausible, but verging on the impossible. She toys with the oxymoron of the male lesbian. She questions what makes a lesbian a lesbian: and asks whether 'she' can be a man. Remember, within a theory of radical discontinuity between sex and gender, new configurations of bodies can emerge.

The notion that identity is performative (and note that this does not imply voluntarism) lends itself to claims about doing gender differently. If a person's gender 'is not simply an aspect of what one is, but, fundamentally, it is something one does, and does recurrently in interaction with others' (West and Zimmerman 1991: 27) – if there is no 'essence' to our gender, sex, 'race' or our desire – then this provides possibilities for radically *redoing* gender. West and Zimmerman (1991: 31), however, go on to suggest that 'if one wishes to be recognised as a lesbian (or heterosexual woman) one must first establish a categorical status as female'. This raises the fundamental question of how we are going to define 'female'. If women are still 'women' despite the surgical removal of breasts, ovaries and wombs (removed in an attempt to reduce the risk of getting cancer; see Hunt 1992) where does that leave our 'biological' truths?

Returning to Zita's essay, we are still left with the question 'does a lesbian have to have, and be, a female body'? And, if so, on what grounds do we demarcate femaleness? If sex is no longer assumed to be fixed and immutable, no longer the 'natural' bedrock upon which layers of gender are built, then where does that leave us when we want to demarcate a 'lesbian community'? Or a 'heterosexual couple'?

Zita's essay questions whether we should use the same genetic–gonadal–anatomical lexicon as the master norms of the dominant culture. Should we really be using the 'master's tools' to 'judge' who is a woman and who is not, and hence who a lesbian could be, and who, categorically, could not? She argues persuasively that lesbians may indeed be women who have had 'mastectomies, vulvectomies, hysterectomies, and other surgical operations removing body parts' (1992: 112). Yet theories of performativity leave open the political needs of forging a community. What grounds do we make alliances on, and do we still need to base our definitions in biology?

Our bodies are used to legitimate our social, sexual and biological truths. Our bodies are said to speak the truth. But what kind of truth do we have and whose is it? Laqueur's work has exposed the 'making of sex'. He has traced the historical traditions in the construction of a heterosexualised body, a body that is either male or female: a body that is no longer allowed to be anything in between. He also highlights the ways in which our bodies are naturalised and legitimised. How often do we hear that homosexuality is not 'natural' because quite simply 'things don't fit'? Well, as Laqueur (1990) eloquently illustrates, the sexual organs that 'fit' are clearly heterosexualised and contingent on a specific time and place.

Recent work such as Laqueur's on the constructions of gender, sex, 'race' and sexuality provide opportunities for relaying our foundations of truth. If we are no longer able to start with 'the body', adding colour to make white, or adding sexuality to make heterosexual, then we need to begin again. We need new ways of thinking about nature and biology: ways of thinking that do not endorse immanence or truth. Corporeal truths are deeply embedded within ideological discourses, and are used to legitimate what people can and cannot do, as well as their place in society. We can no longer accept nature as 'nature intended', but always as intended for some purpose.

ACKNOWLEDGEMENTS

Many thanks to David Bell, Jon Binnie, Laura Cream, Claire Dwyer, Peter Jackson, Nuala Johnson, Larry Knopp, Charlotte Pomery and Gill Valentine. The support of the Economic and Social Science Research Council (ESRC) is gratefully acknowledged.

3

LOCATING BISEXUAL IDENTITIES

discourses of bisexuality and contemporary

feminist theory

•

Clare Hemmings

Is there a way to think outside the patriarchally determined Same/Other, Subject/ Object dichotomies diagnosed as the fact of culture by Simone de Beauvoir thirty years ago, and, in the process, still include women as a presence? In other words, do we want to continue reorganizing the relationship of difference to sameness through a dialectics of valorization, or is there a way to break down the overdetermined metaphors which continue to organize our perceptions of reality?

Jardine 1980: xxvi

Bisexuality appears to be all the rage at the moment. It has been much discussed recently within lesbian and gay politics and feminism. After such a long silence *vis-à-vis* bisexuality as a viable political identity (really since the early days of gay liberation), more and more people appear to be 'coming out' as bisexual, thinking about or writing about bisexuality. There has been a spate of books on bisexuality and its relation to feminism in the last few years, as well as a number of 'personal narrative' volumes (e.g. *Bisexual Lives* (1988) – *Bisexual Lives II* is on its way). Sue George's *Women and Bisexuality* (1993) was the first British book on the subject for more than ten years.[1] Yet Anglo-American feminist theory has still failed to address bisexuality as worthy of theoretical and political attention in its own right. The above-mentioned volumes are all by self-identified bisexuals. Interest in bisexuality has rarely been articulated by non-bisexuals. French feminists have written about bisexuality more consistently,[2] but always within the specific French philosophical and psychological traditions. Anglo-American feminists have tended to keep away from those traditions and so have not been greatly influenced by the French interest in bisexuality. It is the Anglo-American traditions that

I address in this chapter as they have most obviously compelled the recent resurgence of interest in theorising bisexuality.

Within lesbian and gay communities, a significant number of people have talked openly about desire for people of the 'opposite sex', even while maintaining their own political identity as lesbians or gay men. Lisa Power, amongst others, mentions that many lesbians have felt attracted to men, but have felt unable to admit to this, for fear of being ostracised from their communities (Power 1992). With the advent of 'queer', cross-dressing and playing with conventional notions of sex, gender and sexuality has become *de rigueur*. As a result it has become easier to talk about bisexuality without necessarily being seen as apolitical, or, in terms of feminism, a traitor to the sisterhood. I would not go so far as to say that bisexuality is accepted within lesbian and gay communities, but at the very least it has become part of a discourse of differences, and is now often mentioned if only to be forgotten.

Within feminism I perceive that the move from politics of identity (1960s and 1970s) towards politics of difference (late 1980s and 1990s) has facilitated the renewed interest in bisexuality to a great extent. This chapter sketches a partial history of the relationship between bisexuality and feminist theories of identity. My aim is to show how contemporary feminist debates around identity and differences have facilitated the coming out of a bisexual voice. For me to attempt to theorise bisexuality from a feminist perspective without examining these sources would be foolhardy (if not treasonable). Yet bisexuality's position in relation to feminist theories of identity and difference is anything but unproblematic. Through my enquiry I highlight the many difficulties within contemporary feminist debates – difficulties of structure and emphasis that often remain unarticulated, or ignored. My concern is also with the location (or locatability) of bisexuality in relation to existing structures of sexual identity and subjectivity. Or, to put it another way, what or where is my bisexual 'home'?[3]

There are three parts to this chapter. Firstly, I look at the shifting focus within Anglo-American feminism from identity to difference, and ask what relevance this has for an understanding of bisexuality. Secondly, I analyse discourses of bisexuality within radical lesbian feminism and feminist theories of difference (also drawing on queer politics as an intertwined set of theories). Thirdly, I look at the problems and possibilities of articulating a bisexual feminist location, both out of, and in parallel to those theories.

FROM IDENTITY TO DIFFERENCE

The theory and practice of the early feminist movement in Britain and the United States focused on sexual difference – women's difference from men, and women's sameness in relation to each other (their common identity as women).[4] This was, and is still being critiqued by women who have felt and still feel excluded from 'the sisterhood'. The shift in the last decade or so in Britain and the United States has been from a focus on sexual difference to one of difference*s*: differences among and between women have been highlighted, and connections between different forms of oppression – race, class, sexuality, etc., as well as sex and gender – have been brought to the foreground. This has often led to bitter debate between the proponents of opposing feminist theories. The assumptions often made about women's sameness in relation to each other and in opposition to men have been challenged, opening up new spaces for bisexual women to talk about their desires, as well as for heterosexual women and lesbians to challenge monolithic assumptions about identity within rigid sex and gender binaries.[5]

Central to the debates about difference within feminism is the concern about exclusion and marginalisation expressed by black feminists. Black feminists[6] have argued that feminism not only marginalises and excludes them, but that the very premises of feminist thought and practice are white and Western. Early attempts to rectify this situation, such as Michele Barrett's and Mary McIntosh's article, 'Ethnocentrism and socialist-feminist theory' (1984), were heavily criticised for being 'additive', that is, attempting to assimilate black women's experience into an already-established white socialist-feminist framework of oppression. Caroline Ramazanoglu, amongst others, has pointed out that such an additive response to black feminist criticism still pivots on the misguided belief that extending the field of vision to include black women's experience (becoming non-ethnocentric), is the only alteration to Western feminist theory that needs to be made (Ramazanoglu 1986). One of the major criticisms levelled at Western feminism by black feminists is that sex oppression is taken to be the primary oppression, with 'race' and class oppressions being seen as derivative.[7] Within such a framework it is primarily through sexual difference that one's identity is understood to be formed. The problem is that in prioritising *Woman*, feminist theory has been unable to take account, other than nominally, of differences among women. Western feminism has stressed that while we are all different, there is also something that we all share as women – oppression *as women*. Yet black feminists have shown that different strands of identity cannot be separated in such a simple answer. We do not experience our identities as women and *also* as white or black, as lesbian, gay or bisexual. Other forms of identity cannot be added on as an afterthought. As a white bisexual woman, I experience my sexuality and my whiteness as a woman; I experience my femaleness and my sexuality as a white person; and I experience my whiteness and my femaleness as a bisexual. The fact that I do not experience oppression and identity in strands makes the polarisation of debates within white feminism particularly frustrating.

The shift in feminist theories from focusing on identity as women to differences between women has mostly come from critiques of existing feminist theory as replicating hierarchies of patriarchy. Radical feminists have been accused of being patriarchal because of the implicit assumptions made about the 'feminist body' as white, middle-class and able-bodied. These interventions by women who did not recognise themselves in the term 'feminist' as it had come to be understood, pointed towards theories that take a less universal stance, and interrogate instead the convergence of a variety of different discourses ('race', class, age, as well as sex) in specific moments. Feminist questioning of assumed identity categories is characterised by a move from seeing identity as sameness, as recognition, to sexual identity as *location*, as something formed *in relation* to other identity variables. The idea that a 'woman' is a fixed category, that has only a predetermined range of meanings, has 'gone out of fashion'.

If the white, middle-class lesbian body has been privileged as the site of the greatest resistance, and as promising the greatest equality, lesbians have now begun to talk about inequality within their specific communities. Firstly, lesbians have pointed out that the equality promised through same-sex relationships does not exist *per se*. Issues of race, class, disability, age, etc., mean that all lesbians are not 'equals'. And as lesbians have begun to talk about violence within lesbian relationships, for example, it has become clear that power differentials do not cease to operate in relationships between women. Secondly, some lesbians (e.g. SM – sadomasochist – lesbians) have questioned the assumption that 'equality' *per se* is what is desired or desirable in the first place.[8] Their arguments are that being lesbian does not have to mean 'same', or searching for sameness. Power-relationships, they argue, can be acted out in pleasurable ways. Differences are what are homed in on in all these cases: differences, and how these intersect with or are produced through power.

What I have forgotten to mention is that none of this has occurred without difficulty or crisis.[9] There has been no smooth progression from one set of ideas to another. In fact, the debate about what constitutes Woman and women's oppression in terms of these debates is one of the central sticking-points within feminism at the moment. Often this results in simplification of ideas, as a false opposition between identity and difference is set up. Further, the lures of postmodernism, and the individualistic liberatory promise of notions of flux and transitory identity, seem unable to offer an alternative politics. Neither do postmodern ideas of *coalition* rather than community always satisfy the emotional and practical needs of marginalised/oppressed groups to feel a sense of belonging to something other than the dominant discourse (that of white, heterosexual, able-bodied men).

Until recently a feminist's bisexuality was more likely to be considered as a sign that some patriarchal conditioning remained (hence the continued interest in men), than as an identity worthy of theoretical and political attention. While heterosexual and bisexual women have always been part of feminist movements, the emphasis on consciousness-raising (and hence false-consciousness) since the late 1960s, has meant that many women have felt guilty about 'needing' men to survive.[10] Recent interest in bisexuality coincides with the changes in feminist ideas I have outlined above. As a result I would say that bisexuality is worth further enquiry in its own right, as a little-explored and much-excluded sexual identity, and as a way of making sense of the impasse that seems to have occurred between theories (and theorists) of identity and those of difference(s).

A contemporary bisexual identity (and in particular a bisexual *feminist* identity) is intricately bound up with theories, practices and politics of difference versus sameness. Discourses of bisexuality come out of the relationship between these conflicting positions. Bisexuality could almost be seen as the embodiment of those tensions. For example, an emerging British bisexual movement spends much of its time critiquing the notion of identity, and deconstructing the more established discourses of a lesbian and gay identity and movement. Yet at the same time the bisexual movement is also trying to find a *place* for itself, trying to find a home. Coming out as bisexual in the 1990s may be partly to do with embracing postmodern ideas of the multiple self, yet can we afford to lose ourselves in a plethora of free-floating signifiers?[11] At the 11th National Bisexual Conference David Bell asked whether bisexuals really want to spend the rest of this century balanced precariously on the margins of a variety of discourses and meanings, laughing parodically at heterosexual and homosexual communities, swaying between community and dissolution (Bell 1993a)? My argument is that in analysing bisexuality's exclusions and contradictions some insights into the patterns of feminist thought, and possibly some proposals for ways out of the 'feminist impasse',[12] may emerge.

DISCOURSES OF BISEXUALITY

Bisexuality has been understood within feminist theories of identity and difference in particular ways. I now move on to examining those feminist positions in more detail. How are their theories structured? I am particularly interested in how bisexuality is or is not assimilated into those structures. I believe that many of the problems with feminist theories of identity and difference may be highlighted through a focus on bisexuality.

Identity politics

Early radical feminism has often privileged sameness over difference (collectivism over individualism; straight or 'vanilla' sex over SM; non-penetrative sex over simulated penetration; the pre-oedipal over the oedipal). Sameness is female/feminine, which is positively invested. To be a 'good girl' in feminist terms (which is a bad girl in patriarchal-capitalist terms; see Kaloski 1994), is to shun the enemy man/masculine, which is negatively invested and on the side of difference. This is almost a direct reversal of negative/positive sameness/difference paradigms within sexological theory.

The sought-after sameness of late twentieth-century feminism could be seen to replace the sought-after separateness of the masculine sexual subject. Individuation in terms of becoming a feminist sexual subject requires oneness with the mother, rather than a post-oedipal separation. What radical feminism seeks to do is set up sameness as the ideal, in opposition to the differentiation that is negatively associated with masculinity. Radical feminism becomes assumed to be based on equality and mutual support among women. For example, Adrienne Rich's conceptualisation of her 'lesbian continuum' that all women are on because of their connection with their mother, their women friends, etc., means that 'lesbian' becomes figured as a 'return', as the alternative to masculine fragmentation (Rich 1980). It is only a short step before sexual practices among women are also expected to mirror this movement towards non-individualism.[13] Those women who seek to become subjects in other ways, not bound to the embrace by and with the mother, must be held to occupy a position of masculine subjectivity within this model if female sameness is to be maintained.

Within radical feminist theory and politics the bisexual woman causes many problems. In many respects bisexuality is both absent (never discussed, except to dismiss), and abhorrent (a lingering threat – how can you be sure you're not sleeping with one?). The bisexual woman is doubly masculinised. She rejects sameness, in that she has certain desires for men, yet she cannot be contained in the parallel stream of difference either. Within such a structure a bisexual woman's identity is seen to be formed through differentiation from the feminist mother, yet she also attempts to penetrate, and hence contaminate and subvert the vigilantly guarded barracks of female sameness. I do not use such metaphors lightly. The bisexual woman is cast as the double-agent of sexual politics, selling out to the highest bidder. She is seen to be immune to politics, interested only in exchange, and the currency here is pleasure (Hemmings 1993). The limits of the boundaries are simultaneously acknowledged and denied. For example bisexual women are often blamed for bringing an HIV and AIDS risk into the lesbian community, yet I have never been to an HIV and AIDS workshop on same-sex safety for women, or seen one advertised, that addresses the issue of bisexuality to any significant degree. This applies also to those that have addressed other difficult or taboo areas such as SM.[14] The question is not whether lesbians should meet independently of bisexuals (I have no doubt that they should on a whole variety of issues), but what is at stake in maintaining the boundaries around female sameness, that are in the end, arbitrary (and by that I mean that they do not correspond to reality or experience).

It seems to me that such a discourse of invasion is central to understanding the production of bisexuality in relation to a particular form of identity politics. The bisexual is portrayed as ready to transgress the boundaries of lesbian identity (and also heterosexuality of course, but this does not have the same political implications for feminism at present). She is depicted as poised – rather than sitting, I think – on the proverbial fence, deliberately deceiving, delighting in masquerade, out-of-control, apolitical, unable to accept responsibility for same-sex desire and relationships (because, of course, there is no responsibility in opposite-sex relationships),[15] and so the list goes on. What amazes me is the level of power attributed to the 'bisexual body', to a body that is considered to be mythical and abhorrent, that is, not a *real* body. What this highlights for me is the lesbian feminist investment in maintaining sameness/difference boundaries, not because those boundaries are 'true' or static, but because, in fact, they are not. The fear of a 'bisexual invasion' is proof positive that those boundaries are not fixed. It is a fear that bisexual women are not outside of lesbian communities, but within them, that results in attempts to exclude them.

A further argument that is used to reject the necessity for a consideration of bisexuality is one that says that bisexuality is not political. This is seen in reference to the fact that lesbianism *is* political. Lesbianism is political in this context because it stands in opposition to hetero-patriarchy. Bisexuality is said not to be, because it is only in so far as bisexual women experience same-sex desires that they are political. That is, bisexual women are only political when they are 'temporary lesbians'. This is another distancing tactic.[16] I would say that nothing (or everything) is political in and of itself, yet labelling bisexuals as generically apolitical is a way of setting bisexual women *in opposition to* lesbians. It might be more useful and more accurate to think of bisexuals as located close to or nearby lesbians (Kaloski 1994).

Theories of difference(s)

Feminist theories of differences between women have been greatly influenced by post-structuralist and postmodern theories of the fragmentation of the self. It would seem that if early feminism may be accused of reproducing models of masculinity in its emphasis on a univocal identity, this could not be levelled at recent feminist theories of the 'post-structuralist school'. So does a contemporary focus on difference provide enough space for a 'bisexual home'? Here 'queer theory' is also relevant, though this is more often seen as emerging from within the lesbian and gay movement rather than the feminist movement. Queer focuses on fragmentation and difference within lesbian and gay communities. Such a development is interwoven with a feminist progression of ideas about what constitutes identity, in particular through the figure of the lesbian feminist. As a bisexual feminist my 'home' could perhaps be found through either or both sets of theories. Postmodern feminist/queer theorists such as Judith Butler and Teresa de Lauretis challenge the notion of power as repressive. Instead they argue that our identities are formed in and through negotiation with a *network* of power, not opposite to or outside of power (Butler 1990, 1993; de Lauretis 1991). There is much to be said for this approach, not least the extent to which it opens up the possibility of 'non-fixed' deviant identities that are opposed to heterosexism but are not univocal or static in themselves.

The first question is whether or not postmodern theories actually *do* challenge the sameness/difference oppositions that underly fixed notions of identity. If we look at the concept of *transgression*, which is one that I have mentioned as being part of the construction of bisexuality within lesbian feminist discourses – and also one that is frequently used by postmodern theorists as a positive idea – some of the problems with an exclusive focus on difference emerge. Transgression is a mutable term taken up by fascists and left-wing militants alike (E. Wilson 1993). Its primary function, however, seems to be the crossing of existing boundaries, the deliberate reversal of the status quo. Transgression is, of course, also associated with the avant-garde, and often with decadence. But does transgression really challenge dominant discourses? Elizabeth Wilson paraphrases Michel Foucault, who defines transgression as a 'going further' which then sets up new boundaries that need to be transgressed in their turn:

> What you then have is a transgressive spiral which at least in theory is interminable. From that point of view, transgression can define no final goal, and there can never be any final mastery; it is rather a process of continuously shifting boundaries, the boundaries of acceptable behaviour, the boundaries of what may be shown in terms of sexually explicit representations for example.
>
> E. Wilson 1993: 110

Continually shifting boundaries do not necessarily denote new territories, or new discourses. Transgression of the status quo can, in fact, *consolidate* the dominant discourse, rather than undermining it. Dominant discourses rely on the presence of an 'other', defining what is dominant through what is not. There is no guarantee that a postmodern focus on difference within sexual politics (queer, SM, etc.) is not simply setting up an alternative opposition that equates difference with the post-oedipal, the rejection of the mother – and hence sameness/difference dichotomies are maintained. Difference can end up being privileged for its own sake, and the necessity for analyses of power and possibilities of community or coalition may frequently be ignored. Unless transgression actually disrupts the underlying *forms* of the discourses being challenged, the attempt runs the risk of becoming yet another partner in the endless spiral of binary oppositions.

It does not seem accidental that bisexuality is occasionally mentioned within queer and postmodern theories, but never engaged with in a serious theoretical way. For example, Teresa de Lauretis questions the boundaries of the category 'lesbians and gay men' but merely in an additive way, noting rather scornfully that the trend on her campus is to speak of ' "Lesbian, Gay, Bisexual and Questioning" ' (de Lauretis 1991: vi). Yet de Lauretis never actually considers the implications for lesbian and gay studies of engaging with bisexuality, and in fact she refrains from mentioning it again. Within queer *politics* the role of the bisexual in Britain and the US has been similarly marginalised, partly because many of the meetings have taken place in Lesbian and Gay Centres that do not extend access to bisexuals. Cherry Smyth, in *Lesbians Talk Queer Notions* (1992) – a bold attempt to link feminism and queer politics by tracing the reinvention of the 'lesbian' – mentions the importance of bisexuality in contemporary queer politics. Yet like de Lauretis, Smyth does not take bisexuality seriously enough to discuss it in any depth. Bisexuality, or the presence in one body of same-sex desire and opposite-sex desire, might be said to be the epitome of identity as temporary and shifting location. Yet attempts to deconstruct the univocal identity of the lesbian are, it appears, only possible if we retain the fixed categories of lesbian and gay: difference from a position of sameness in other words.

It would seem that binary oppositions structure both identity politics and politics of difference within feminism. The 'other' in question may change, but the paradigm does not. Otherisation is a profoundly complex process: you need the very thing that you are unable to accept. It seems to me that this necessity for the Other also manifests itself within the *terms* of the debates that constitute the 'crisis' in feminism. For example, the lesbian SM and pornography debates are represented as clear-cut issues of pornography and violence against women on the one hand, and as censorship of material, behaviour and fantasy on the other. Again, the issue of 'race' cannot be adequately discussed within such frameworks. Jasbir Puar, in a paper presented at a conference in Utrecht in June 1993, discussed the taboo within both white Western and black feminism against black women considering the effects of their own 'whiteness'. Puar argued that 'whiteness' is not simply a category of 'race' or being. She used the example of second-generation Asian women born and brought up in Britain who may 'use' whiteness strategically – e.g. adopting 'white' clothing, attitudes, lovers, education – as a means of self-defence against racism. Hence the notion that 'whiteness' can be analysed only as an external or oppositional category of oppression by black feminists is problematised. She also raised the point that western feminism has a vested interest in understanding South Asian cultures as different (but equal) in order to maintain the relationship between sameness and difference within an overall structure of female sameness (Puar 1993). At my most pessimistic I would be tempted to say that theories of difference are often only new and mutable forms of the old argument (the more things change the more they stay the same).

BISEXUAL POSITIONING

I have suggested that bisexuality cannot be understood through existing feminist structures, and that, in fact, an analysis of bisexuality in relation to feminist models highlights the very difficulties that result in a bisexual exclusion. So what of my own positioning in relation to the structures I have described? In critiquing feminist structures of sameness and difference, am I trying to create myself as somehow not implicated in those structures? If bisexuality is not adequately accounted for, where could it be located? From what position(s) could a bisexual feminist theory be explored?

Claiming outsider status

I realised as I was writing that while I profess not to be prioritising bisexual identities over and above lesbian and gay identities, I often am, and not just in terms of voicing what has been silent either. Of course this is something unmentionable, given the extent to which bisexuals have had to defend themselves against charges from the lesbian and gay communities that they are fragmenting lesbian and gay communities and detracting from 'the real issue' – homophobia. I certainly don't believe that bisexuals are freer or better than lesbians or gay men, yet in using my positioning as the entry point into a critique of, in my case, lesbianism and feminism, is this implicitly what I *am* saying?

Using bisexuality (*my* bisexuality) as a way of highlighting the binarisms of sameness and difference within theories of identity can be a way of privileging outsider status. Yet can being 'outside' of something be used automatically as a mark of having 'inside' information about or on something? Such status seems to have replaced status through power (or lack of it); a hierarchy of suffering, replaced by a hierarchy of exclusion.[17] To maintain a sense of my (privileged) outsider position, I must invest heavily in reproducing those binarisms, particularly as having 'nothing to do with me'. So I rail against the dualisms that I claim are 'keeping me down', preventing an adequate theory of my own *marvellous fluidity* from emerging triumphant. But of course, those 'dreadful binaries' are scarcely somewhere 'out there', they inform and produce my identity as much as anyone else's. The conversations I have with myself, the operation of binaries within my psyche, the way I see the world, etc., all reconstruct what I claim to deconstruct.

According to Elizabeth Wilson, bisexuality is either the same as homosexuality, but weaker, or different from it, in which case it must in the sphere of heterosexuality (E. Wilson 1993). Within this framework it is not difficult to see why bisexuals have embraced notions of 'outsider' status, or entered into the competition for exclusionary honours. Traditional identity politics have to go out of the window to be replaced with notions of transgression and gender-play. But in fact, I do *not* ally myself with Elizabeth

Wilson's indictment of bisexuality as just another apolitical fuck. The attempt to exclude bisexuality often occurs *because* of the structures of power, of opposition politics. One response is to claim that it is better to be outside and visible, than inside and invisible. In that sense bisexuality's exclusion by others, *and* its self-conscious exclusion, are both immensely political.

Bisexual theorising

One of the major difficulties – as well as pleasures – of theorising bisexuality is precisely the lack of foundational categories to work with. Whatever the shortcomings (and there are many) of structures of lesbian or gay male desire, there are at least assumed meanings and identities to kick against.[18] So while a bisexual theory may be critiquing sameness and difference classifications, there are at present no alternative structures that have been fully theorised – no home other than the ones I carry on my back.[19] Bisexuality, then, is both produced and not produced within sameness and difference. It is given meaning through those structures (as they are the only ones we have), yet there is no sense of an identity from which bisexual subjects might position themselves. Writing as a bisexual would seem to be a contradiction in terms. Yet paradoxically this 'writing of oneself' is one of the things that marks out contemporary theorising about bisexuality, or bisexual theorising. There is a tension between the bisexual self one knows oneself to be at a given time, and the positive desire not to label bisexuality as one particular set of desires, choices or behaviours.

More work needs to be done on examining the differences between bisexual as an ontological category and bisexuality as an empirical category.[20] The differences between sexuality as a set of acts and sexuality as identity is, of course, a central issue here. This has been discussed in relation to homosexuality, but not in relation to bisexuality. The fact that bisexuality has not been pathologised as a sexual identity *per se* may be one reason for the contemporary claims that bisexuality *does not exist*. It is still considered (problematically) as a set of acts. Yet there is a danger that in claiming an identity *per se*, bisexuals will be categorised and contained in a similar way to homosexuals at the end of the nineteenth century. Does recognition of other sexual subjectivities outside of homosexuality and heterosexuality necessitate the assumption of a particular identity? Yet if I reject the notion of 'identity', I cannot ignore my desire to articulate positions from which bisexuality might be theorised.

Perhaps a way of ensuring against (i) the privileging of a specific bisexual identity, or (ii) the privileging of difference for its own sake in the search for methodologies and homes, is to emphasise the relationships between particular locations at particular times (e.g. lesbian–bisexual; bisexual–bisexual, *ad infinitum*). In this way different bisexual acts or subjectivities might be theorised in conjunction, not as if in a vacuum. Perhaps we might try and understand location in terms of the ways in which people's individuality is formed through power (so that we are both unique and similar to others): a move towards a politics of location that actually does take into account the relationships *between* individuals.

What particularly interests me is how individuals makes sense of their own locations. For example I would say that I am closer to a lesbian feminist than to a male bisexual 'swinger' in many cases, yet at times I might ally myself with that swinger in response to biphobia from lesbian and gay communities. I am simultaneously located in terms of class, 'race', education and age. Hence I am able to speak in less dangerous places, within the academy for example, where the risks of declaring oneself bisexual and feminist are minimised. Postmodernism has, of course, addressed these issues in terms of 'specificity', but it can still feel terribly lonely. The difficulty is whether one can form any sense of belonging on the basis of temporary identifications and alliances. The burning question is how one can become a *subject of dislocation* that is able to recognise other such subjects.

The problem may also be the way forward, may be the impetus to explore new ways of theorising not just bisexuality, but all forms of sexual location. Maybe reading the personal is about finding new ways to talk about yourself, re-examining the relationship between insider and outsider status. The problems are worked with not before engaging or writing, but *in the process of writing*. Becoming a subject of dislocation is a two-fold enterprise. Firstly, it involves the use of the personal – the bisexual – in highlighting the difficulties of existing structures. Secondly, it may lie in reading the contradictions within oneself, as well as within 'the world'. To read oneself may be to read culture, from within.

NOTES

1 Interestingly, there has been more recent work on bisexuality in women than in men. Perhaps this is because of the shifts in feminist debate that I outline in this paper (Daumer 1992; Cantarella 1992; George 1993; Hemmings 1993; Kaahumanu and Hutchins 1991; Weise 1992).

2 French feminists such as Cixous and Irigaray have used the metaphor of bisexuality as 'possibility', as a way of breaking through existing binary oppositions, but bisexuality as a viable political and sexual identity is not taken on board (Irigaray 1981, 1985a; Cixous 1975). In the work of Hélène Cixous, for example, bisexuality is conceived of as a 'bridge', as a way of connecting heterosexuality and homosexuality, but not always as an active identity or location in itself (Cixous 1975: 63–130).

3 David Bell and Ann Kaloski have both recently discussed the idea or possibility of a 'bisexual home', Bell in relation to lesbian and gay identities, and Kaloski more specifically in relation to lesbian/feminism (Bell 1993a; Kaloski 1994).

4 For overviews of the theories espoused by the US feminists of the late 1960s/early 1970s, see Eisenstein (1980); Gallop (1992); Tong (1989).

5 The body of work by lesbians who are aiming to broaden the category 'lesbian', or challenge the range of its meanings is really quite vast. Those texts that come to mind most strongly and in no particular order are: Lorde (1984); Pratt (1984); Butler (1990, 1993); de Lauretis (1991); Fuss (1991); Bristow and Wilson (1993); Boffin and Gupta (1990). Writings by lesbian sadomasochists such as Pat Califia (1988, 1993a), within the context of the feminist debates around SM and pornography, have also challenged the notion of a univocal feminist or lesbian identity. Califia's short stories highlight differences between lesbians, by looking at SM erotics, desire and fantasy. As yet, not a great deal of feminist work exploring the boundaries of heterosexuality has been published. Useful exceptions are: Hamblin (1983); Hollway (1983); Kitzinger and Wilkinson (1993); Valverde (1985).

6 I use the words 'black feminists' here, although in many respects I prefer the American words 'women of color', which has not yet become current in British theory or politics. The term 'women of color' perhaps denotes the diversity of women critiquing 'white feminism' better than the British equivalent 'black feminists'. The term 'black' is a disputed one that, for example, Asian women do not always find adequate to their experiences. As a British

woman I have decided to use the British term, bearing in mind that it is highly problematic.

7 See, for example, hooks (1984, 1989). White/Anglo feminists who have addressed this issue are Spelman (1990) and Ware (1993), who begins to problematise the term 'whiteness'.

8 Many articles have been written on sadomasochism, pornography, lesbianism and feminism in the last ten years, some good, some appallingly bad. One recurrent problem within these (seemingly endless) debates is the way in which a number of different debates have been grafted onto one another. Hence, to be pro-sadomasochism – in principle or in practice – is to be pro-pornography; to be anti-pornography is to be anti-sadomasochism. The do's and don'ts of feminism and lesbianism are usually fought within this mutually reinforcing and exclusive terrain. Exceptions to this are: Ardill and O'Sullivan (1986); Echols (1985); Eisenstein (1984); Ferguson *et al.* (1984).

9 By 'crisis' I do not mean to imply that such a feminist impasse is irredeemable. Rather a crisis could be seen as a point in time (or in theory) where contradictions can no longer be hidden, or where tensions between ideas or actions can no longer remain an undercurrent. Crisis can be defined as 'a crucial stage or turning point in the course of anything', *or* 'a time of extreme trouble or danger'. In the context of this introduction (and within feminism at this time), I would say that the *two* meanings of crisis are present, and that the crisis must be analysed in order to facilitate the first rather than that the second meaning holds sway.

10 If, as radical feminists have argued, women are prevented from realising their lesbian

potential by patriarchal conditioning which oppresses them, then heterosexual and bisexual feminists have not managed to rid themselves of that conditioning (yet) – in other words, they are suffering from false consciousness.

11 This is not to suggest that postmodern theories do not include an analysis of power, but that it is easy to interpret such theories as to do with individuals and nothing more.

12 Caroline Ramazanoglu takes up the issue of the feminist 'impasse' in *Feminism and the Contradictions of Oppression* (1989).

13 Julia Creet argues that feminism has reversed the symbolic order, but has not changed the actual structure, by creating instead the law of the Mother. She says that the maternal law works *in tension* with the law of the Father, creating 'a new set of structures which function symbolically although, unlike the law of the Father, they cannot be accurately located in a legal, institutional, or other discursive system' (Creet 1991: 145–6).

14 For example, 'Let's Talk About Sex', a Nottingham day-school 'for lesbians to talk about sex and safer sex issues' in May 1993, included an SM workshop with a note that stated, 'the SM workshop is for SM lesbians only, and is not a political discussion meeting. All lesbians have a right to non-judgmental safer sex advice'. Bisexuality, or bisexual behaviour, was not mentioned at all in the advertising. It appears that all lesbians except those that have sex with bisexuals, or those that have sex with men (yes, some lesbians do) have a right to 'non-judgemental safer sex advice'.

15 The idea that relationships with opposite-sex partners are a way of avoiding the responsibility of relationships with same-sex partners is one that also surrounds men's bisexuality. See, for example, the portrayal of the indecisive bisexual man, who is finally able to commit himself to his male lover at the end of the narrative in *Torch Song Trilogy*.

16 Recent articles that present this view of bisexuality are, notoriously, Ara Wilson, 'Just add water: searching for a bisexual politics' (1992), and Elizabeth Wilson, 'Is transgression transgressive?' (1993). Both present bisexuality as a wishy-washy version of lesbian or gay identities.

17 Simply being 'outside' of a particular identity does not necessarily mean that that position is oppressive. For example, black and white women could be said to be 'outside' one another's experiences, yet, as Maria C. Lugones and Elizabeth Spelman (1984) argue, those positions are not equal. Black women actually have extensive knowledge about white women and their communities as they have been exposed to white education, theories, lifestyles, yet white women do not automatically have that knowledge about black women's lives.

18 It might be possible to argue that the meanings of bisexuality articulated by nineteenth-century sexologists and psychoanalysts serve as 'foundational categories' to kick against. I would argue, however, that such categories do not function in their own right, but as a 'fall-out zone' for those cases that cannot be understood as heterosexual or homosexual. A bisexual *identity* is not a possibility, unlike a homosexual identity – however pathologised.

I must add that it would of course be wrong to suggest that theorists have not looked at other ways of understanding sameness/

difference relationships, while not necessarily speaking of a bisexual subject. Melanie Klein (1988), for example, develops her concept of the mother's good and bad breast, that the child has ambivalent feelings towards before differentiation through the oedipus complex is said to occur. Hence pre-oedipal sameness is challenged. Jessica Benjamin's work on intersubjectivity foregrounds the need for the differentiated subject to acknowledge others as subjects in their own right, again challenging the assumption that a subject needs its other to survive (Benjamin 1980, 1986). Yet notably these theorists are hardly part of the dominant canon of psychoanalytic or feminist criticism.

19 I am drawing on Ann Kaloski's use of Gloria Anzaldua's term in *Borderlands/La Frontera: The new mestiza* (1987) (Kaloski 1994).

20 Thanks to Derek McKiernan (Trinity and All Saints' College, University of Leeds), for suggesting this difference to me.

4

OF MOFFIES[1], KAFFIRS[2] AND PERVERTS

male homosexuality and the discourse of moral

order in the apartheid state

•

Glen Elder

Discourses of sexuality in South Africa were central to the creation, support and final collapse of the apartheid state. Sexuality, unlike other state-regulated social relations that characterised South African life between 1948 and 1991, transcended the public and private spaces of life. The control of sexuality was accordingly an important (although seldom noted) tool of the apartheid government. One of the ways in which such control was exercised was through the public debate surrounding sexuality that sought to codify and shape the private actions of individual South Africans.

Overt efforts to control sexuality by apartheid legislators included the legal enforcement of a racially-based sexual segregation. The *Immorality Act (1957)* prohibited 'carnal intercourse between white and coloured persons'.[3] However, it was not only legal recourse within the public courts that the authorities drew on for support. A well-established masculine order in South Africa also underpinned the smooth operation of the regime. Within recent accounts, the way in which a 'patchwork of patriarchies' (Bozzoli 1983: 3) shaped the evolution of South African society, and particularly the role of women within that evolutionary process, has received attention. Unclear at this point is the extent to which the apartheid state regulated and constructed masculine identity through sexuality. The intention of this chapter is to examine the way in which the construction of one aspect of male sexuality in South Africa was subject to an 'apartheid-style' regulation: the way in which male homosexual activity in particular was publicly articulated and acted upon by the state, will form the focus of this discussion.

LESBIAN AND GAY STUDIES: PINK SHADES ON A POST-COLONIAL GAZE?

An emergent 'gay culture' in South Africa has resulted in the development of a literature around the topic of homosexuality. For a variety of historical reasons which are not unrelated to the history of apartheid itself, it is gay white men who have framed the debate thus far. The structural marginalisation of black South African and lesbian voices has created a rift in the literature between the constructions of so-called 'situational male homosexuality' and 'gay male identity'. The 'situational homosexuality' refers to the documented sexual encounters between (otherwise) straight males living in migrant worker hostels, prisons and military barracks. Work illuminating gay male identity, on the other hand, has found expression in a growing popular and academic literature which focuses on 'coming out' in South African society, the increasing numbers of openly gay venues, pride parades, heightened levels of public visibility, as well as some significant political gains in South Africa's interim constitution (see for example Isaacs and McKendrick 1992).

Internationally, the emergence of a literature focusing on the social development or construction of sexualities has with few exceptions examined these questions within the Anglo-industrialised nations of the world. Within geographical studies in particular, questions of sexuality have without exception emerged from the United Kingdom and North America (Weightman 1980; Castells and Murphy 1982; McNee 1984; Lauria and Knopp 1985; Holcomb 1986; Knopp 1990b; Bell 1991; Davis 1991; Jackson 1991; Adler and Brenner 1992). The social relations of the South African spatial economy shaped by apartheid pose a challenge to the prevailing geographies of sexuality. How are sexualities constructed and negotiated in peripheral or semi-peripheral economies, and how do these spatial processes feed into the emergence of, amongst other things, 'gay and lesbian culture' in 'First World' settings?

In a similar vein of thought, Fuss (1991) argues that certain ideas and understandings can only be articulated through what she calls an 'indispensable interior exclusion', by which she means binary opposites like masculine/feminine or heterosexual/homosexual. By extending Fuss' point concerning the interdependence of meaning through a geographical imagination, the meanings of sexuality in different global locales emerge. Thus, the construction of sexuality becomes part of a global process of local sexual discourse, whereby one meaning becomes defined in terms of another. For our purposes this questions the extent to which the emergence of studies of 'essential' sexualities, and 'lesbian and gay' culture in general occurred through a silencing of 'other' homosexual experiences around the globe.

Despite insightful advances made in the vibrant (albeit marginal) literature of sexuality in South Africa, there has unfortunately emerged a tendency to conceptualise questions within sexuality through essentialist frameworks, imported from a prevailing sexuality literature in the United Kingdom and the United States. Accordingly, the essentialist 'gay/straight' divide which characterises most accounts of sexuality internationally emerges in the South African context too. It is the divide in turn which has led to the emergence of a bifurcated understanding of male homosexuality in South Africa and 'othered' homosexual experiences that do not fit the 'lesbian and gay identity' model.

As South Africa moves towards a more open society, the challenges of the moment are painfully clear. Perhaps most clear of all is that apartheid was not only a policy of racial discrimination. More than a racial order, apartheid was also an essentialising process of state control and regulation of daily life. Women and men, black and white South Africans all felt the consequences and reaped the benefits of apartheid in significantly different ways. These differential experiences were not unforeseen consequences. Rather, they were painstakingly laid out and codified in the apartheid statutes through essentialised understandings of identity. To simply embrace these understandings as part of our analysis does little to further the understanding of apartheid. Similarly, as geographies of sexuality emerge, attention must focus on sexual relations outside of traditional 'First World' settings, and in so doing seek to move beyond essential categorisations informed by Anglo-industrialised experience.

Sexuality and discourses surrounding it are controlled in distinctly different ways. Sexual encounters between men under apartheid were differentially articulated and interpreted depending upon who they were, where they acted out their sexual intimacy and where it came under public scrutiny. The differential consequences of homosexual activity, depending on racial classification and class position amongst other things, has made itself felt in the present imposed essential notions that inform current sexuality studies in South Africa. To break down the division is to move the terms of the debate away from a 'ghettoisation' of gay issues towards a more integrated analysis which sees discrimination against same-sex encounters as part of the discrimination of apartheid. Also, an anti-essentialist understanding will inform current debates within the geography of sexuality. What follows, then, is an investigation of two instances that clearly demonstrate the somewhat fascinating attempt to define an essential male homosexuality in South Africa along racial lines.

BLACK MALE HOMOSEXUALITY AND THE MINED COMPOUND SYSTEM

On the mines there were compounds which consisted of houses, each of which had a *xibonda*[4] inside. The *xibonda*'s job was as head of the living quarters, he had authority and was known as a counsellor. Each of these *xibondas* would propose a boy for himself, not only for the sake of washing his dishes, because in the evening the boy would have to go and join the *xibonda* on his bed. In that way he had become a wife. He, the husband, would double his join on the mines because of his boy. He would make love with him. The husband would penetrate his manhood between the boy's thighs. You would find a man buying a bicycle for his boy. He would buy him many pairs of trousers, shirts and many blankets. Eventually the young miners would go back home to their parents or wives with many things, after having been substitute wives on the mines. The old ones did this because by experience, they knew that they were not allowed to go and have fun outside the compounds.

Wa Sibuyi 1993: 54

The existence of homosexuality within the mine compounds of southern Africa was openly acknowledged by employers and employees alike. The preceding quotation serves to illustrate the extent to which relationships between men in the mines were institutionalised and formed part of hostel life. It is argued by Moodie (1983, 1986, 1988) in particular that a material basis shaped the nature of these liaisons. The demand for domestic service within hostels, as well as the lack of women, resulted in some of the homosexual activity that ensued. Also, the activity that was precipitated by these circumstances did very little to challenge the existing social order in the mines. In fact, homosexual activity was seen by many as a 'necessary evil' in order to sustain the highly exploitative relationships that existed in the form of migrant labour.

Concerns about the long-term consequences of these relationships did filter through into liberal accounts that advocated the abolition of the migrant labour system:

In their plight to satisfy their sexual needs, [migrants] indulge in terrible practices such as homosexualism [*sic*] which is an outside practice and is now beginning to reach broader extents. Young men reaching the mines get involved in this practice. There are even men, (I am not exaggerating), who move around the compounds and their sole business is to entice men in the compound to sleep with them. Some men would even divorce their wives afterwards because of this practice that has become important in their lives. All kinds of atrocious vices take place in these hostels such as sodomy and the like. A close investigation of this problem would unearth quite a number of vices which are unknown to the public but common talk to the inmates of a hostel.

Mohlabe 1970: quoted in Wilson 1972: 114

Despite claims by liberals and religious groups alike, little action was undertaken by the state or mining authorities to discourage the practice. Not even seedy descriptions of male prostitution and sodomy moved those who controlled the apartheid order to take action.

The existence of a black male sexuality that was at once threatening but also necessary in urban areas was constructed and controlled within the spaces of the hostel. The debates that ensued came from the liberal establishment, and more often than not were totally ignored by the apartheid architects. Homosexual activity and the discourse that surrounded it helped to contain the threat of unbridled black male sexuality. The hostel space provided the all-important spatial context in which this activity took place, and therefore served to shape the discourse that surrounded the issue of black male homosexuality.

A control of black male sexuality that was contained within the hostel system served to contain the homoerotic threat and grudging respect that mine officials and the apartheid state held for the strength and power of black labour. The closed system of mining life in South Africa saw the emergence of sprawling white family residential estates alongside hostels accommodating thousands of black male workers. The presence of white miners' wives, daughters and mothers alongside thousands of black men who had no sexual outlet not only poignantly demonstrates some of the internal contra-dictions that racked racial capitalism, but also the root of homosexual tolerance.

By tolerating and at points encouraging homosexual encounters between men within the mine compound and hostel system more generally, a public discourse emerged that served to contain the threat that a perceived black sexuality and virulence posed. By constructing black male sexuality in this way, mining houses and the apartheid state also managed to contain growing demands on the part of workers to bring their wives and children to the mines on a permanent basis, thereby increasing the pressure for higher wages. Obviously loving sustained relationships between men did develop within the ugly edifices of racial capitalism like the mine compounds. In turn lovers' beds became niches of resistance against the alienation that black mining life presented. These actual encounters, however, did not detract from the way in which black homosexuality, or the discourse that surrounded it, was used to quell white male fears and contain the threat that the occupants of worker compounds presented.

Further instances of homosexual activity abound within other structural edifices of apartheid. Obviously the prison system has produced its own series of accounts concerning homosexual encounters, as has the all-white South African defence force. In both instances, however, there is a suggestion of violence and sexual assault which in fact distances these events from the hostel homosexual relationships. The highly racialised and sexual way in which the apartheid state sought to regulate the public discussion of private acts between men, however, was also well demonstrated through the way in which the apartheid and post-apartheid courts have sought to frame discussions about male homosexuality. Interestingly, a bulk of the discussion in this realm has revolved around the control and regulation of white male homosexuality exclusively.

WHITE MALE HOMOSEXUALITY AND THE LEGAL SYSTEM

Homo sex is not in black culture.[5]

The preceding quotation was taken from a protestor's placard outside the Johannesburg Supreme Court. The demonstrator sought to give support to Winnie Mandela who was in the process of being tried for the kidnapping of young Stompie Moeketsi Seipei. During the trial of the African National Congress matriarch and three others on kidnapping and assault charges, a case was built by the defence attorney around Mrs Mandela's protection of the youths against the alleged homosexual molestation of the young men by a white Methodist minister. As the court case unfolded it appeared that an assumed level of societal homophobia was being used by the defence council to clear Mandela's involvement in the kidnapping and ultimate murder of the youths. As local gay activist Simon Nkoli said at the time: 'linking homosexuality to sexual abuse is as ludicrous as equating heterosexuality and rape'.[6] Despite these objections, the theme of the defence's case continued. Winnie Mandela was eventually convicted on charges of kidnapping and assault. On appeal the assault charge was overturned, but the charge of being an accessory to kidnapping Stompie from the Methodist manse was upheld.

The outcome of the trial is not what interests us at this point. Rather the prevailing discourse in and around male homosexuality as used in South Africa's legal system – a context that permitted a level of homophobia to prevail – shapes the discussion below. Historically the South African legal system has not exercised sympathy to the cause of homosexual rights. In examining the historical accounts as reflected in law, parliamentary and legal practice, it is only the control of male homosexuality that is discussed and debated. Furthermore, it is also evinced that the laws, and legal precedents, involved only homosexual activities between white males. Although the Mandela trial involved the kidnap and murder of a black youth, the bigoted remarks were levelled at the white minister accused of child molestation and homosexuality – two terms that were used interchangeably. Within this context, disclaiming the existence of homosexuality in black South African culture by the protestor outside the Supreme Court is understandable.

Based upon Roman-Dutch and English law, South African common law legislates against so-called 'unnatural offences'. Given the developments in English law around the mid-1960s, that saw the decriminalising of private consensual sex between men, it is noteworthy that at the same time the apartheid regime set about investigating and criminalising the same sexual acts (Hunt and Milton 1982). The extent to which these actions were an attempt by the apartheid state to defiantly define its own moral code, and thus distance itself from the previous coloniser, is open to debate. What is clear though is that towards the latter half of the 1960s an unprecedented amount of attention was focused on the issue of white male homosexuality. A police raid on a party in one of

Johannesburg's more salubrious white suburbs in 1968 precipitated a request from the all-white parliament to investigate the issue of male homosexuality.

In the opening paragraph of the parliamentary report,[7] the committee chairman ironically stated that between July 1966 and June 1967 there were fifteen cases of sodomy before the court involving white men, while 147 cases involving sodomy between black men came before the courts during the same time period. Despite the disparity, the report went on to document only one very particular kind of homosexuality. In a revealing paragraph, a police major stated:

> Although the South African Police has [sic] dealt with various forms of homosexuality over the years, the circumstances were such that it was regarded as isolated and not really constituting a threat to the moral basis of the populace. The seriousness of the situation came pertinently to the notice when in January, 1968, a police raid was carried out on a double-story residence in the suburb of Forest Town, Johannesburg.[8]

The quotation reveals the kind of bias that later informed much of the legislation against homosexual activities between South African men. Whereas a history of black male homosexuality was openly acknowledged in compounds and elsewhere, it was white middle- and upper middle-class homosexual encounters that came under extreme scrutiny by the state. Despite a debate around whether homosexual men were in fact sexually gratified or satisfied,[9] the threat to 'the moral basis of the populace' eventually precipitated an amendment to the Immorality Act. The enactment guaranteed criminal consequences for any 'male person who commits with another male person at a party any act which is calculated to stimulate sexual passion or to give sexual gratification'.[10] Beyond the bizarre nature of this amendment, as noted by Cameron (1993),[11] there is also an intended race and class bias to the amended legislation which also demonstrated the area of the state's concern.

The reasons for the clear bias that informed much of the debate and legal precedent around male homosexuality lay in an attempt by the state to contain a perceived threat. Unlike the spatially-containable threat of black masculinity in the mines, white male homosexuality threatened the very existence of a patriarchal apartheid system. The predominantly white masculine parliament and legal system of apartheid, which had sought to categorise (and in postmodern terminology 'other') the South African population as a form of control found itself under threat. The idea of white male homosexuality in turn objectified the apartheid architects and practitioners. The idea sent panic down through the trenches of an unassailable order. The panic-stricken commission and parliament set about trying to understand and contain the 'transferable' condition of homosexuality between white men. A conspiratorial intent to overthrow the 'moral order' of apartheid was attributed to homosexual men. In an attempt to control and contain the assault, amongst other characteristics, the following were noted:

> Homosexuals have no difficulty in identifying one another and know precisely how to approach one another when they find themselves in a strange area ... Queer[s] ... sometimes have sham marriages. The marriage, however, is never consummated and each of the parties goes his or her own way ... The older members of the queers derive pleasure in getting an attractive young man dressed as a female. The latter then performs a vulgar 'strip tease', this satisfying the onlookers sexually ... A queer is 'just ripe' for homosexuality from the age of 18 years. His 'life span' is approximately to the age of 30. After that he is 'over his youth'. He still practices it thereafter but he is introverted and he has acquired a mate and they are satisfied together ... The facts embodied herein were obtained by discussing the matter with queers, as well as from persons who associate with the latter without practicing the cult.[12]

The clandestine and dishonest disposition of male homosexuals, as outlined above, served to strengthen the need for strict legislation. The metamorphosis of a homosexual also revealed an effort on the part of the state to create a model in which to locate an undefinable threat to white masculinity. It is clear that the act of sexual intimacy between men, in itself, was not sufficient to sway opinion. Instead, the activity was constructed within a 'twilight zone of lust' and thereby marginalised. Such efforts once again served to contain a growing threat in the middle-class white suburbs of residential South Africa.

YOUR 'PLACE' OR MINE?

From a brief and pointed history of homosexual activity between men in South Africa, it is clear that the state's response has been at best inconsistent. A legacy of written documents and spoken memories assists in revealing the invisible past of South African male homosexuality. From these accounts it is clear that, within the public political life of South Africa, the act of male homosexuality was discussed with varying degrees of intensity, and within several fora. More important than the homosexual act, however, was the 'race' of the perpetrators and the locale in which their intimate encounter took place. The question of 'race' and place served to shape the state's response and that of the public in general.

Apartheid's control and regulation was piloted by an overwhelmingly male-dominated, Calvinist-inspired order. The case of black male workers living inside hostels reveals the absolute need to control and define those caught up in life under apartheid.

What followed then was a subtle shaping of the public's understandings of life within the hostel. In so doing, the state and mining authorities managed to assuage fears that challenged the smooth operation of apartheid. In defining the hostel as a veritable Sodom of mining life, the idea of sex occurring between black men somewhat lubricated the advance of mining-related capital accumulation.

The challenge that white middle-class men practising homosexuality posed to the apartheid state was their seemingly undefinable character. Unlike black miners who were contained and controlled within compounds, white male homosexuals had no outward signifiers, and thus could invisibly infiltrate the comforts of white middle-class suburbia. Accordingly much effort on the part of the state went into the defining and studying the homosexual 'cult'. As late as 1987, an attempt by the state to grasp the nature of homosexuality was once again revealed in a President's Council Report that tacitly accepted the innate character of homosexuality.[13] Unfortunately the fact that the discussion fell under a section on promiscuity, and is listed along with extra-marital intercourse and prostitution as a threat to the white masculine youth of the country once again located the practice outside of the 'natural moral order'.

CONCLUSION

Homosexual acts between men have occurred throughout South Africa's apartheid past. The way in which they have been interpreted and acted upon by the state, however, has remained hidden through an essentialist framework that sought to separate the so-called 'situational homosexuality' from a wider practice which has more recently been called 'gay identity'. It is clear that the 'race' of men caught up in the act, as well as the spatial confines in which they chose to become intimate, shaped the South African state's response.

Understandings of both acts as reflected in the varied public responses all concurred on the idea that homosexuality was in fact a transferable condition. Examples show how migrant workers left their wives permanently for deceptive and beguiling male prostitutes, while 'butterfly-like' young white men 'ripened' into full-grown homosexuals, after performing the dance of the seven veils for lecherous old men. The possibility of an innate compulsion as part of a masculine sexuality was never considered, for fear of the long-term implications this might hold for the future of the male-dominated apartheid order. The lessons which we can draw from the control and discourse of sexuality under apartheid go beyond a dismantling of South Africa's past. Rather, these instances evince the shortcomings of an essentialist understanding of sexuality, as informed and created by experience in 'other' contexts like the Anglo-industrialised nations of the 'First World'.

ACKNOWLEDGEMENTS

My grateful thanks to Gordon Pirie and Edwin Cameron for their insight, to Tim Davis and Larry Knopp for the inspiration to write this piece, and to Lydia Savage for helpful comments and editing.

NOTES

1 Derogatory Afrikaans name used for male homosexuals.

2 Derogatory name used for black South Africans during apartheid, stemming from a colonial term referring to a non-believer.

3 Republic of South Africa (1957) *The Immorality Act* No. 23(s1).

4 The *xibonda* was an elected member of each hostel dormitory who could take the inmates' grievances to management. One must note that the *xibonda* system was also used by management to find out what was happening in every hostel dormitory.

5 *Weekly Mail*, Johannesburg, 15 March 1991.

6 Simon Nkoli, Chairperson of Gays and Lesbians of the Witwatersrand. Reported in *Weekly Mail*, ibid.

7 Thanks to Professor Edwin Cameron of the Centre for Applied Legal Studies, University of the Witwatersrand, Johannesburg for making me aware of the report and for referring me to: Retief, G. (1994) 'Keeping Sodom out of the Laager: State repression of homosexuality in apartheid South Africa', in E. Cameron and M. Gevisser (eds), *Defiant Desire*, London: Routledge.

8 Republic of South Africa (1968) *Report of the Select Committee on the Immorality Amendment Bill*, p. 9.

9 Republic of South Africa (1968) *Report of the Select Committee on the Immorality Amendment Bill* [S.C.7 – '68.]

10 Republic of South Africa (1975) *Sexual Offenses Act* No. 23, as amended by *Immorality Amendment Act* No. 57 of 1969 (ss20).

11 Referring to S V C 1987 2 SA 76 (W) 81I–J, Cameron states: 'The critical jurisprudence this provision has evoked includes a solemn decision by two judges of the Supreme Court that "a party" was not constituted when a police major, visiting a well-known gay sauna in Johannesburg for entrapment purposes, barged in on a cubicle where two men were engaging in a sexual act and turned the light on. The court held no doubt, properly and fairly that the two men's jumping apart when the major switched on the light prevented a "party" from being constituted' (p. 34).

12 Republic of South Africa (1968) *Report of the Select Committee on the Immorality Amendment Bill*, p. 14.

13 Republic of South Africa (1987) *Report of the Committee for Social Affairs of the President's Council on the Youth of South Africa*.

5

FEMME ON THE STREETS, BUTCH IN THE SHEETS

(a play on whores)

•

Alison Murray

Being both a lesbian and a sex worker involves conflicts and convergences, and acting differently in different situations. Sex work as a job is still generally stigmatised (including by many lesbians). At the same time a lot of women who aren't lesbians act like they are for the price of a male fantasy. Working and pimping have their place in working-class butch-femme subcultures, although they were denied and drowned by some streams of 1970s feminism. Meanwhile, lesbian and sadomasochistic (SM) chic shows that you can still make up the rules of the game as you go along.

PRO-LOG

this knowledge makes me dirty.
pro found knowledge.
Fallon 1989: 201

Scene: Sydney

You are visiting Dick Loony in a motel room (he says his name is not pronounced like that but you just laugh). You are wearing a short dress and lacy underwear. He asks what a nice girl like you is doing in a job like this. He asks what time you knock off and if you want to meet him after work. He asks if he can fuck you up the arse. You ask him if he is gay. He is insulted and doesn't pursue the issue.

Scene: Jakarta

You are drinking in a bar with a friend. You are wearing jeans and a leather jacket. One of the bar girls warns your friend not to talk to you because you are likely to start cracking a few heads. When the bar closes one of the bar girls invites you home. She wants to fuck in the bed where her husband is sleeping, but you refuse so she takes you upstairs. In the morning she gives you money for a taxi fare even though your motorbike is outside.

Scene: Queensland

You are shooting up junk in a bathroom in a tropical town where you came to dry out. Your girlfriend's mother is watching soap operas on TV. She didn't say anything two hours ago when you and your girlfriend got dressed up in black dresses and stilettoes, leather and studs and went out. You performed a humiliation scene in a motel room for a client who just sat in the corner watching: he was a businessman and had brought all his own equipment with him in a special suitcase.

As Valentine (1993b) notes, lesbians develop strategies of time-space and appearance to adapt to different contexts, such as home, a lesbian bar or public (ambient heterosexual and implicitly homophobic) space. Lesbians establish their identity through their images and how they perform them (cf. Butler 1990), both for themselves and for a changing audience. An audience can use these performances to make rigid categorisations of exclusion and inclusion, making stereotypes something to avoid: 'I had these fears that I might have to have cropped hair, an earring through my nose and wear a pair of army boots. And so now I think it's laughable' (quoted in Valentine 1993b: 240). Or, stereotyped images can be exploited, played with and used to confront: you can use an image to look the part (of a dyke), but increasingly you could be straight, or a sex worker, or an academic, or all of these, or something else – and still wear cropped hair and army boots. Lipstick lesbians of the 1990s have adopted a hyperfeminine image which makes it harder for homophobes to point out lezzos to each other, but as Bell *et al.* argue (1994: 42), the image is not designed to destabilise heterosexuality but to confront stereotypical lesbians. It is 'a political backlash against the ideological rigidity of lesbian feminism and androgynous style'.

While dykes can choose to adapt to or resist both the mainstream and each other, dykes who are also sex workers are more likely than most to make changes in their performance according to the space they are in (the client's space, the girls' room, the street, the dyke bar, the prison). Being a sex worker is first and foremost an act, and this audience is paying. Changing after the show makes a personal distinction between work and not-work, and is a strategy to avoid labelling and stigma.

Joan Nestle has written about the 'historical sisterhood' of lesbians and sex workers, where both have a similar experience of being subjected to surveillence, official and unofficial policing. Both know a history of fear, loss and hiding, and talking about 'going straight': 'whore and queer are the two accusations that symbolize lost womanhood – and a lost woman is open to direct control by the state' (Nestle 1992: 245, see also 1987b). Labels of deviance like 'lesbian' and 'prostitute' are used to keep 'good girls' in line, but it's not only the state who controls – lesbians do it to themselves:

> It would be great to think that lesbians have gone beyond the tired old stereotypes of sex worker as sad (junkie/victim), bad (immoral nympho slut) or mad (acting out unresolved childhood abuse), but unfortunately this is not so.
>
> O'Sullivan 1994: 40

ACT ONE: THE ACT

> sexuality is onanism
> playing for keeps
> fellating for advancement & profit
> Fallon 1989: 208

Sex work encompasses many acts and people, and the 'sex' (as work or labour) which is exchanged for money is not always about male penetration, 'high-risk activity' or the invasion of a passive female body. Workers who perform 'straight sex' have their own strategies to minimise the amount of actual fucking that goes on, while clients are limited by their sexual vocabulary for the 'pleasure' that they pay for: this means there's always the potential for talking them into something else.

The traditional menu of the western sex industry includes lesbian performances and 'doubles', which Pheterson (1989: 155) explains: ' "Trio" is the Dutch prostitute's term for two working women and one male customer. In the States we call it a "double" because we don't count the customer.' Plenty of female sex workers who don't identify themselves as lesbians are fucking each other for money, and plenty do it for fun. You don't have to tell male clients about your personal life. They· are usually happy with whatever story you think they might like to hear, and especially happy if you tell them you're a dyke – it makes them feel like they've made a sexual conquest. They might want you to bring a girlfriend along, too. They are just as confused as some feminists about the differences between *sex as work* and *sexuality as identity*:

we used to do shows together and we thought this was hysterical, because we were both stone butches, so we would never have anything to do with each other sexually; but we would just put on these great shows that we'd make a lot of money for – and I can just remember laughing in her cunt while all these guys would think we were sexually excited.

> Doris Lunden, interviewed in Nestle 1987b: 114[1]

Lesbians are also involved in the sex industry as owners, managers and a small but growing number of clients (such as wealthy women paying for an escort service in Sydney, or the commercial lesbian scene in Jakarta: see Sunindyo and Sabaroedin 1989). It's not only men who can make money out of women, and it seems like only the matter of economics is stopping more women from paying for sex, or for whatever takes their fancy. That doesn't mean that lesbian workers would necessarily prefer female clients, as they can be a lot more demanding, but people still make assumptions that desire is part of the sex worker's job. In another area, Mathew (1988) criticises the assumption that most male prostitutes are gay. These assumptions ignore the complicated links and dissonances between sex and desire, identity and sexual practice: it's just as dense to assume that a self-identified lesbian would not get some kicks out of sex work or the power relations of commercial sexual practice:

> [M]any lesbians hate anything other than vanilla sex or mutual masturbation, they cannot handle diversity . . . Regardless I am proud to be a pervert and have always been attracted to the power play of dominance and submission and that is why I am in this profession . . .
>
> We rarely see female clients which is unfortunate, but on the whole gender is irrelevant. Just because we identify with the lesbian community doesn't mean we can't play with men.
>
> Castel 1994: 9

ACT TWO: THE SCENE

> if you see me with my girlfriend
> please don't say hello
> she's very jealous
> Fallon 1989: 201

The 'lesbian community' can seem a sorry splintered thing. Since the arrival of lesbian feminism and the sex wars, to some factions, '[w]hores, and women who looked like whores, became the enemy ... Lesbian prostitutes have suffered the totality of their two histories as deviant women – they have been called sinful, sick, unnatural and a social pollution. In the decade of lesbian feminism, they have not been labelled because they are invisible' (Nestle 1987b: 232, 243).

Lesbian communities are still fairly small, and experience would suggest that they include a disproportionate number of sex workers, but it's not something you usually hear about. Kimberly O'Sullivan (1994: 40) makes a plea for a lesbian sex worker float in the Sydney Mardi Gras parade, but given that she also mentions the 'collective dyke cold shoulder' for whores, it's hard to imagine many dykes getting up there without bags on their heads.

Being in the business of sex, workers are among the best informed people about sexual health and the most accepting of any kind of sexual practices, including those they would not choose to do personally. Coming out as a dyke to a bunch of workers is a lot easier than telling most dykes that you work. Disclosing about your sex work experience, for instance because you work as an HIV/AIDS peer educator, can be social suicide outside the industry and almost as bad as being out as an unrepentant junkie. Many sex workers have lovers who are also working, as this avoids having to deal with unsympathetic attitudes after a hard night.

Lesbians are divided and workers are stigmatised by those feminists who argue that the sex industry supports patriarchy. They emphasise coercion, sex slavery and child prostitution to stir an emotive response against the whole industry, and they reinforce the anti-pornography campaigns of Andrea Dworkin and Catherine Mackinnon, who can't see that there's nothing inherently wrong with the commodification of sex. Lesbian separatists with utopian fantasies can't come at the idea of any sex with men at all. Unfortunately, as Biddy Martin suggests (1992: 118), some academic studies have legitimated these ideas by institutionalising them. The academy has progressed from women's studies to gender to sexuality, getting closer to the cunt of the matter while continuing to marginalise class, race and alternative subject-voices.

Scene: a conference on 'feminism in the 1990s'

In a paper on sex work, the speaker has discovered in her research that some workers actually enjoy the job: 'a worker said to me that more women should realise they have a goldmine between their legs (but don't all go rushing out the door!)'. A complaint is heard from the floor: 'we're going back to the days of selling our looks and not all of us can do it!'. The speaker's response: 'Oh, you'd be surprised – there's even a demand for pregnant women, I know that sounds squeamish, but ...' (gasps and squeaks come from the floor).

ACT THREE: PIMPS AND CHICKS

I think there are a lot of lesbians who are really hostile to butch and femme because they have this incredibly literal reading of what butch and femme means, and they simply equate it with heterosexuality and with sort of mimicking heterosexuality, I think, so that's how they'd explain their hostility to butch and femme, I actually think there's sort of another layer of explanation, which is that I think a lot of lesbians are actually really scared of sex, and really scared of sexuality, and particularly lesbians who became lesbians through feminism and the rejection of men, actually get very freaked out at the idea of women being sexually powerful.

in Carr 1993

'Political' lesbians didn't believe there should be butch and femme, and so everyone was androgynous, which really meant everyone with their short hair and diesel dyke overalls ended up looking butch – but not the aesthetic butch of the 1950s and 1960s: 'Skillfully, seemingly carelessly, the butch fingers her pompadour and casual curl into place. Then, with a flourish, using a comb followed by the fingertips of her other hand, she creases the duck's ass down the middle of the back' (Mushroom, 'How the butch does it: 1959', in Nestle 1992: 134).

This subcultural style and what butch-femme stood for was an embarrassment to lesbian theorists of the 1970s:

Indeed, because it is a gender system, butch-femme has come to occupy the position of 'whore' relative to lesbian feminist 'marriage', not only in a literal sense, where the whore is the woman with whom sex is illegitimate and unspoken, but in a more symbolic one, wherein butch-femme, particularly because of its class and race associations, has become another manifestation of the 'whore stigma': that portrait of uncontrolled sexualness given groups deemed 'other' by a dominant culture.

MacCowan 1992: 327

The history of butch-femme is working class, or at least it is imagined and imaged as such – and the dykes in those scenes are not only symbolically 'whores' but are also doing sex work. The bottom line is economics and the class-gender system: you might dream of an allowance and a room of your own to write in, but as a working-class woman you have few choices about the type of work you can do, and you still need to pay the rent.

In Sydney, some of the old-time workers from Kings Cross will tell you how some butches were pimps, but many butches also had some experience of working themselves when they needed the money. The scene carries on among street and drug-injecting workers, and extends into the prison system which crushes women into the deviant stereotype of dyke/whore/junkie/criminal – whatever you were like or were put in jail for in the first place. In the 'community' of middle-class political and/or lifestyle 'scene' lesbians, these women are either ignored or fantasised:

> The typical stereotype is that the butch is a working class woman, she's a diesel dyke . . . the real pimp, gigolo type of butch woman, and the thing that attracts me to that image is not just the clothing or the image but the fantasy of like, the working class girl who's sort of like built herself up by pimping off the earnings of a femme.
>
> in Carr 1993

Scene: the Philippines

You are drinking in a karaoke bar with some butches who have formed a working-class lesbian group: 'We are not ready for femmes in the group yet. If they came in now there would just be a fight'. Two nights ago they trashed this bar, but tonight they are mellowing out. One leans over: 'There are some beautiful chicks in here, but you can't trust them. They will leave you for a customer'. A moral is being implied: you can take the girl out of the bar but you can't take the bar out of the girl.

Scene: Sydney

You are on outreach, giving out condoms and pamphlets, and talking about immigration raids. The Vietnamese brothel manager wears overalls and drives a black BMW. One of the Thai workers is her girlfriend; she has powdered her face whiter than white. The other workers are not lesbians and they giggle about doing lesbian doubles. One of them went out on a date with a dyke but she didn't know what to do: 'When we got home and she touched me I got goosebumps'.

Practices and identities vary so much that it's difficult and probably pointless to try and define who is a dyke and who is a worker. People enter and leave sex work and lesbian relationships, and may deny they ever existed. Butch-femme is just one set of possible images which can be adopted at particular times and places, places which can be found in Asia as well as the West. The cultural contexts vary, but like America in the 1950s the butch-femme scene may be currently the only choice available.

When they make that choice, these lesbians don't have their own space to perform in. People marginalised and stigmatised by rigid laws and social attitudes claim spaces in the city where they can, and they share them. Another article (Murray 1993) describes a night-time scene under a concrete flyover in Jakarta, where amplifiers are set up to blare the music of a gamelan band, and singers in traditional costume are also available for sex. The crowd includes sex workers, butches, transsexuals, unemployed youths and street fighters, and the police frequently clear them all away.

For alienated people to create an alternative scene, finding your own space helps. It might be alienated land such as under a bridge or a derelict building, or a temporary use of space like a street at night. It's inevitably an urban scene – not a lesbian utopia but the streets of the city. 'I stand knowing in my bones this city of tired workers. I have enough to cherish in just the courage of these days and nights. This is my land, my ancient totems, this tenacious grip on life' (Nestle 1987a: 15).

ACT FOUR: MASSIVE AMOUNTS OF RUBBER

> (thinking) so camp, so S&M, can't
> stand the heat but will she get out of
> the kitchen? Nope
> Fallon 1989: 138

It seems like all the dykes are now SM dykes: 'For self-respecting lesbian sadomasochists, the ante has been upped considerably in terms of what it might take to be a s/m lesbian' (B. Martin 1992: 99). Dykes of the 1990s like to play around with categories like gender, butch and femme, dominance and submission – playing with morality and the 'straightness' of many lesbians: 'the desire to practice the dangers of sex needs protection and the massive supply of rubber/lubrication/imagination' (Munster in Bashford *et al.* 1993: 27). At a lesbian SM wedding in Sydney, the mistress wears a wedding dress and has her slave on a chain. The slave wears an antique military uniform. As well as the shake-up of styles and stereotypes, the time of HIV/AIDS has also brought some convergences of interest between sex workers and lesbians, and has sat them both under the umbrella of a predominantly gay male AIDS bureaucracy. AIDS and the politics around the disease has brought the invisible and the deviant to the surface for inspection. HIV is transmitted through intimate exchanges of blood and cum – surveys have been done, and it is no longer a secret that lesbians sleep with men, inject drugs and play with a variety of sharp and blunt instruments.

Dykes are among the most politicised sex workers, especially around AIDS issues. As mentioned already, being a 'peer educator' for an HIV/AIDS agency can be a hassle when it means disclosing as a worker to people you don't know, when there is still a lot of stigma attached to working. It means changing how you perform your identity in different contexts: it can be a nakedly stressful experience to say your lines without the costume and the backdrop.

The choices people make about sexuality and sexual practice depend on what choices and attitudes are available at the time: the increasing range of queer sexualities and identities has widened the choice and blurred the boundaries. Prostitute performance artists have led the way in demolishing dyke antipathies and dichotomies like good and bad, abused and empowered (S. Bell 1993; Juno 1991). As SM has become more mainstream in the lesbian community, more dykes are working as mistresses. The sex industry has its own system of class and status distinctions, with some bondage and discipline (B&D) mistresses believing themselves above and beyond it, something which is acknowledged and admired in some lesbian scenes and texts. They are perceived as an elite who can choose not to have sex with their clients: they can piss and shit on men, grind their stilletoes into men's balls, and get paid for it.

Some 'political' lesbians argue that power games and genderfucking ignore the struggles of the past to make lesbians visible; but this can be seen as political in itself: dispersing power relations and breaking down the state's capacity to control by splitting up convenient categorisations of deviance (such as lesbian and sex worker). When the margins become illusory, the centre can no longer define itself:

> Straight is not heterosexual or gay. Straight crosses into both these worlds. Straight is the fact that a lesbian tells me I'm straight because I fucked a man. Because I fucked a man for money or sport or whatever, that makes me straight, does it? … I'm straight because I cream my pants when I see a woman shove, yeah I said shove, her black cock up another woman's arse who's loving it.
>
> Munster in Bashford *et al.* 1993: 11

Dyke whores are no longer double deviants, in some parts of the West at least. After being invisiblised by some feminisms, dyke whores have come out in a babble of trendy deviances, though the working-class junkie dyke whores are still invisiblised. There are new games to play, where the referee is not the only one with a whistle.

NOTE

1 Stone butches in the 1950s were so-called because they would pleasure their feminine lovers but would not be touched sexually themselves.

6

BODY WORK

heterosexual gender performances

in city workplaces

•

Linda McDowell

BALLS STREET

Talking of talking dirty brings us to the City, where journos vied to come up with the most demeaning examples of City Gent at work. Paul Delaire Staines, a 26-year-old with Yasuda Europe, explained: 'If a person is market-making, and they are asked to show too much of the size of their position, a common response would be, "What do you want me to do? Lift up my skirt and show you the lot?" Men doing well are known as "big swinging dicks", and women dealers are "honorary big swinging dicks"'.

Guardian Weekend 30 April 94: 2

This chapter is concerned with the links between power relations, heterosexuality, identity and the body in the workplace. As the quotation above illustrates, City workplaces and practices are saturated with heterosexist imagery and behaviour, demanding of women a physically impossible performance. Their embodiment as female clearly raises vexed issues for their recruitment to and progress within the City environment, if conformity to particular social practices is a condition of success. In an empirical investigation of changes in the social relations in merchant banks in the City of London during the years of rapid expansion in the 1980s, I began to explore the ways in which the language and practices in merchant banks conformed to this heterosexist stereotype of aggressive masculinity. I began by investigating whether women in particular, but also men from different social backgrounds, had been able to capitalise on the growth of employment opportunities in the City. Who was recruited in these years? What types of employment opportunities opened up? Was the image of a merchant banker, especially the association with a certain type of class-based masculinity, being challenged, and if so, how and by whom? Although the majority of merchant banks had put in place equal opportunities programmes during these years, it quickly became clear that conventional equal opportunities policies based on disembodied liberal notions of individual merit had relatively little impact on the culture of banking which operated to produce an atmosphere in which certain attributes of heterosexual masculinity were valorised. In these boom years of the mid-1980s, women certainly were recruited in growing numbers by merchant banks in the City, but my examination of personnel and survey data revealed that they were not making as much progress up the occupational hierarchy as their male counterparts recruited at the same time. As the banks were committed to expanding opportunities for women and were careful in their selection procedures, it seemed evident that something else was happening that made banks an inimical working environment for many women. In my investigation of why women failed to progress as fast as men I began to examine in detail the everyday working environment of three banks,[1] looking at the ways in which women were made to feel 'out of place' in the City. In this chapter I focus in particular on questions about the (hetero)sexed body and its significance in the shaping of power relations in the workplace.

As Fiske (1993: 57, emphasis added) argues:

> The body is the primary site of social experience. It is where social life is turned into lived experience. To understand the body we have to know who controls it as it moves through the spaces and times of our daily routines, who shapes its sensuous experiences, its sexualities, its pleasures in eating and exercise, *who controls its performance at work*, its behaviour at home and school and also influences how it is dressed and made to appear in its function of presenting us to others. The body is the core of our social experience.

That work, in the sense of paid employment, is a performance undertaken by embodied, gendered and sexed individuals has become increasingly clear as Britain shifts towards a service-based economy. An increasing proportion of jobs in contemporary Britain involve the marketing of personal attributes, including sexuality, as part of the product. They are jobs which depend on what Hochschild (1983) termed a 'managed heart' – jobs in which seemingly spontaneous forms of personal interaction are in fact carefully managed. The smile of the airline attendant is a classic example; the performance of the sex worker the most obvious example. But, as I shall argue here, increasing numbers of professional jobs, once seen as the epitome of disembodied, rational workers – as mind work rather than body work – are also characterised by sexualised performances. Service sector employment has been defined by Leidner (1991, 1993) as 'interactive' work in which 'distinctions among product, work process, and worker are blurred or non-existent, since the quality of the interaction may itself be part of the service offered' (Leidner 1991: 155). As she explains:

> workers' identities are not incidental to the work but are an integral part of it. Interactive jobs make use of their workers' looks, personalities, and emotions, as well as their physical and intellectual capacities, sometimes forcing them to manipulate their identities more self-consciously than do workers in other kinds of jobs.
>
> Leidner 1991: 155–6

One of the ways in which interactive service workers are selected and controlled is through emphasis on and careful surveillance and disciplining of their bodies; explicit rules about weight, permitted hirsuteness, and style of dress and implicit rules about sexual identity, or at least its transfer into workplace performances, are enforced to produce a particular corporate image of the worker. The following brutally honest statement from the supermarket chain Asda illustrates the bodily requirements of their potential employees. The chain looks for people with a 'healthy, well-groomed appearance and pleasant expression, clear complexion and near ideal weight for their height'. Nobody should be hired who is 'significantly overweight, ungainly, not sociable'. Also disqualified are candidates with 'a thick accent or a speech impediment' and (perhaps surprisingly) those who are 'concerned with their own status and image' (Atkinson 1993: 14). So, although looks and interaction with customers are important, for Asda the performance of selling one's self clearly must not be too self-conscious.

The Disney Corporation is perhaps the best-known example of an explicit encoding of the notion of performance. Each employee of the Disney theme parks, for example, is referred to as a cast member and the following rules apply to their appearance: the first and the fourth to all members of the cast, the middle two to men only.

As a condition of employment you are responsible for maintaining an appropriate weight and size.

A neat, natural haircut and a clean shave are essential.

Sideburns should be neatly trimmed and may be permitted to extend to the bottom of the earlobe, following their natural contour.

Single earrings are not permitted. Women cast members may wear two, men may not.[2]

What these two examples reveal is the establishment of corporate norms to which the employees of these capitalist corporations must conform. Although sexuality and sexual attractiveness are not explicitly considered in these rules, it is a clearly taken-for-granted assumption that workers conform to a conventional, heterosexual image of masculinity and femininity. The gaze of both the employer and of the clients and customers of interactive service workers ensures this 'normalisation'.

In investigating the ways in which bodily and sexual norms are constructed and imposed in the workplace, albeit in the different context of merchant banking, the work of Foucault is a helpful theoretical framework. Drawing on his notions of social normalisation and social resistance in the analysis of my empirical data, I examined the ways in which women and men in merchant banks enacted an appropriate sexed gender performance at work – a performance that might, or might not, conflict with their non-workplace identities. These performances are clearly enforced by a number of mechanisms, including the establishment of workplace rules, everyday social practices and through self-discipline. As Foucault argued, power, rather than being a totalising system, is diffused throughout the whole social order, and exists at all levels, from the micro-scale of the body, the home and the workplace to the structural institutions of society. The control, discipline and surveillance of bodies is particularly important in the production of what Foucault terms 'docile bodies', which conform to historically- and spatially-specific ideas of what is normal and appropriate forms of the presentation of self and daily behaviour in particular spaces; in this example in City workplaces.

The dominance of appropriate norms is achieved not solely through decree or power imposed from above but through multiple 'processes, of different origin and scattered location, regulating the most intimate and minute elements of the construction of space, time, desire, embodiment' (Foucault 1979: 138). These norms, the prevailing notions of self, ideals of physical appearance, sexual identity and acceptable behaviour, are maintained not through physical violence or coercion, but through self-surveillance and self-correction. Thus,

> there is no need for arms, physical violence, material constraints. Just a gaze, an inspecting gaze, a gaze which each individual under its weight will end by interiorising to the point that he is his own overseer, each individual thus exercising this surveillance over, and against himself
>
> Foucault 1977a: 155

The use of the masculine pronoun in this quotation is marked. In his own work, Foucault failed to distinguish between bodies that are marked as either masculine or feminine. He treated the body as if it were not differentiated by gender, and he was blind to the specific practices that produce particular feminine versions of docile bodies. For this reason, the relationship between feminism and Foucault's work has been an uneasy one (Ramazanoglu

1993). However, there are many points of mutual interest. Women have, historically within Western culture, long been synonymous with the body. Many feminists have identified women's bodies as the sites and expressions of power relations, recognising that female bodies and the attributes of femininity mapped on to them, including variants of passive heterosexual desire, are social constructions or productions, constituted and produced by the effects of power and self-surveillance. The multiple ways in which women discipline themselves in order to conform to idealised notions of feminine beauty, of compulsory (hetero)sexual attractiveness and particular expectations of feminine behaviour have been well documented (Bordo 1993; Coward 1984; Rich 1980; Wolf 1990).

Men, as well as women, adopt various strategies in their efforts to make their bodies conform to historically specific ideas of femininity and masculinity and a hegemonic heterosexuality. Thus, there is no transhistorical male or female body or essence but rather 'what is called "the body" is a site and expression of different, interested power relations in various times and places' (Bailey 1993: 106). As Strathern has pointed out, this argument has long been recognised by social anthropologists, whose work on initiation rites, for example, reveals the ways in which bodies receive the imprint of cultural norms: 'mutilated or forced into unnatural positions, the body presents a vivid image of "construction" ' (Strathern 1989: 51).

Recognising that ideas of masculinity and femininity, their embodiment and engagement in particular sexual identities and practices are culturally specific does not imply, however, that there is an absolute porousness of materiality – bodies have a material existence which is undeniable. As Bailey (1993: 104) argues 'ideas of the "feminine" are the result of the interplay of previous historical understandings of femininity and the bodies these have produced', and Grosz (1990: 72) has emphasised that 'what is mapped on to the body is not unaffected by the body on to which it is projected'. Masculine characteristics and attributes have different meanings depending on their embodiment in male or female bodies. Bodies are more than the effects of power: 'there is a complex interaction between grounded embodiment, the discourses of sexuality and institutionalised power' (Ramazanoglu and Holland 1993: 260) which is as yet incompletely understood and needs to be examined in particular locations at different times.

In a series of detailed interviews in three merchant banks, I began to examine some of the specific ways in which this complex interaction between embodiment, discourses of sexuality and institutionalised power work out in a particular workplace. Merchant banking is an interesting location for such an examination as it is pervaded by images of power and desire and by sexualised images and language. As Thrift *et al.* (1987) intimated, the world of international finance is a 'sexy, greedy' world. The interplay of glamorised images of youth, masculinity, sexual desire and (ultimately) corruption are central elements of a series of representations of this world. The film *Wall Street* (see Denzin 1990), Carol Churchill's play *Serious Money*, the British TV series *Capital City* and David Lodge's novel *Nice Work* all illustrate this interplay.

In the next sections of this chapter, I turn to the ways in which ideas about masculine and feminine bodies, sexual attractiveness, sexual desire and sexuality position men and women differently in the workplace. I focus on the micro-politics of power at work, examining how mechanisms of normalisation and surveillance and strategies of resistance operate within the everyday social relations in the world of merchant banking. The argument focuses on three specific areas of micro-politics – the body, sexualised performances as expressed through dress and client–worker relations. In each case I show how assumptions about an idealised heterosexual identity affect everyday social interactions in the banks.

BODIES/SEX AT WORK

The relationship of women's bodies to the workplace raises a number of challenging questions. The first and most difficult issue that women in the workplace face is establishing their right to be there at all. In contemporary society the body is represented as the location where nature meets culture and, as feminist scholars have long argued, woman or femininity is associated with the nature side of this dichotomy. The aesthetic 'disembodied' bodily presentation recommended to men, especially in the workplace, is out of reach for women. Women's bodies are, by very definition (and in contrast to those of men), grotesque, incomplete, fertile and changing. Like nature, women too are natural, marked by sexuality, fecundity and growth and so apparently uncontrolled and uncontrollable. And while culture is appropriately found in the public or civic sphere, nature is located in, and should be confined to, the private or domestic arena.

Thus, women's sexed bodies are threatening in the workplace for the very reason that they are not meant to be there. They challenge the order of things. As Martin argues,

> Numerous contrasts dominate postindustrial capitalist society: home versus work, sex versus money, love versus contract, women versus men. Because of the nature of their bodies, women far more than men cannot help but confound these distinctions every day. For the majority of women, menstruation, pregnancy, and menopause cannot any longer be kept at home. Women interpenetrate what were never really separate realms. They literally embody the opposition, or contra-diction, between the worlds.
>
> Martin 1987: 197

Women's embodiment of these contradictions forces men in their everyday workplace lives to face their preconceptions and privileged position and to deal with issues which they define as 'private'. It also results in diverse strategies both to deny the contradictions and to exacerbate them in order to exclude women from workplace privileges. This is a particularly important issue in extremely masculinist workplaces where women are a small minority. Rodgers (1981), drawing on historical commentary about women in the House of Commons, has argued that through a range of behaviours from joking, excessive courtesy to more unpleasant forms of personal comment, women are made to feel out of place in Parliament. She documents the particularly difficult time that women who are pregnant have when they attend the House. A woman who is visibly pregnant or breastfeeding is at her most explicitly female and cannot be classified as anything but a woman. Interestingly, Rodgers documented women's, as well as men's, overt hostility to the (very few) women MPs who attend the House through this stage of their lives. She explained their reaction as follows:

> Women whose success has been geared to the male construct, have discarded the symbols by which they would be anchored into the traditional female domain of domesticity and nature. They fear that if one of their women colleagues openly combines the public symbols with the female and domestic ones, they themselves will be at risk of being seen as the women, which, on some levels they, of course, are. Their position in the dominant category is after all a tenuous one.
>
> Rodgers 1981: 60–1

Similar forms of behaviour, attitudes and everyday interactions are also common in the world of merchant banking. Jokes, personal comments, excessive and ironic courtesy and other types of overt behaviour which constitute sexual harassment are used to draw attention to women's female embodiment and to construct them as out of place at work, as the 'other' to a disembodied masculine norm. These behaviours constitute a form of oppression that Iris Marion Young (1990a) has termed 'cultural imperialism'. She has argued that 'cultural imperialism' is one of several key structures in the enforcement of oppressive and dominant social relations. Cultural imperialism refers to the ways in which the dominant meanings of a society render the particular perspectives of a group invisible at the same time as paradoxically stereotyping the group and marking it out as 'other'. Thus, it involves the universalisation of the dominant group's perspective and its establishment as the norm. What is particularly significant for the arguments here is that Young suggests that 'the culturally imperialised are stamped with an essence. The stereotypes confine them to a nature which is often attached in some ways to their bodies, and which cannot thus be easily denied' (Young 1990a: 59). And, as she points out, one of the most significant groups who are equated with or reduced to their bodies is, of course, women.

Young, like Foucault, emphasises the everyday establishment of power relations through discursive strategies. 'Group oppressions are enacted ... not primarily in official laws and policies but in informal, often unnoticed and unreflective speech, bodily reactions to others, conventional judgements, and the jokes, images and stereotypes pervading the mass media' (Young 1990a: 148).

In the three banks in which I undertook detailed interviews with male and female employees in similar positions in the occupational structure, it was clear that a range of everyday behaviours from comments on clothes and appearance through what most respondents reported as relatively innocuous jokes to more overt and troubling forms of sexual harassment were used to reinforce women's embodiment and to remind them of their 'otherness' in the workplace. With only one or two exceptions, all the women whom I interviewed were under 40 years of age and so assumed to be both sexually active and either potential or actual mothers (in fact very few of them had children) and it seemed to be universally assumed by their male colleagues that they were heterosexual and so potential receivers of their sexual favours.

The following quotations are all from women talking about the ways in which they were continually confirmed as (hetero)sexualised and embodied beings in workplace interactions. Whereas the first six comments focus on comments and behaviour directed at particular women colleagues, the second four illustrate the ways in which women in general are regarded and so reinforce these particular women's feelings of being 'the other' in the workplace. Although many of the comments and behaviour seem juvenile, they combine to make women's daily lives in the workplace uncomfortable. It is noticeable that they are based on assumptions that men hold about women as sexually active and their place in society as wives and mothers rather than as men's equals in the workplace.

'Men, but especially the young men here, are very open about where they think a woman's place is – in the kitchen or in the bedroom ... They think it's a man's prerogative to work at the bank and women shouldn't be here.'

'A whole series of sexist faxes ... very derogatory to women, went round the office and they generally landed up on my desk. I try to let them wash over me. It's not really worth bothering about them.'

'I asked someone to do something for me, a young male graduate, and he hadn't got a computer password, so he asked me what mine was and he came back about half an hour later. Two or three of them were giggling and I said what is it? And he was very sorry but he'd changed my password and I looked at him and he said "It's nipple".'

'If you see two or three of the girls[3] in the department standing together having a chat, somebody will always comment "Oh, mothers' meeting". If you see four blokes together nobody even bats an eyelid.'

'Men think crude jokes are funny rather than offensive.'

'I hate it when people make personal comments. I've even had people say things about my underwear.'

On attitudes to women more generally:

'They had a blow-up woman that they used to kick around the floor.'

'When I was a dealer, they had strippers on the trading floor – for birthdays and so on – that sort of thing.'

'For most of the time they [male colleagues] treat me as an honorary male, and that's fine, I much prefer it but it also means that I see the way they look on women. If I go out for a drink with them then they will comment on anything that walks past in a short skirt … But I guess I'd rather be the honorary male and then not have all the comments than be on the other side.'

'I think there is harassment against women as a group but not of individual women.'

For men too, however, embodiment/sexuality is an issue, and conformity to an idealised bodily image was a significant mechanism of discrimination among male applicants to positions within merchant banking.

The following comment, albeit accompanied by a certain shamefaced justification, illustrated the way in which the striking physical uniformity among the male employees was achieved:

'We had a very nice chap in for interview but he was very overweight … we sat and talked about it very seriously, about whether the fact that he was very large was going to weigh on the client's mind.'

The pun may have been unintentional, but the candidate was not employed.

The respondent, a male assistant director, went on to elucidate in greater detail, clearly feeling somewhat defensive:

'We are not all clones, we don't all look the same; we don't recruit physical stereotypes at all. But if someone was very ugly, it would make a difference, because we are selling a service and if people don't want to buy that service from that person, it's difficult to say that that person has a future in something that is very much a frontline selling service.'

In fact, there was a striking uniformity in the physical appearance of the men whom I interviewed that explains the respondent's comment about clones. All the interviewees were white, slim and above average height, with a clearly cared-for body and style, including a good haircut and expensive clothes (style is the focus of the next section). While it has been suggested that this attention to the body verges on the homoerotic, the culture of these banks is intensely heterosexual. The relations between men and women, as the quotations indicate, are premised on sexual attraction between the sexes. In the overwhelmingly heterosexual atmosphere and emphasis on sexuality in language and social interactions, especially in trading, dealing and sales, women who do not respond are dismissed as lesbian:

'Any woman my colleague can't get on with he thinks "she's a lesbian anyway" ... If they don't respond to him treating them as a sexual object they must be a lesbian.'

However, this respondent was quick to reassure me that her women colleagues were not, in fact, lesbians. Sexual attraction between women was barely mentioned as an issue in the interviews, other than as complaints about anti-lesbian repartee and offensive comments about all-women networking. One group, for example, was dubbed the 'dykes dinners'. Disapproval of homoerotic relationships, however, was expressed more strongly. As a woman respondent remarked of her co-workers: 'most of them are completely homophobic'. Consequently, a number of male respondents indicated that they had decided to conceal their sexual preferences while at work and participate in the construction of an overwhelmingly heterosexist atmosphere and workplace practices. The only out gay man who was interviewed revealed that the homophobic atmosphere also made him the subject of unacceptable sexist jokes and behaviour, which, in common with many of his women colleagues, he endeavoured to treat as unimportant. The following statement from one of his colleagues reveals commonly-held, if somewhat astonishing, attitudes:

'There's this guy who's completely queer, makes no bones about it, thinks it's hilarious. I mean he's always making jokes, and I mean once upon a time we sent him up in the lift. We stuck him to his chair, taped him up with Sellotape, and sent him to the twelfth floor in the lift, and this kind of stuff. I mean it goes on all the time; it's really very funny. So it's a very nice place to work; it's very friendly.'

Clearly, the mechanisms of cultural (and sexual) imperialism identified by Young lead to discriminatory practices that affect men and women alike in banks. Once recruited, however, for most men the workplace is a more comfortable place to be than it is for women. Men's physical presence at work is unremarkable, that is as long as they conform to a version of heterosexual masculinity. Embodiment as a heterosexual male is taken for granted in the workplace in ways that construct women workers, not only as different, but as an inferior 'other', particularly in extremely masculinised spaces like these merchant banks. As a woman respondent remarked, 'the demand of the culture to "be one of the boys" makes it so much harder for women to fit in'. And a male respondent was clear that 'in a sense, being a male is almost like being normal and – we are talking about this particular environment – you know it's expected, so there's no particular advantage or disadvantage'.

While women – defined as 'natural' bodies – are out of place in the disembodied world of corporate finance, where a rational mind is the prized attribute, they are equally out of place in other departments in merchant banks, where, rather than a disembodied masculine rationality, notions of a sexualised masculinity pervade the discourse and the daily practices of the workplace. In the physical culture of trading and dealing, a heroic or 'macho' heterosexual masculinity is counterposed to the cool rationality of the world of corporate finance. As an interviewee working in a dealing room remarked, 'the whole place is incredibly macho. There's an extremely macho kind of culture'. Here, male embodiment is valorised through sexualised discourses that emphasise the possession of 'iron balls' or 'big dicks' (Lewis 1989; McDowell and Court 1994) by successful traders and dealers as they 'consummate' deals. And not only deals. One of the few women traders explains that one of her colleagues 'used to spend most of his mornings explaining to his clients which woman he was pursuing at the moment or whether he got laid last night'.

Another dealer reported that

'there's a lot of sexist banter … the guys talking about how many women they screwed the night before or something … there's always a lot of reading of *The Sun* and all that sort of stuff.'

And, as another respondent explained, 'down on the floors, it's a very laddy atmosphere … the language can be quite blue'. Asked to sum up her colleagues, a woman dealer replied that they are 'Jack-the-lad basically'. For male dealers, social interaction in the workplace is also often based around typically masculine interests. Thus 'they talk about their fast cars, they talk about the sports they do . . . There's quite a lot of showing off that goes on'. A male foreign exchange dealer commented, 'as you are generally working with other men, you have interests in common . . . from football, to cricket, to sport'.

The environment in this side of banks 'is noisy, aggressive and very pushy'. One respondent commented with surprise on the exceptional behaviour of her boss:

'He didn't seem to feel the incredible need to bite everybody's head off and knock them out of the way and trample on their heads where other people did.'

Whereas other dealers and traders working on the trading floor emphasised the aggressive environment:

'You don't write memos to somebody to sort things out; you just go and shout at them.'

The work also involves 'bellowing down phones'.

As this type of behaviour is socially constructed as masculine, women are at a disadvantage. The following comment is a clear expression of how women are perceived:

'Aggression from a woman is seen as trying to compete too well. It's probably – this sounds awful – but it's not natural, certainly from a dealer's point of view. An aggressive male dealer, well, that's how dealers are. An aggressive female dealer is . . . isn't she just trying to sort of show something, isn't it a put-on sort of thing?'

It seems, however, that the degree of aggressive, masculinist behaviour varies between markets. As a respondent who worked as a salesman in gilts explained, 'foreign exchange fits the media images – it tends to be work hard and play hard. Gilts doesn't have the same sort of foreign exchange tag put on it – which is the type of dealer that appears on the television. Gilts is less aggressive'.

Despite these variations, however, it is clear that a hegemonic idealised notion of heterosexual masculinity is the dominant image in the world of merchant banking. For women, and for many men, everyday social relations in the workplace demand a sexualised gender performance that contradicts their bodily appearance and continually constructs them as an inferior other at work.

THE POLITICS OF APPEARANCE: DRESSING FOR WORK

If the body *per se* is one of the main sites where discipline is exerted in the workplace, then a closely related area where power relations and self-discipline are exerted is personal appearance, what we might term the body's extension into dress. The appearance of a worker makes it possible to assess the extent of bodily discipline or docility, the adherence to corporate norms, or, alternatively, to assess visible strategies of resistance. Here too hegemonic notions of masculine and feminine appearance are based on norms of heterosexual attractiveness. Whereas in the previous example, alternative masculinities were important, in the arena of dress, alternative femininities are more significant. A single dress code for men contrasts with a contradictory concern among women in professional occupations to differentiate themselves from junior workers on the basis of occupational status at the same time as minimising their difference from their male colleagues.

In professional interactive service occupations, by comparison with routine service jobs, disciplining the body through explicit dress codes or corporate uniforms is not common. However, informal norms are clearly important and a limited set of possible dress styles are available for workers in particular occupations and positions within these occupations. Appearance is crucial to potential employment prospects. For example, Fiske has described the advice given to male candidates for professional employment by a consultant with an outplacement firm in the USA:

> A blue suit and a white shirt are common to all male candidates: what will . . . really attract the corporate interest is the details. The lower tip of the tie should come to the top or center of the belt buckle and the back of the tie should go through the label so it cannot escape control and reveal its undisciplined self to the interviewer. The belt should not only be new, but should show no sign of weight loss or gain. The body of the candidate should be totally disciplined, and should indicate that it is always controllable. Weight gain or loss are signs of a body breaking out of control and having to be redisciplined. The tie relates to the belt symmetrically, producing the body as aesthetically balanced around both vertical and horizontal axes. The aesthetics of symmetry, of the repetition and balance of form, represents human control over nature. Nature is asymmetrical, ever changing and growing. Aesthetic form is static, completed, controlled.
>
> Fiske 1993: 59

Here is a clear illustration of the relationship between an idealised male body (and the emphasis on discipline, symmetry and form is again almost homoerotic) and clothing. The particular emphasis on the tie is interesting. As Angela Carter pointed out, the tie is a metaphor for the phallus, marking the masculinity of its wearer.

In the world of merchant banking, discipline is similarly asserted on/by male employees by their standard dress of dark suit, shirt and tie and black shoes. It is the cut, style and quality of these items that are the crucial distinguishing characteristics which are read by clients and co-workers alike to place and evaluate both the adviser and the advice he or she is giving. Careful attention to appearance is common. For example, male senior executives routinely advise their male subordinates to 'buy good quality shirts, Marks and Spencer won't do. Get a decent haircut, take care with your personal hygiene'.

The significance of appearance to City workers is recognised in the recent launch of a brand of make up for men that is, according to the women's magazine, *Elle* (July 1994), aimed specifically at City workers. However, although many of the respondents admitted to using hair gels and mousses, there was no other visible evidence of the use of cosmetics, few signs of bodily decoration other than wedding rings and little experimentation with dress. As respondents explained, 'only dark suits are permissible', and 'the most men can do is wear tartan socks, and nobody can really see them, or a large tie or braces. Some of the braces are amazing'. So sartorial experimentation for men is limited in the world of merchant banking, at least in Britain. The single image is that of the sober, besuited and preferably heterosexual family man.

For the women respondents, dress raised more difficult issues. Already positioned as the 'other' by their female embodiment, some women chose to try and minimise their difference from the masculine norm through forms of disguise of their femininity and sexual attractiveness. Many women whom I interviewed adopted a style that aped the appearance of their male colleagues, as the following comments illustrate:

'I wear these men's shirts; I mean they are ladies', they are made for ladies at a men's tailors.'

'To look professional you must wear a suit.'

'I always wear a suit. I like to look as male as I can or at least neutral.'

'I tend to wear sort of rather boring frumpy things because I felt initially that there was less of a distinction between me and the men and I wouldn't be noticed as a girl. I don't know why I do it now.'

It is clear that both a desire to look 'professional' (for which read masculine) and a desire to minimise conventionally defined sexual allure combine to produce a particular uniform look for women.

The second dress code or norm for these women, however, depended on making status distinctions plain rather than gender differences indistinct. Thus, they must not be confused with women working as secretaries whose femininity, accentuated by dress, is part of a congruent gender performance in which 'natural' attributes of a sexualised femininity are accentuated. Thus many respondents emphasised that 'women should wear jackets unless they want to be associated with a secretary' and recognised that careful attention to dress was essential to establish status differentials between women. Thus an older woman commented that

> 'Young women associates [a professional position] really change their style in the first three months. You have got to make it obvious that you are not a secretary, because everyone assumes that a woman of 25, say blonde with a fringe, is a secretary.'

While most of the women respondents in professional positions dressed exceedingly carefully and conformed to the norms that insisted that they did not mark themselves out as 'the other', as overtly feminine or sexual through their dress codes and appearance and so out of place in the professional echelons of sober, besuited merchant bankers, a number of respondents deliberately flouted the distinctions or blurred the boundaries. The following discussion with a senior woman illustrates her opposition:

> 'I try and choose my clothes to do several things at once. I don't fancy the female version of male work dress, you know shapeless grey serge suits and a floppy tie. One of my colleagues always wears a sort of black skirt and black stockings. I have red clothes, I have yellow clothes, I would have orange clothes. I have a colleague on the trading floor who says "here you come like the traffic lights today". They would be absolutely unremarkable in the West End. In the City they probably mark me out more as an executive secretary than an executive, in that secretaries dress a little more wildly. Most women professionals wear smart tailored things. I don't like that sense of uniform; I just don't like the kind of voluntary uniform aspect of the City.'

For many more senior women, with greater self-confidence won through their success in the workplace, dress had become a more pleasurable performance that could be used to create or subvert a particular image. They were aware of the codes and the norms, the limits to acceptability that constructed the City uniform for women, and enjoyed playing with the rules and pushing the acceptable limits. This confirms Bordo's observation that ' "feminine" decorativeness may function "subversively" in professional contexts which are dominated by highly masculinist norms (such as academia)' (Bordo

1993: 193) and parallels conclusions about dress codes among senior women in the British civil service (Watson 1993) and among hospital doctors (Pringle 1993). The same respondent who enjoyed dressing as a traffic light was clearly aware of the subversive effects of an overtly sexualised style for senior women from whom such provocativeness is not expected. Thus she reports that 'sometimes I come to work in a leather skirt just to confuse them'. The association of leather with dominant sexuality is well known. As Judith Williamson has remarked, 'the black leather skirt rather rules out girlish innocence' (Williamson 1992: 222). But the same respondent is equally clear that overtly provocative forms of dress are also inappropriate in certain circumstances when the purpose of dress, image and style is to maximise business. Thus 'when I make a cold call, I dress like this [a plain blue tailored dress]. I want to blend into the background so the client will listen and have no distractions', by which she meant sexual distractions. However, even then, she confesses to a certain provocativeness: 'I wear high heels, so I'm 6 feet tall when I stand up. And I think that commands some small sense of "well, I'd probably better listen to her, at least for a little while".'

Another respondent was equally frank about women's ability to confuse and divert men through an overt display of sexualised femininity: 'It sometimes helps to go disguised as an executive bimbo.'

INTERACTING WITH CLIENTS

The purpose of bodily discipline and the controlled presentation of self in interactive service occupations is, of course, to create an acceptable image in the interaction with clients and bring a transaction to a successful conclusion. In many selling occupations, this interaction is carefully scripted. The emotions, attitudes, sexual performance and behaviour brought into play are regulated through intensive training programmes, standardised routines and managerial control of daily transactions. Many aspects of the face-to-face interactions with customers and clients are carefully scripted, regulated and routinised. Workers are trained, for example, to stand in particular ways, to use certain gestures, phrases or jokes and to follow a set routine in verbal transactions. As Leidner (1993: 87) argues, these employees 'are asked to cede to the company the right to reshape many aspects of their selves, including their emotions, values and ways of thinking'.

In selling financial advice and making deals in merchant banks the element of discretion is considerably greater than that open to routine service sector workers. Explicit scripting of client interactions is not appropriate at this level in the financial world. The product on sale is a speciality service, rather than a routine interaction with interchangeable clients – adapting a manufacturing analogy, it is a post-Fordist niche product rather than a Fordist mass good. The aim is to make clients feel that they are receiving a specialist service, tailored to their unique demands.

A number of different ways of ensuring satisfactory client relations and a positive outcome are in place, however, in merchant banks, although they are implicit mechanisms of control rather than the explicit structures adopted in the fast-food and life insurance industries. These include careful initial selection of employees, induction programmes on recruitment and particular ways of doing business that produce the illusion of a personal service for each client. In each case the significance of gender and of sexuality in these sets of relations varies. Here I focus solely on the ways in which the illusion of a personal service is created, illustrating the significance of sexuality and gender in these interactions.

There are two main ways in which a personal relationship is built up with clients. The first is by matching the client and adviser, ensuring that they feel comfortable with each other, have interests in common and are able to sustain a relationship throughout their association. A range of behaviours that draw on aspects of the social construction of femininity are used to make clients feel 'special', including flirtation. Many women whom I interviewed recognised that the maintenance of a 'special' relationship is at least in part based on sexual attraction between their clients, who almost without exception were men, and themselves. The following statements are representative of the ways in which the senior women whom I interviewed explained their dealings with clients:

'You need flair, a feeling for people. I'm no good with a client I'm not interested in . . . if there is no spark in the relationship, you just can't turn it on.'

'One's clients have fairly sizeable egos, which they have needed to get where they are and it's just much easier to play along with that if you are a woman.'

'Women work very hard and they are good at getting on with people. People tell them things; people tell me things.'[4]

'You flirt a bit with them and you take them out for a drink and they just think you are wonderful.'

'I mean, some of my clients that are men, I have no doubt deal with me because they like my legs.'

'Women seduce their clients, not literally. I'm quite certain it's done that way.'

And as a young male corporate financier remarked with some wonderment in his voice:

'I never really thought of this, but she [a female colleague] said, "sometimes I can use my female skills to get things that you couldn't get".'

Another, more alert, male respondent, recognised that women have certain advantages: 'If you are resilient, slightly sugar coated and not unfeminine, you have an advantage. Women seem able to strike up an instant rapport with a client.'

The clear implication of all these statements is that these 'female skills' include heterosexual attraction between a client and a woman banker. Some respondents were more open about this:

'Initial communication is easier because you are the opposite sex. From then on you have to be careful – as a single girl [*sic*] you can get yourself into hot water – that can be a disadvantage. The job is to be friendly and open with people and it can get difficult. For men that will never happen because they'll just become big buddies.'

The second way of establishing a special relationship with a client is through involvement in a range of social and semi-social activities, in which pleasure and business are mixed. In the world of corporate finance, especially at the more senior levels, and at all levels in trading and dealing, the boundaries between work and leisure are often difficult to draw. Indeed, as Budd and Whimster (1992: 2–3) have argued, merchant banking is a prime example of those occupations in which 'elements of life style and culture as well as conceptions of personality become implicated with the worlds of work and finance' leading to 'the interpenetration of areas of life previously separated by hierarchies and boundaries'. Many of the negotiations with clients take place in what are ostensibly leisure environments – in clubs, in restaurants or on the golf course. In addition, contacts are made and cemented at some of the key social events or arenas of 'high culture' – at Glyndebourne or the London opera for example – or at key sporting events such as Wimbledon and international rugby matches. Although these events are not necessarily exclusively masculine affairs, many women have neither the interest (in rugby for example) nor the skills (golf is a good example here, even assuming that particular clubs and courses are open to women) necessary to ensure their participation. A number of respondents also remarked on the difficulty of entertaining clients as a single woman. Many of their male colleagues and their clients involved their wives or female partners in such events, but women found that their own partners (if they had one), who were often high-powered professionals themselves, were reluctant to become involved in these forms of work/leisure activities.

Other ways of making clients feel that they are receiving a unique and personalised service do tend to exclude women. For example, many respondents, especially the 'big dicks' from dealing and trading, mentioned the value of taking clients drinking, or to a range of somewhat risqué events including strip clubs to cement their masculine bonding. As a gilts salesman explained, 'I'll go anywhere with a client, from drinking to go-karting.'

Although many of the women interviewed rued their exclusion from these methods of cementing social interaction, they emphasised that the different feminised strategies available to women in establishing a personal relationship were just as likely to be successful:

> 'I have the ability to listen and to make polite noises. I gain clients' confidence and friendship rather than being the chap that they take off with to watch rugby. It's just a different way of doing things.'

But most men still agreed with the following statement of one of our male respondents:

> 'Women may have a natural advantage, as the majority of clients are men, and clearly their PR skills and general warmth of approach is much better than a man's, but at the end of the day I think that clients will be looking for somewhat sober advice and probably men are perhaps more able to give that advice sincerely, even if it isn't actually sincere, than women would.'

This quotation embodies not only a set of assumptions about women's 'natural' attributes, but, rather disarmingly, makes transparent the way in which selling financial advice is a gendered performance.

Although the methods of interacting with clients are far more implicit in merchant banking than in routine service jobs, and do not involve the overt scripting of interactions, it is quite clear that merchant banking, like other interactive service jobs, involves 'emotional work' which is also 'sex work'. Client/banker interactions depend on what Stinchcombe (1990) has termed 'ethnomethodological competence', that is the capacity to make use of unspoken norms of behaviour, in this case norms based on assumptions of heterosexual attraction between men and women and heterosexual 'buddiness' between men, to control interactions. For bankers, too, selling oneself – one's body, sexuality and gender performance is part of the job. As one respondent explained, 'we rely on people putting together what is necessary to meet a client's needs, whatever that is, that's the over-riding principle'. It took a woman trader, working in and struggling against the aggressive masculinised environment dominant in her part of the bank, to admit that this 'whatever it is', for women, involves commodified sexual exchanges:

'If you are an attractive woman in this environment it can help on the male side of things; and frankly you have to learn to use all your assets and swallow your pride sometimes because in some form or other, obviously not in the literal sense, but in some form or other, it can be a form of prostitution of your sex … and you, hmm, and you have to learn to use that.'

CONCLUSIONS: MULTIPLE MASCULINITIES AND FEMININITIES

That the affluent world of merchant banking has anything in common with the fast-food outlet or the massage parlour initially may seem unlikely. And yet, as I have shown here, for workers in these different types of 'interactive' services, the everyday world of work involves the construction of a gender performance, in which attributes of masculinity and femininity, including a more or less authentic presentation of sexual identity, are an integral part of selling a particular product, be it financial advice, hamburgers and fries or sexual services. Selling these products also involves selling oneself. Hochschild (1983) has argued that this requirement to sell oneself as part of a product has significant consequences for notions of subjectivity. She suggests that workers whose emotions are managed by employers become alienated from their feelings in a process similar to the alienation of manual workers from the actions of their bodies and the products of their labour. Certain service workers, she suggests, have difficulty experiencing themselves as authentic even when they are not at work because they are unable to distinguish which feelings are their own. This distinction, based on an essentialist notion of self, has been questioned more recently in postmodern work on the fragmented and partial construction of identity but it is still a commonsense notion that is widely held.

In interviews in the merchant banks significant numbers of respondents referred to a distinction between their 'real self' or the 'real me' and the persona they felt that they adopted at work. It was clear from many conversations that the workplace persona was constructed as false in opposition to a real self that had to be hidden in the workplace. It was particularly noticeable that it was the women respondents who were far more likely to experience, or at least give voice to, this dichotomy between a 'real' and a constructed self, an inner core and outer shell, explicitly recognising that they were constructing a gender performance to meet workplace demands. Whereas junior women were likely to adopt a masculinist version of self in the workplace – what Acker (1990) has termed the 'honorary male' strategy – more senior women tended to exaggerate their feminine characteristics, and indeed either parody them or use them to confuse (McDowell and Court 1994). For the junior women, the construction of a masculinised identity seems particularly likely to give rise to conflicts as client interactions are predicated upon the manipulation of a heterosexualised femininity.

While the emphasis here, in the main, has been on the consequences of interactive 'sex' work for women and the construction of female subjectivities and identities, the ways in which men also 'do gender' (West and Zimmerman 1987) on the job and rely on variants of heterosexualised masculinity are of growing significance in an economy in which service occupations are expanding. Merchant banking, in particular, makes demands on men too to maintain a particular regulatory fiction or sex/gender performance. As I have shown, it demands a particular heterosexual performance of men, exacerbated by excessive 'macho' behaviour in dealing and trading and a peculiar paternalism in corporate finance. While the consequences of these demands for the construction of alternative masculinities have begun to be explored in banking (McDowell and Court 1994) a great deal of further work on both masculinity and femininity in a range of service occupations remains to be done. This chapter has begun the exploration in a particular subsector of the economy. Comparative studies of other professional occupations will confirm or challenge the specificity of these results.

NOTES

1 This chapter draws on an ESRC-funded study (grant number 23 0006) carried out with Gill Court between 1991 and 1993. We administered a questionnaire survey to all the merchant banks in the City of London (some 360 in total), undertook detailed interviews with seventy-eight men and women in three merchant banks, as well as a smaller number of interviews in four other banks, and analysed the personnel data of the three banks in which we did the detailed case studies.

2 Information from the British daily press on the opening of EuroDisney.

3 Interestingly, almost all the women respondents referred to themselves and their women colleagues as 'girls', implicitly confirming their sexualised status.

4 Notice the ungendered term 'people' – almost certainly being used to refer to male clients.

2

sexualised spaces
global/local

SEXUALISED SPACES: GLOBAL/LOCAL

The relationships between sexualities and space are made clear when we begin thinking about the power of particular landscapes as either liberatory or oppressive sites for the performance of our sexed selves. The contradictory meanings of 'home' or 'city' as places to be in – but also to dream about – have a resonance which stretches far and wide. This section explores both real and imagined landscapes – the city, the home, the desert island – and thereby reflects on the relationship between sexual identity and these generic spaces. The lesbian adrift in the city or the male castaway both show how we shape and are shaped by our location in space: searching for a sense of place to match our sense of self, and a sense of self to match our sense of place.

Greg Woods' meditations on his 'fantasy island' as a stage upon which dramas of (homo- and hetero-)social and (homo- and hetero-)erotic are enacted indicates the power of a particular fictional place as a metonym of masculinity which shifts between prelasparian 'wilderness' and the forces of civilising modernity. Another symbolic landscape wherein identities are constituted is the home, and Lynda Johnston and Gill Valentine explore the complex meanings of 'home' to lesbians, basing their theories on empirical material from New Zealand and the UK.

The relationship between the city and sexuality is discussed by both Sally Munt and Larry Knopp. Sally's chapter *dérives* through the urban landscape – a landscape of books as much as of streets and skyscrapers – in the company of the lesbian *flâneur*, whose body in movement rewrites the city as it rewrites her. Meanwhile, Larry theorises the spatial dynamics of the exchange between urban spaces and urban sexualities, arguing that the public spaces of the city and the private lives of its inhabitants must be understood within a complex web of power, space and difference.

7

WHEREVER I LAY MY GIRLFRIEND, THAT'S MY HOME

the performance and surveillance of lesbian identities in domestic environments

•

Lynda Johnston and Gill Valentine

WANTED: LESBIANS TO LIVE IN AN ALL-DYKE HOUSE. VEGETARIAN AND NON-SMOKING PREFERRED. 40 DOLLARS PER WEEK. FREE SUZANNE CLIPS TO THE FIRST THREE APPLICANTS.

Home is a word that positively drips with associations – according to various academic literatures it's a private, secure location, a sanctuary, a locus of identity and a place where inhabitants can escape the disciplinary practices that regulate our bodies in everyday life (Allan and Crow 1989; Saunders 1989). Above all the home is often presented as being synonymous with the heterosexual 'family' and the ideal of family life (Allan 1989; Madigan *et al.* 1990; Oakley 1976; Saunders 1989). But not all homes are exclusively occupied by heterosexuals. 'Home' can take on very different and contradictory meanings for sexual dissidents who share a house with heterosexual family members (Bell 1991; Valentine 1993a) and for those, like the lesbians who placed the advertisements above, who create their own domestic space. 'At home' sexual identities are both performed and come under surveillance. Whilst 'the home' may be taken for granted or appropriated as the terrain of heterosexual family life and therefore be regarded as normative, it is also a possible site of challenge and subversion. This chapter draws on research carried out in New Zealand and the UK[1] to explore the experiences of lesbians in the parental home; and to examine how lesbians create and manage their own domestic environments.[2] The chapter does not seek to reify the idea of 'lesbian identity' or 'lesbian homes' in a universal sense. Rather, it suggests that lesbians' experiences from different cultures be part of the wider debate of geographies of difference. The local politics of lesbians in NZ and the UK are brought together to highlight the fragmented nature of difference.

HAPPY FAMILIES: LESBIANS IN THE PARENTAL HOME

The word 'home' has multiple meanings. In an attempt to clarify the concept, Somerville (1992) has picked out seven key dimensions: shelter, hearth (i.e. emotional and physical well-being), heart (loving and caring social relations), privacy, roots (source of identity and meaningfulness), abode and paradise ('ideal home' as distinct from everyday life). This is, he claims, a classification that can be supported by Watson and Austerberry's (1986) empirical findings. Of these seven meanings, it is the notions of privacy and heart that appear to have received most academic attention.

Being in a private space is at the heart of what it means to be 'at home' according to Graham Allan and Graham Crow. They argue that 'A home of one's own is ... valued as a place in which members of a family can live in private, away from the scrutiny of others, and exercise control over outsiders' involvement in domestic affairs' (Allan and Crow 1989: 4). This ability 'to relax' and 'to be yourself' away from the gaze of others, was also identified as one of the most important meanings of home by participants in Peter Saunders' (1989) research. As one of his respondents explains:

'I can dress how I like and do what I like. The kids always brought home who they liked. It's not like other people's place where you have to take your shoes off when you go in.'

in Saunders 1989: 181

Peter Saunders summarises such sentiments when he states: 'The home is where people are offstage, free from surveillance, in control of their immediate environment. It is their castle. It is where they feel they belong' (ibid.: 184).

But although the home may be a more or less private place for 'the family' it doesn't necessarily guarantee freedom for individuals from the watchful gaze of other household members: 'the public world does not begin and end at the front door' (Allan and Crow 1989: 5). Rather, the ideology of 'the family' actually emphasises a form of togetherness, intimacy and interest in each others' business that can actually deny this privacy. Linda McDowell (1983) is one of many authors to have argued that women have little access to private space within the family home. Likewise, children's space (usually a bedroom), is often subject to intrusion and violation by parents (Hunt and Frankenberg 1981) and young people usually have less power than other members of the household to make decisions that determine the 'family lifestyle' (Madigan et al. 1990). The privacy of a place is not therefore necessarily the same as having privacy in a place. In this sense the distinction between public and private is complex and hard to draw, being simultaneously articulated at a multiplicity of levels.

Lesbians living in (or returning to) the 'family' house, who haven't 'come out' to their parents can find that a lack of privacy from the parental gaze constrains their freedom to perform a 'lesbian' identity 'at home'. Home is not for them, the place where they can, in Peter Saunders' words, establish the 'core' (Saunders 1989: 187) of their lives. It does not, to use Somerville's (1992: 533) classification, have any meaning as a source of identity or 'roots'. Rather it is a location where their sexuality must often take a back seat. The most obvious expression of their identity – lesbian sex – is definitely off limits (at least when parents have them under surveillance) as Janice and Sharon, a New Zealand lesbian couple, explain:

'Well it makes me sad 'cos I can't take Sharon home, that's my problem 'cos I never came out to my parents.'

Janice, New Zealand lesbian

'She took me home, and it was really uncomfortable. We didn't do anything. We slept in the same room but in separate, single beds. And your mother sounded confused 'cos Janice was going "Oh we'll just use the double bed, we'll just sleep in there, that's all right". Your mother [Janice's] was going "Um, are you sure? Look we've got the two single beds, how about you sleep in the single beds, come on Janice?" '

Sharon, New Zealand lesbian

One option to try and get round these moments is to 'come out' to the family. But this means running the risk of taking on parental pain, anger, disgust and even rejection. And so fear of being 'found out' or of giving themselves away drives many women to use time/space strategies to separate the performance of their lesbian identity from the performance of their identity as a daughter (Valentine 1993b).

'My sister knows, my parents don't . . . I moved away so there didn't seem any point in saying anything. I mean I got a job here away from them and they were back in Cardiff, so there was no need for them to find out. But now they've moved to Redcar [a few miles away] which is a source of irritation to me.'

Sandra, English lesbian

Unlike Janice and Sandra, Julie is 'out' to her family, but in practice it makes little difference to her experience of the asymmetrical family home, as her sexual activity is still policed by her vigilant parents.

'When I came out to my parents, my mother said "there's only one stipulation, you can bring your girlfriends home but they can't sleep in the same room with you" ... That was it. When I have taken a lover home there she has just been really different. It's felt really uncomfortable.'

Julie, New Zealand lesbian

Whilst some parents may also feel squeamish or prudish about their daughter having sex with a male partner under the family roof, within the discourse of heterosexism, a male partner is at least the established 'norm' and although 'sex' may be banned, kissing, holding hands and other expressions of (hetero)sexuality are usually accepted as part and parcel of 'normal' relationships. For many lesbian couples, the expression of anything beyond 'friendship' is tantamount to 'flaunting it' and so they modify their behaviour to such an extent that their relationship is virtually invisible.

It is not only sexual activity that is inhibited by the hegemony of heterosexuality; there are many other moments when a lesbian identity cuts across the grain of the heterosexual 'nature' of the 'family' home. The home is supposed to be a place where family members participate in communal activities, socialise and share their feelings. These basic patterns of social relations are often underlain with a heterosexual ethos. At the kitchen table and round the TV the asymmetrical family can serve up a relentless diet of heterosexism and homophobia – 'Have you got a boyfriend?' 'Don't you fancy him?' 'Letting those poofs on telly, it's bloody disgusting.' Not surprisingly this cultural web of heterosexual norms can inhibit the performative aspects of a woman's lesbian identity.

'the comments Dad makes about queers and lezzies. I mean, he said it in front of me. Michelle [her partner] and I have been sitting there and I'd feel sick.'

Catherine, English lesbian

According to James Duncan (1981: 2–4), the home is a medium for the expression of individual identity; a site of creativity; a symbol of the self. Such that Mary Douglas and Baron Isherwood (1979) describe the contents of the house and garden as 'the visible bit of the iceberg' (quoted in Duncan 1981: 175). These semi-fixed domestic items, from curtains and wallpaper to pictures and books, are all supposed to help inhabitants to communicate an identity and outsiders to read it (Rapoport 1981). Many asymmetrical family homes are impregnated with 'heterosexuality'. Its overwhelming presence seeps out of everything from photograph albums to record collections. But the love that dare not speak its name in the family house can hardly cover the walls and smile down from the picture frames. And so lesbians restrict the performance of their sexual identity in their own physical surroundings, hiding pictures of lesbian icon kd lang under the mattress and gay fiction behind the bookcase, ever cautious that the privacy of their bedroom may be subject to the gaze of brothers, sisters and parents.

The constraints on the performance of a lesbian identity don't stop at the bedroom door. Judith Butler (1990) has critiqued gender, sex and the body as categories, suggesting that they are discursively produced by the effects of various institutional practices and discourses. She argues that the body is not a ready surface awaiting signification but a set of boundaries, 'a surface whose permeability is politically regulated and established' (1990: 139). As these women describe, the parental home can inscribe the lesbian body. While still 'be-ing' a lesbian, for these women there is not the repetition or redoubling of the role that is necessary for the lesbian category to be expressed in a heterosexual environment.

> '[I] cover my tattoos up when I go home, especially if mum and dad have company coming over. I do that, it doesn't worry me, that's it.'
>
> Jackie, New Zealand lesbian

> '[I dress more] conservatively . . . kind of straight and less scruffy.'
>
> Hayley, New Zealand lesbian

The home can therefore be a site of tension for women who identify as lesbians – a place where the ideal of the home as a place of security, freedom and control meets the reality of the home as site where heterosexual family relations act on and restrict the performance of a lesbian identity. Rather than being 'where above all one feels "in place"' (Eyles 1984: 425), 'at home' is where many lesbians feel 'out of place' and that they don't belong or fit in. In Somerville's (1992) terms, home may have meaning as a 'shelter' and an 'abode' but not as 'roots' or 'paradise'.

> 'I mean, as much as I love my family I always feel I don't fit in. The only place I feel at ease is with gay people . . . I feel I sit in a room full of my family and I feel I'm just not part of this, I don't fit in.'
>
> Jane, English lesbian

This lack of ontological security can also be accompanied by a lack of actual physical security. Research shows that whilst lesbians experience less abuse at the hands of strangers than gay men, they are at the receiving end of more domestic violence perpetrated by family members 'disgusted' by their sexuality (Berrill 1992; Comstock 1989). This violence, which can range from physical assault to verbal intimidation and harassment, contributes to shattering the myth of 'home as a haven'.

But the heterosexual family home isn't only a site of oppression, but also a site of subversion – a place where a lesbian identity can sometimes be discreetly performed so that it is not read as such by other family members. For example, by dressing in a way

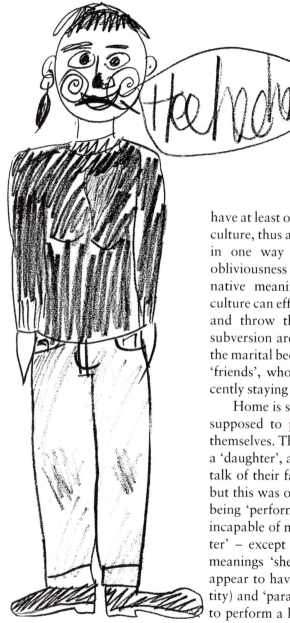

that has lesbian meaning for them, wearing discreet lesbian jewellery, or more subtly by listening to the music of lesbian icons like kd lang (Bradby 1993), women can eke out a lesbian identity in a home environment that is constraining and repressive. This is most easily done when women appropriate musicians, athletes, TV programmes and so on who have at least one foot planted squarely in heterosexual culture, thus allowing a lesbian audience to read them in one way whilst having a quiet laugh at the obliviousness of friends and relatives to their alternative meanings (Plate 7.1). In this way lesbian culture can effectively take symbols of heterosexuality and throw them back in its face. Other acts of subversion are less subtle with women having sex in the marital bed when their parents are out or sneaking 'friends', who are to all intents and purposes innocently staying over for the night, into their own bed.

Home is supposed to be where your heart is. It is supposed to provide a space for individuals to be themselves. The parental home may meet the needs of a 'daughter', and most of the women interviewed did talk of their family home as 'loving and supportive', but this was only when their lesbian identity was not being 'performed'. The parental home seems largely incapable of meeting the needs of the 'lesbian daughter' – except in a material sense. It may have the meanings 'shelter', 'abode', 'hearth', but it doesn't appear to have the meanings 'privacy', 'roots' (identity) and 'paradise' (ideal home). Rather the freedom to perform a lesbian identity (Plate 7.2), to relax, be in control and to enjoy the ontological security of being 'at home' appears to be best met when lesbians can create and manage their 'own homes'.

Plate 7.1 A self-portrait by a
New Zealand lesbian
Source: Lynda Johnston

AT HOME: MAKING LESBIAN SPACE

Housing in nineteenth- and twentieth-century Britain has been and is 'primarily designed, built, financed and intended for nuclear families – reinforcing a cultural norm of family life with heterosexuality and patriarchy high on the agenda' (Bell 1991: 325). As Louise Johnson argues:

> What is being offered to both women and men is a set of recognisable cultural symbols (chief of which is the suburban home and its ownership) ... [but housing] also allows the subversion of dominant social relations. For there is no reason why 'Bedroom 1' should not be occupied by children or a gay couple.
>
> Johnson 1992: 44

Lesbians occupying a home built on these traditional cultural symbols often do subvert them by making structural changes to the house to express a non-heterosexual identity or lifestyle, as this woman explains.

> 'I've made my room ... I've built a mezzanine bed, and I have lots of things of comfort around me in my room ... Things which reflect me and reflect the things that I've done in my life, or the people that are important to me and my lifestyle.'
>
> Mary, New Zealand lesbian

Louise Dooley (1985) argues that there may be connections between the self which is invested in private and personal things and the home, because the home as the house of these things automatically assumes a sacred nature. Notwithstanding this, women also make more conscious efforts to produce

Plate 7.2 'Being oneself' – a New Zealand lesbian 'at home'
Source: Lynda Johnston

a space within which they feel 'at home'. Posters depicting famous lesbians, pictures, personal photographs, music collections and colour schemes (and the compulsory cat and/or dog)[3] are used to make the house a 'lesbian' space. Some women living in shared houses, particularly those influenced by lesbian feminist politics, also attempt to create different ways of living within the framework of the house, for example by producing combined living spaces and by organising daily activities on a collective basis (Egerton 1990; Ettorre 1978).

But our identities are not singular; they are multiple and often contradictory. Identities performed by lesbians in their homes may produce discordant spaces and odd juxtapositions; on the one hand, spaces may resonate with lesbian identities, on the other hand they may resonate with childhood things that reflect the identities of the 'child', 'the daughter' and 'the biological family'. This interface of needs and desires between the lesbian home and the parental home is captured by this woman, who expresses a need to have artistic objects around her that remind her of her upbringing, and her attachment to her 'family' home:

'Pictures, colours and comfortable things from my family, like, my mum, my grandmother and my aunty all paint and they've always had paintings around them. That's been their hobby, and I pick up from that.'

Elaine, New Zealand lesbian

Tensions between parents and a daughter's lesbian identity can resurface even when she has fled the heterosexual nest. Having a home of one's own may allow a woman enough control over the space to express her sexuality in the physical environment but it doesn't necessarily guarantee freedom from the prying eyes of parents, relatives and neighbours. Discouraging people from popping in and trying to arrange planned rather than spontaneous visits can buy enough time for the home to be 'prepared' for visitors. Alternatively, visitors can be limited to one or two rooms that are 'produced' for public scrutiny to symbolise the whole home (Allan 1989; Mason 1989). For example, the living room and dining room provide a formal statement about the home for outsiders whereas the bedroom has a greater aura of privacy and is an easier space in which to perform a lesbian identity. One New Zealander explained that she restricted her parents' movements within her home in order to stop them entering rooms covered with lesbian posters, but unfortunately the off-limit rooms included the toilet. In the event, her parents didn't need to use the bathroom during their visit. If they had she would have been forced 'to come out' or be found out. Her flatmate recalled the experience as being 'quite nerve racking'.

One way to take the tension out of these fraught occasions is to change the performance of the home according to the identity of the visitor. Whilst some women 'de-dyke' the house completely, others make more subtle changes depending on the level of discomfort likely to be expressed by visitors or experienced by the occupants.

'Things like *Macho Sluts* [a book] would go down like a cup of cold sick. I freak out and run round and de-dyke the whole place.'

Joanne, New Zealand lesbian

'[If her mother was feeling] really uncomfortable ... I mean there would probably be some things that I'd remove so that it [lesbianism] wasn't so blatantly obvious.'

Sarah, New Zealand lesbian

'My parents came to stay and they slept in my room, and I was really aware of the poster I had on the wall which has got a picture of myself and [the word]

Plate 7.3 'My home'
Source: Lynda Johnston

"lesbian", and it's got heaps of positive words for lesbians around it. And I was really aware of that picture all the time and I really wanted to see what their reaction was . . . I wondered if they even think about why I had it there.'

Janice, New Zealand Lesbian

Despite these moments the lesbian home can be more than a place of arrival and departure and a location for shuffling a pack of identities and laying out a different hand. It can also be a focal point of lesbian activities, a place of support, a sanctuary and a secure area where a lesbian identity can be maintained usually without threat from other occupants. In Somerville's terms it can embrace 'roots' and 'paradise'. This is reflected in Plate 7.3, a drawing of 'home' by a New Zealand lesbian; and by these quotations:

'It's comfortable for me as a lesbian, 'cos I know I can come home after a day . . . and it could be a really challenging day, and I'm able to talk it out. I can just be me, I don't have to face challenges unless it comes on the televison or the radio. And then we [lesbian flatmates] can all stand round and yell at the TV.'

Sue, New Zealand lesbian

'It's like when with my parents, I have to live a separate life and with work it has to be a separate life and it gets sort of hard work at times, keeping everybody happy, everybody in their place, being one thing to one person, not ever being your *whole self* except obviously *in the home*.'

Stacey, English lesbian

As this last quotation explains, juggling multiple identities in public space in this way, so that a lesbian identity is not performed to the 'wrong' audience, can be nerve-racking. To be seen by family, friends or colleagues going into a gay bar or holding hands in the street with another woman can rupture a carefully cultivated illusion of heterosexuality. 'The home', particularly for those who are very wary about the personal and employment consequences of being 'outed', can therefore take on a vital role as a lesbian social venue and meeting place. Indeed, in many provincial towns and rural areas, informal networks of private homes fill the entertainment gap created by a complete absence of lesbian institutional spaces. And, in other places, homes become alternative focal points for groups of women alienated from gay bars and institutional spaces because of political or personality clashes.

'Most of the lesbian bit of my life is home-based I suppose, with supper parties and things.'

Sara, English lesbian

But this is not to suggest that the lesbian home is anymore the idyllic romanticised haven than the heterosexual nuclear family home was before it fell from grace under the weight of feminist critiques of domestic violence and so on. Like the heterosexual home, the lesbian home is also a site of conflict and disagreement. It is a site where a lesbian identity must be performed, but it is also a site where this identity comes under surveillance from other lesbians. 'Political correctness', which has come to haunt the lesbian feminist landscape, or other 'orthodoxies', can be invoked by some women to regulate the performative aspects of others' lesbian identities within the domestic environment.

Children can also be at the heart of domestic conflict. Anyone replacing 'Dad' in mother's affections is liable to run the gauntlet of children's anger and jealousy but this hostility can take on a new edge when Mum's new partner is another woman. Children's aggression and rejection can also be accompanied by overt attempts to exaggerate their own performance of heterosexuality. Not surprisingly, being constantly under surveillance in this climate can limit the performance of lesbian sexuality in a 'lesbian home', by inhibiting women from being affectionate to each other, sharing a bed and so on. Ultimately the clash of identities under the same roof can come to a head with either the 'child' deserting the 'lesbian home', often to take up residence in the heterosexual environment of a grandparents' home; or the mother's lover being driven out as the children struggle to reproduce the space as a 'normal family' home.

'She [daughter, aged 6] gets very upset that we live together. She gets very jealous. She heard me saying to Pat that I love her. And she said "Don't say that, I don't want to hear it". She's very set that she's gonna get married ... So it's very difficult to be a couple and relax together when Tracey's so hostile.'

Maria, English lesbian

'when the relationship before this one, was breaking up, I did have a conversation with them both and she [daughter] did say "Does that mean we are going to be back to normal?" and what she meant was that would I no longer be a lesbian. So of course I found that quite upsetting.'

Lesley, English lesbian

The privacy of the home is not always the same thing as privacy from the neighbours. Prying eyes over the garden fence, eavesdropping through badly soundproofed walls, and the efficiency of local gossip networks can expose the most 'closeted' of couples to neighbourhood surveillance. Usually, this evokes nothing more than a few snide or petty remarks but occasionally lesbian and gay homes can become the target of hate campaigns or vicious attempts to restore the 'respectability' of the neighbourhood by driving the occupants out of the street (Harry 1992). In studies of anti-gay attacks in Pennsylvania State, Anthony D'Augelli (1989) found that of 125 incidents of victimisation recorded by participants, 17 per cent involved property being damaged or destroyed. The 'home' is not always therefore a place of emotional and physical well-being.

> 'When we first moved in the person who lives the first house round the corner said "Are you sisters?" And I said "no, just friends" and I think we got known as the couple of dykes on the corner . . . One night . . . we were in bed and it woke me up and one of them said "Queers live here". But I didn't hear anything else and there was a bit of giggling and I felt like hanging out the window and saying "Yes, and they're trying to get some sleep!" '
>
> Chris, English lesbian

> 'There's a neighbour who's a bit of a worry . . . I walked past them [neighbour and his son] and they said "fucking dyke". I just ignored it and carried on walking. When they started giving Mike [another neighbour who the two men have accused of being gay] trouble and wrecking his car and things [by pouring acid on it], I got more careful about kissing Emma goodbye at the door and whatever because I don't want to be victimised.'
>
> Jo, English lesbian

Ironically, the more lesbians withdraw from local and family life and put up barriers against outsiders, the more this privacy can become an isolation that suffocates the relationship or facilitates one partner's ability to emotionally manipulate or physically abuse the other (Hall 1992; Mann 1993). This isolation from heterosexual friends, neighbours and family can often be compounded by a lack of contact with other lesbians, particularly in provincial towns and rural areas where there are few places and opportunities for gay women to meet. Thus two women in a lesbian relationship can often become very dependent on one another. This dependence can give one the power to control or dominate the other, especially if one woman is just 'coming out' or has less experience of a lesbian lifestyle than the other. NiCarthy (1982: 234) argues that: 'An abusive lesbian might insist she knows what is "correct" about the lesbian lifestyle, as if there is only one.' Emotional abuse, about how a woman dresses, behaves and what she should do to be a lesbian, is, according to Lezli Mann (1993), often the prelude to actual physical harm. For lesbians trapped in these relationships the difficulties of telling a friend, colleague or relative about the abuse and seeking help are often compounded by the trauma of having to simultaneously 'come out', concern about how this information will be received and a fear that not only one's relationship but also one's sexuality will be judged negatively (Mann 1993).

While a lesbian home may become almost a prison or a very static and stifling place to be, lesbian homes can also be very fluid and unstable environments. Because of the limited opportunities lesbians have to meet one another in everyday environments, lesbian social networks, particularly in provincial towns, can be very incestuous (Valentine 1993c).

> 'It would be fascinating to do a genogram, not of parents but of ex-partners. It would be like this [hand action] – so jammed and intertwined. It's like a family. It's a lot more interwoven than a comparable heterosexual set-up. We used to have a saying "sisters-in-lust" because so-and-so's slept with so-and-so, so we're all indirectly related.'
>
> Vicky, English lesbian

As this quotation implies, as women shuttle between relationships so the ownership and occupation of lesbian homes can become fluid and complex as women hop from living in one 'lesbian home' to another. This movement can also produce different sorts of living arrangements and alternative conceptions of what constitutes 'home' that may have greater meaning in terms of 'identity' or 'privacy' but less in terms of material security as 'shelter' or 'abode'.

CONCLUSION

We all have a multiplicity of subject positions and identities. 'Home' is one site where our identities are performed and come under surveillance and where we struggle to reconcile conflicting and contradictory performances of the self. 'Home' itself is also a term laden down with a baggage of multiple meanings: shelter, abode, hearth, heart, privacy, roots, paradise and so on. For women who identify as lesbians, the parental (or 'family') home is often a site where they have to manage the clash of their identity as a lesbian with their identity as 'daughter' from a heterosexual family. The struggle to control how their identity is read and received under the surveillance of vigilant parents can rob the parental home of its meaning as a place of 'privacy', 'roots' and 'paradise'. Whilst being a place of material and emotional comfort ('shelter', 'abode', 'hearth' and even 'heart') that can meet the needs and desires of the 'daughter', the parental home does not appear to meet the needs and desires of the 'lesbian'. It is a location where lesbianism and heterosexuality do battle. The heterosexuality of the home can inscribe the lesbian body by restricting the performative aspects of a lesbian identity but it can also be subverted itself by covert acts of resistance.

The 'lesbian home' is one site of lesbian identity construction and maintenance.

Constituted to meet the needs and desires of lesbians, it appears to be a place of significance, of 'roots' and even 'paradise', for many women. But despite the greater freedom to perform a lesbian identity within the boundaries of a 'lesbian home', it is still a location where this identity comes under the surveillance of others, especially close family, friends and neighbours. It is not necessarily a place of 'privacy'. In some cases the physical site of the home is actually altered depending on the relationship of the visitor to the occupants so that a lesbian identity is not performed in the physical environment to the 'wrong audience', thereby disguising the identity of the occupants. Alternatively, in an attempt to create the privacy necessary to conceal a lesbian relationship, couples can often withdraw from family, friends and the local neighbourhood and become isolated. This isolation can become stiflingly claustrophobic, smothering relationships and enabling abusive domestic situations to develop unnoticed under this cloak of privacy. Thus a lesbian home is not necessarily a place of emotional and physical well-being ('hearth' and 'heart'). Neither is it always a stable 'shelter' or 'abode' – domestic conflicts between women and their children and the usual ebb and flow of sexual relationships can all contribute to a fluidity in the membership and constitution of lesbian households.

The meanings of 'home' to the lesbians involved in this research are numerous and beset with contradictions. They are perhaps most neatly summed up by Massey when she writes about the home (in a different context): 'each home-place is itself ... a complex product of the ever-shifting geography of social relations present and past' (Massey 1992: 15).

ACKNOWLEDGEMENTS

Lynda Johnston would like to acknowledge the invaluable ongoing encouragement, support and advice from Robyn Longhurst and Robin Pearce at the University of Waikato. Thanks are due to Lou Englefield for suggesting the chapter title.

NOTES

1 This paper is based on findings generated by focus group discussions exploring the meaning of home conducted by Lynda Johnston with lesbians in New Zealand and by forty in-depth taped interviews about lesbians' perceptions of everyday places that were carried out in the UK by Gill Valentine. The drawings used in this text are self-portraits and sketches of the lesbian home made by participants in Lynda Johnston's research.

2 Gay women obviously also live in many other forms of 'home', for example with male partners in marital homes, in rented accommodation and so on, and are of course also 'homeless'. These issues, however, lie beyond the scope of this chapter.

3 In the UK within lesbian culture it is joked that the stereotypical lesbian has at least one cat. In New Zealand 'she' owns dogs!

8

THE LESBIAN *FLÂNEUR*

•

Sally Munt

I haven't been doing much *flâneuring* recently. Six months ago I moved from the British coastal town of Brighton, where I'd lived for eight years, to the Midlands city of Nottingham, chasing a job. A four-hour drive separates the two, but in terms of my lesbian identity, I'm in another country.

Geographically, Nottingham is located in the exact centre of England: the land of Robin Hood. This local hero is mythologised in the region's heritage entertainment – next to the (fake, nineteenth-century) castle, one can purchase a ticket for *The Robin Hood Experience*. Nottingham, formerly a hub of urban industry, is nostalgic for a time when men were men, and codes of honour echoed from the heart of the oak, to the hearth, to the pit. D.H. Lawrence is this city's other famous son. English national identity is thus distilled into a rugged romanticised masculinity, an essence of virile populism which is potently enhanced by its attachment to the core, the fulcrum, of England. Its interiority is endemic to the boundaries which entrap it; in its corporeality it is the heart, the breast, the bosom, and to each tourist is offered the metaphoricity of home.

Brighton is on the edge. Thirty miles from France, this hotel town is proud of its decaying Regency grandeur, its camp, excessive, effeminate façades. It loves the eccentricity of Englishness, but laughs at the pomposity of England. Brighton looks to Europe for its model of bohemia, for it is just warm enough to provide a pavement culture to sit out and watch the girls go by. Brighton, the gay capital of the South, the location of the dirty weekend, has historically embodied the genitals, rather than the heart. Its sexual ambiguity is present on the street, in its architecture, from the orbicular tits of King George's Pavilion onion domes, to the gigantic plastic dancer's legs which extrude invitingly above the entrance to the alternative cinema, the Duke of York's. Aristocratic associations imbue the town with a former glory. Its faded past, its sexual history, is a memory cathecting contemporary erotic identifications as decadent, degenerative and whorelike.

The stained window of nineteenth-century permissiveness filters my view of Brighton. Promenading on a Sunday afternoon on the pier, loitering in the Lanes, or taking a long coffee on the seafront, ostensibly reading the British broadsheet *The Observer*, the gaze is gay. Brighton introduced me to the dyke stare, it gave me permission *to* stare. It made me feel I was worth staring at, and I learned to dress for the occasion. Brighton constructed my lesbian identity, one that was given to me by the glance of others, exchanged by the looks I gave them, passing – or not passing – in the street.

It's colder in Nottingham. There's nothing like being contained in its two large shopping malls on a Saturday morning to make one feel queer. Inside again, this pseudo-public space is sexualised as privately heterosexual. Displays of intimacy over the purchase of family-sized commodities are exchanges of gazes calculated to exclude. When the gaze turns, its intent is hostile: visual and verbal harassment make me avert my eyes. I don't loiter, ever, the surveillance is turned upon myself, as the panopticon imposes self-vigilance. One night last week, I asked two straight women to walk me from the cinema to my car. The humiliation comes in acknowledging that my butch drag is not black enough, not leather enough, to hide my fear.

As I become a victim to, rather than a perpetrator of, the gaze, my fantasies of lesbian mobility/eroticism return to haunt me. As 'home' recedes, taking my butch sexual confidence with it, my exiled wanderings in bed at night have become literary expeditions. As I pursue myself through novels, the figure of the *flâneur* has imaginatively refigured the mobility of my desire. These fictional voyages offer me a dream-like spectacle which returns as a memory I have in fact never lived. Strolling has never been so easy, as a new spatial zone, the lesbian city, opens to me.

The *flâneur* is a hero of Modernity. He appeared in mid-nineteenth-century France, and is primarily associated with the writing of poet Charles Baudelaire; he appears successively in the criticism of the German Marxist and follower of the Frankfurt School, Walter Benjamin, in the 1930s. The economic conditions of rising capitalism that stimulated his appearance resulted in the rise of the boulevards, cafés and arcades, new spaces for his consumption of the city-spectacle. Neither completely public, nor completely private, these voyeuristic zones were home to the *flâneur*, engaged in his detached, ironic and somewhat melancholic gazing. He was also a sometime journalist, his writings on the city being commodified as short tableaux in the new markets for leisure reading. His origin, in Paris, that most sexualised of cities, traditionally genders his objectivication as masculine, his canvas, or ground, as feminine.

Elizabeth Wilson (1992) has taken issue with the predominant feminist opinion that this *flâneur* is essentially male. She writes in the presence of women as subjects in this urban narrative. She also directs us to acknowledge the figure's insecurity, marginality and ambiguity, rejecting the preferred version of the *flâneur*'s voyeuristic mastery:

Benjamin's critique identifies the 'phantasmagoria', the dream world of the urban spectacle, as the false consciousness generated by capitalism. We may look but not touch, yet this tantalising falsity – and even the very visible misery of tramps and prostitutes – is aestheticised, 'cathected' (in Freudian terms), until we are overcome as by a narcotic dream. Benjamin thus expresses a utopian longing for something *other than* this urban labyrinth. This utopianism is a key theme of nineteenth- and twentieth-century writings about 'modern life'. In Max Weber, in Marxist discourse, in the writings of postmodernism, the same theme is found: the melancholy, the longing for 'the world we have lost' – although precisely what we have lost is no longer clear, and curiously, the urban scene comes to represent utopia and dystopia *simultaneously*.

E. Wilson 1992: 108

The *flâneur* is fascinated, transfixed and thus trapped into representing wishes, without fulfilment:

The flaneur represented not the triumph of masculine power, but its attenuation . . . In the labyrinth, the flaneur effaces himself, becomes passive, feminine. In the writing of fragmentary pieces, he makes of himself a blank page upon which the city writes itself. It is a feminine, placatory gesture . . .

E. Wilson 1992: 110

Is the *flâneur* someone to be appropriated for our *post*modern times? I don't wish to rehearse the arguments concerning whether the *flâneur* is a good or bad figure, partly because they tend to be articulated within a heterosexual paradigm, reliant upon heterosexual discourses of the city. I'm interested in this observer as a metaphor, who offers at once a symbolic hero and anti-hero, a borderline personality in a parable of urban uncertainty, of angst and anomie. Within the labyrinth, the process of making up meaning in movement becomes the point, and perversely too the pleasure, as we become lost among the flowing images. This act of performative interpretation is crystallised in this early urban tale of lesbian cross-dressing:

So I had made for myself a *redingote-guérite* in heavy gray cloth, pants and vest to match. With a gray hat and large woollen cravat, I was a perfect first-year student. I can't express the pleasure my boots gave me: I would gladly have slept with them . . .

quoted in Moers 1977: 12

What happens if the *flâneur* is cross-dressed not just in actuality, here as George Sand vogueing in her butch drag dandy suit, but symbolically too? Writing in 1831, she claimed 'my clothes feared nothing' (ibid.: 12). When she is dressed *as a* boy, she is all-image, a spectacle of auto-eroticism, desired only by herself – 'No one knew me, no one looked at me, no one found fault with me ...' (ibid.). *As* such she is a simulcrum, if, as Wilson (1992: 109) continues on to argue 'the flaneur himself never really existed', then there is no material ground of maleness *or* femaleness to be invoked. Is the *flâneur* a transvestite? Can s/he be a cross-dressed lesbian? It's possible the *flâneur* is a borderline case, an example of a roving signifier, a transient wild-card of potential, indeterminate sexuality, trapped in transliteration, caught in desire.

One crucial problem with the conventional line on the *flâneur* is the idea that he roams the streets *untouched*. As pure male essence his visual trajectory-projectile is uncorrupted – he sees windows, not mirrors. To stretch the analogy, even the clearest window will frame the picture, and reflect back the tiniest reflection of self. I'm simplifying, condensing, extracting and probably bowdlerising the *flâneur* here, as a vessel to be filled by the lesbian narrative, in order that I can contribute to the unfixing of the supremacy of the heterosexual male gaze in urban spatial theory.

Preliminary writers to procure the form included Renée Vivien and Djuna Barnes. The poet and traveller Renée Vivien imagined a visionary lesbian city, Mytilène, as an escape from early twentieth-century Paris. The lesbian voyager's imagination is freed from cultural constraints to wander at will, for in this Sapphic paradise all temporal and spatial barriers are excised. The fantasised map of Lesbos has no restrictions, but critic Elyse Blankley (1984: 59) has noted how the *real* island of Lesbos turned out to be Erewhon: Vivien, on her frequent visits, refused to leave her villa, finding the native women 'unattractive and disappointing'.

Both Djuna Barnes' descriptions of the 1920s' Paris salon culture in her novel *Ladies Almanack* (1928), and particularly the character of Dr O'Connor in *Nightwood* (1936), retain elements of the Modernist *flâneur* (Tyler Bennett 1993). Ur-*flâneur*ing is also evident in her journalistic sketches collected together in *Djuna Barnes in New York* (1990), which combine to form a panorama of city life from 1913 to 1919. Predominantly, Barnes is remembered as an expatriot in Paris, thus a traveller, and an outsider ideally located to comment on an alien, European, culture. Her positioning in the New York text as an exile is particularly revealing. She returns to the city not as a native, but retains the inside/outside dichotomy of the alienated raconteuse, rendering snapshots of a foreign territory. She is the first to emigrate the *flâneur*, taking a European-derived model and appropriating it for US culture.

During the 1920s homosexuality was located in New York in two identifiable spaces, Greenwich Village and Harlem. Homosexuality was made permissible by journeying to a time-zone happening: one *experienced* a present event, rather than took one's pre-formed sexual identity, intact and inviolate, to the party. Social mobility was a prerequisite for sexual experimentation – the bourgeois white *flâneurs* who went 'slumming' in Harlem paid to see in the exoticised black drag acts and strip-shows, a voyeuristic legitimation of their own forbidden fantasies (Faderman 1992).

Margins and centres shift with subjectivities constantly in motion. At the beginning of the twentieth century there was a massive migration of black people from the south to the north of the USA, and many of them came to New York, specifically to Harlem,[1] to make home (Mulvey 1990). Writer James Weldon Johnson dated the beginning of black Harlem to 1900, calling it 'the greatest Negro city in the world . . . located in the heart of Manhattan' (quoted in Locke 1975: 301). A character in a magazine story, 'The City of Refuge', printed in *Atlantic Monthly* in February 1925, exclaims 'In Harlem, black was white' (quoted in Locke 1975: 57). This was (and is) black space, not white space. Art and literature has mythologised the migrant's arrival in Harlem into the making of a new black identity, stimulating the emergence of a new consciousness. It is a continuous happening, endlessly repeated with the arrival of each new traveller from the south, emerging from the subway station. Can we read Ralph Ellison's *Invisible Man* (1952) as another alienated and invisible *flâneur*?

> This really was Harlem, and now all the stories which I had heard of the city-within-a-city leaped alive in my mind . . . For me this was not a city of realities, but of dreams . . . I moved wide-eyed, trying to take in the bombardment of impressions.
>
> Ellison 1982 [1952]: 132

The utopian/dystopian paradox of hope for the city is that more pleasure is taken in the journeying towards it, as a process of desire and transformation, than in the (deferred) arrival. Models of the labyrinth, in which the journey is represented as circular, make this explicit. The boundaries of physical geographies are rebuilt in mental images. 'Harlem' operates as a symbol of black consciousness rather like 'Africa' does – as 'a self-created ontology of blackness' (De Jongh 1990: 145), a myth of 'home' which makes home bearable.[2]

Small groups of lesbians congregated in both Harlem and Greenwich Village during the 1920s. These were different worlds of homosexual identification, divided by race and class. Greenwich bohemian life tolerated a degree of sexual experimentation which conferred upon the area an embryonic stature as erotica unbound, a construction much enhanced during the 1950s and 1960s. As Harlem had functioned as the mecca for black people, now Greenwich Village became the Promised Land for (mainly) white homosexuals. Resisting the conformity of 1950s' small-town suburbia, men and women in the post-war USA were drawn to cities as a place to express their 'deviant' sexuality. Their newly-acquired gay and lesbian identities were predominantly urban, emanating from the social geographies of the streets. The anonymity of the city made a gay life realizable in a repressive era. This odyssey is well represented in the lesbian novels of the period (see Weir and Wilson 1992).

Nightclubs were a visible site for women interested in 'seeing' other women, and it is in this literature of the 1950s and 1960s that the bar becomes consolidated as the symbol of home (K. King 1992). Lesbian/whore became a compacted image of sexual consumption in the dime novel of the period, read by straight men and lesbians alike. The lesbian adventurer inhabited a twilight world where sexual encounters were acts of romanticised outlawry initiated in some backstreet bar, and consummated in the narrative penetration of the depths of maze-like apartment buildings. She is the carnival queen of the city: 'Dominating men, she ground them beneath her skyscraper heels' (Keene 1964: back cover), a public/private figure whose excess sensuality wishfully transcends spatial and bodily enclosures. This Modernist nightmare of urban sexual degeneracy is crystallized in the identification of the city with homosexuality. Lesbian-authored fictions of the period, like the *Beebo Brinker* series (1957–62), are less sensationalist syntheses of the available discursive constructions of 'lesbian', but still depend on a myth of the eroticised urban explorer (see Hamer 1990). Transmuting in more liberal times into the lesbian sexual adventurer, this figure can be recognised in diverse texts, from Rita Mae Brown's post-sexual revolution *Rubyfruit Jungle* (1973) to the San Franciscan postmodernist porn parody *Bizarro in Love* (1986) by Jan Stafford.

Within contemporary lesbian writing we encounter a specific, even nostalgic, image of the stroller as a self-conscious lesbian voyeur. The years of feminist debate engrossed with the political acceptability of looking are the background to these lesbian vindications of the right to cruise:

> New words swirl around us
> and still I see you in the street
> loafers, chinos, shades.
> You dare to look too long
> and I return your gaze,
> feel the pull of old worlds
> and then like a femme
> drop my eyes.
> But behind my broken look
> you live
> and walk deeper into me
> as the distance grows between us.

Joan Nestle's first stanza from 'Stone Butch, Drag Butch, Baby Butch' (1987a) ends with the comment 'Shame is the first betrayer'. The extract epitomises the mechanisms of a necessarily coded visual exchange, in a potentially violent, dangerous and sexualised arena – the street. The pun of the title of the anthology is *A Restricted Country* and the spatial penetration of the poem recalls this analogy between the streets and the lesbian body. Inside/outside dichotomies break down, both becoming colonised. A subculture made invisible by its parent culture logically resorts to space-making in its collective imagination. Mobility within that space is essential, because motion continually stamps new ground with a symbol of ownership.

Is the butch dandy strolling through the doors of the bar just a romanticised inversion of heterosexual occupation? The *flâneur* may not have to be biologically male for the gaze to enact masculine visual privilege. The politics of butch/femme and their relation to dominant systems of organising gender relations have been bloodily fought over (see, for example, Hollibaugh and Moran 1992), and whilst I am sympathetic to claims that butch/femme constitute new gender configurations which must be understood within their own terms, they are not intrinsically radical forms springing perfect from the homosexual body. Nor are they naive forms in the sense that they express a naturally good, pure and primitive desire. Nestle's poem is interesting in that it represents the push/pull, utopian/dystopian contrariety of the ambivalent *flâneur*, balancing the temptation and lust for the city (embodied as a woman), with the fear of connection and belonging. Note that the narrator of the poem initiates the glance, then returns the gaze and then becomes the owner of a 'broken look' (line 9). The butch penetrates with her gaze ('walk deeper into me' (line 11)) an assumed femme who is only 'like a femme' (line 7). Evading categorisation, this 'almost femme' narrator is the one whose closing comment of the stanza rebukes invisibility and averted eyes. Who is claiming the gaze here? All we can assume is that it is a woman.

The poem describes movement: both characters are in motion on the street, and the looks which they exchange have their own dynamic rotation. Images of mobility are particularly important to lesbians *as women* inhabiting the urban environment. Feminist struggles to occupy spheres traditionally antipathetic to women go back to the imposition of post-industrial revolution bourgeois family divisions into male–public/female–private spaces, an ideological construction disguising the fact that the domestic space, the 'home', as Mark Wigley (1992: 335) has written, is also built for the man, to house his woman:

> The woman on the outside is implicitly sexually mobile. Her sexuality is no longer controlled by the house. In Greek thought women lack the internal self-control credited to men as the very mark of masculinity. This self-control is no more than the maintenance of secure boundaries. These internal boundaries, or rather boundaries that define the interior of the person, the identity of the self, cannot be maintained by a woman because her fluid sexuality endlessly overflows and disrupts them. And more than this, she endlessly disrupts the boundaries of others, that is, men, disturbing their identity, if not calling it into question.

The familiar construction of woman as excess has radical potential when appropriated by the lesbian *flâneur*. The image of the sexualised woman is double-edged, a recuperable fantasy. Swaggering down the street in her butch drag casting her roving eye left and right, the lesbian *flâneur* signifies a mobilised female sexuality *in control*, not out of control. As a fantasy she transcends the limitations of the reader's personal circumstances. In her urban circumlocutions, her affectionality, her connections, she breaks down the boundary between Self and Other. She collapses the inviolate distinction between masculinity and femininity. Her threat to heteropatriarchal definitions is recognised by hegemonic voices, hence the jeering shout 'Is it a man or is it a woman?' is a cry of anxiety, as much as aggression. The answer is neither and both: as a Not-Woman, she slips between, beyond and around the linear landscape. The physiology of this *flâneur*'s city is a woman's body constantly in motion, her lips in conversation (Irigaray 1985b).

Although the lesbian *flâneur* appears as a shadow character or a minor theme in a number of recent novels, I want briefly to offer examples of her appearance as a structuring principle in three New York fictions: a short story, 'The Swashbuckler', by Lee Lynch (1990), *Don Juan in the Village* by Jane de Lynn (1990) and *Girls, Visions and Everything* by Sarah Schulman (1986).

> Frenchy, jaw thrust forward, legs pumping to the beat of the rock-and-roll song in her head, shoulders dipping left and right with every step, emerged from the subway at 14th Street and disappeared into a cigar store. Moments later, flicking a speck of nothing from the shoulder of her black denim jacket, then rolling its collar up behind her neck, she set out through the blueness and bustle of a New York Saturday night.
>
> Lynch 1990: 241

Perhaps the name 'Frenchy' gives it away – this short passage previews a parodic portrait of the bulldagger as Parisian *flâneur*, complete with portable Freudian phallus (the cigar), given a sexualised ('blue') city to penetrate. The fetished butch drag, the black denims, blue button-down shirt, sharply pointed black boots, garrison belt buckle and jet-black hair slicked back into a bladelike DA[3] constitute the image of the perfect dag. The text foregrounds the plasticity of the role by camping up Frenchy's casanova, gay-dog, libertine diddy-bopping cruising. The sex-scene takes place next to some deserted train tracks, a symbol of transience, travelling and the moment. This generic butch then catches the subway home.

On the journey towards home this *flâneur undresses*. In a classic scene of transformation she then makes herself 'old maidish, like a girl who'd never had a date and went to church regularly to pray for one' (260). In a classic conclusive twist the short story ends with a revelation – she goes home to mother. Fearful of her detecting the sex smell still on her, Frenchy slips quickly into 'the little girl's room' (261) to sluice away her adult self. In the metaphors of change which structure this story, both the closet and the street are zones of masquerade.

The lesbian *flâneur* appears in a more extended narrative as the main protagonist in Jane de Lynn's episodic novel *Don Juan in the Village* (1990). Thirteen short scenes of conquest and submission structure this narrator's sexual odyssey. Kathy Acker has called the book, on its back cover, 'a powerful metaphor of our intense alienation from society and each other. An intriguing portrayal of that strange and trance-like locus where lust and disgust become indistinguishable', a comment which both recalls the *flâneur*'s anomie and highlights the way in which her space is so sexualised. As in 'The Swashbuckler', this novel problematises the predatory erotics of the stroller using irony. In *Don Juan in the Village*, although the protagonist is ostensibly writing from Iowa, Ibiza, Padova, Puerto Rico, or wherever, her actual location is immaterial. The text employs the American literary convention of the traveller in search of (her)self. Delivered with irony, she is a manifest tourist whose every foreign nook temporarily begets a colony of New York City, specifically a Greenwich Village bar, the topos of urbane lesbian identity. Her butch diffidence and boredom unsuccessfully screens a deluded, tragi-comic, self-conscious sexual desperation. Her targets invariably fail to be compliant, and each escapade is a testimonial to her perpetual frustration. This is one moment of supposed sexual triumph:

> As I slid down the bed I saw the World Trade Center out the window, winking at me with its red light. I was Gatsby, Eugène Rastignac, Norman Mailer, Donald Trump . . . anyone who had ever conquered a city with the sheer force of longing and desire.
>
> De Lynn 1990: 186

She is going down on that most evasive of spectacles, the gay Hollywood film star. The star, very politely, but very succinctly, fucks her and dumps her. *Don Juan in the Village* is the solitary *flâneur* stalking the city with the torment of Tantalus in her cunt. Although the narrator confers upon herself the gaze, she is unable to see it through, or through it.

Finally, Sarah Schulman's second novel *Girls, Visions and Everything* (1986) recalls the quest of the American hero/traveller Sal Paradise in Kerouac's *On the Road*:

> Somewhere along the line I knew there'd be girls, visions, everything; somewhere along the line the pearl would be handed to me.
>
> Kerouac 1972: 14

The pearl, a symbol of female sexuality, is something the active masculine narrator seeks to own. This predatory macho role is located historically in the *flâneur*, it is the story of an alienated, solitary sexuality voyeuristically consuming the female body as a ri(gh)te of passage. Modelling herself as *On the Road* with Kerouac, protagonist Lila Futuransky's adventure is similarly self-exploratory, but based on the *female* experiences urban travel offers. Her comparison with Jack is the dream of being an outlaw, reconstructed by a feminist consciousness. Lila's trip is a constant circling between compatriots. Set in Lower East Side New York, she walks the streets, marking out the geography of an urban landscape punctuated by a city mapped out with emotional happenings. Locations are symbols of connection, and constant references to criss-crossing streets remind the reader of the systematic patterns of neighbourhood, in antithesis to the standard early Modernist images of alienation. *Girls, Visions and Everything* is about Lila Futuransky's New York, 'the most beautiful woman she had ever known' (177).[4]

A sardonic wit suffuses *Girls, Visions and Everything*, but there is also melancholic sadness; a sense of decaying nostalgia for a mythical 'home', for streets filled with sisters and brothers sitting languid on the stoop, swopping stories and cementing *communitas*. This is the feminisation of the street, the underworld with a human face, with its own moral and family code. It is rich kids who beat the gays and harass the poor, the prostitutes and the pushers. The lesbians are on the streets, working the burger bar, cruising the ice-cream parlour and clubbing it at the Kitsch-Inn, currently showing a lesbian version of *A Streetcar Named Desire*. Lila meets Emily here, performing as Stella Kowalski. The romance between Lila and Emily is the main plot development in the novel, structuring its five parts. The final chapter sees Lila torn between the 'masculine' desire trajectory of *On the Road* individualism, and the 'feminine' circularity and disruption of affective liaisons. Her friend Isobel urges Lila not to pause:

> 'you can't stop walking the streets and trying to get under the city's skin because if you settle in your own little hole, she'll change so fast that by the time you wake up, she won't be yours anymore ... Don't do it buddy.'
>
> Schulman 1986: 178

The text's constant engagement/disengagement with change and transformation is signified by the urban landscape, which is out of control. Even the protective zones are folding, and yet there are pockets of resistance which pierce the city's metaphoric paralysis with parody: Gay Pride is one such representation, fifty thousand homosexuals parading through the city streets, of every type, presenting the Other of heterosexuality, from Gay Bankers to the Gay Men's Chorus singing 'It's Raining Men', a carnival image of space being permeated by its antithesis. The text tries to juxtapose a jumble of readerly responses, almost jerking the reader into some consciousness of its activity of forming new imaginative space. Lila re-invents New York from her position of other as a heterotopia of cultural intertextuality; she *is* Jack Kerouac, the character not the author, claiming, even as a Jewish lesbian, that '. . . the road is the only image of freedom that an American can understand' (164).

The street is an image of freedom and paradoxically of violence. The female *flâneur* is vulnerable – Lila walks unmolested until the final part of the book whence she is sexually harassed by Hispanics, and saved from serious injury from potential queer-bashers by the black and sick drug dealer Ray. Lila's zone is breaking down: 'People's minds were splitting open right there on the sidewalk' (14).

The fictional worlds start clashing together: Blanche DuBois appears to Emily aged 85 and begging for a dollar. Lila resorts to Emily with a resignation that can only be anti-romance, knowing it is the wrong decision, and nostalgically lamenting the end of the road of selfhood: '*I don't know who I am right now*, she thought. *I want to go back to the old way*' (178).

This whimsical nostalgia also highlights some disillusionment with the post-modernist models of space – wherein the public and private are collapsed onto the street, and the same space is being used by different people in different ways. Hierarchies still exist. Being part of a bigger spectacle, being visible as one subculture among many, may not necessarily create empowerment, only more competition over a diminishing resource.

Three *flâneurs*: Frenchy, *Don Juan* and Lila Futuransky. Each a descendant of eager European voyagers who migrated with their ticket to utopia; each with their separate, feminised, vulnerabilities; each a sexualised itinerant travelling through urban time and space towards a mythical selfhood; none with the sex/gender/class privileges (fixities) of the Modernist *flâneur*. Temporary, simultaneous, multiple identifications mapped out in moments, in the margins, masquerading as the male (and thus *un*dressing him), makes these *flâneurs* engage with the politics of *dis*location:

> And the crucial moment is that brutal instant which reveals that the journey has no end, that there is no longer any reason for it to come to an end. Beyond a certain point, it is movement itself that changes. Movement which moves through space of its own volition changes into an absorption by space itself – end of resistance, end of the scene of the journey as such . . .
>
> Baudrillard 1988: 10

Baudrillard's extended road-poem *America* (1988) is spoken as a man. His narrative of dystopian exhaustion is from the point of view of something being lost. But spatial reconstruction occurs in the moment of presence, however brief. The vacuum sucks us further in, but we need our fictions of consciousness or we will disappear. Lesbian identity is constructed in the temporal and linguistic mobilisation of space, and as we move *through* space we imprint utopian and dystopian moments upon urban life. Our bodies are vital signs of this temporality and intersubjective location. In an instant, a freeze-frame, a lesbian is occupying space as it occupies her. Space teems with 'possibilities, positions, intersections, passages, detours, u-turns, dead-ends, [and] one-way streets' (Sontag 1979: 13); it is never still. Briefly returning to Brighton for the summer, my eye follows a woman wearing a wide-shouldered linen suit. Down the street, she starts to decelerate. I zip up my jacket, put my best boot forward, and tell myself that 'home' is just around the corner.

NOTES

1 Only one of the twelve chapters in Mulvey and Simons (eds) (1990) *New York: City as text* is written by a woman. Perhaps the urban gaze is male after all.

2 In Leslie Feinberg's *Stone Butch Blues* (1993) the Jewish protagonist Jess Goldberg is a he/she, a passing woman, who journeys to New York City to consolidate and make safe her emerging identity. Significantly, as her train travels through the outer urban detritus of NYC, it is seeing Harlem which symbolizes her arrival.

3 'The DA – the letters stand for duck's ass – was a popular hairdo for working-class men and butches during the 1950s. All side hair was combed back in a manner resembling the layered feathers of a duck's tail, hence the name. Pomade was used to hold the hair in place and give a sleek appearance' (Kennedy and Davis 1993: 78).

4 I am aware that I am in danger of entrenching the discourse of 'American exceptionalism'; concentrating my examples in New York encourages the view that it is a 'special' place. It is and it isn't; the myth of New York has a political and cultural specificity in world culture and I am curious about that manifestation. For lesbian and gay people it has a particular set of meanings and associations, and to resist mythologising New York is a difficult practice to perform.

9

FANTASY ISLANDS
popular topographies of marooned masculinity

•

Gregory Woods

Ever since the eighteenth century, and throughout the era of Empire, Western culture has been imposing its values on tropical 'desert islands' while persuading itself it wants to throw off the trappings of 'civilisation'. This trend continues in modern high and popular cultural forms. From UK radio's *Desert Island Discs* to newspaper cartoons and from sentimental romance to gay soft porn, desert islands provide contemporary culture with fertile landscapes for fantasies of pre-industrial peacefulness and prelapsarian sexuality. It is here that the human body can resume its 'natural' condition.

Several main types of island narrative have emerged. In one, marooned children of both sexes grow through adolescent rites of passage into a 'natural' heterosexuality and division of gender roles. In another, isolated males form a relationship with landscape and the elements, then relate homosocially and homoerotically to each other, in febrile renegotiations of their masculinity, before returning to white heterosexual civilisation. In yet a third, an ideal community is conjured up in order to recommend the author's own political theories.

It would not be stretching reality too far to suggest that the 'Orient' is as much a space in the Western sexual imagination as a cultural entity or a segment of the globe. Nor is it confined to the 'East'. As Edward Said has pointed out, while at first 'the Orient was a place where one could look for sexual experience unobtainable in Europe', eventually '"Oriental sex" was as standard a commodity as any other available in the mass culture [of nineteenth-century Europe], with the result that readers and writers could have it if they wished without necessarily going to the Orient' (Said 1978: 190). In the 1850s, Sir Richard Burton developed an elaborate, climatic theory of 'pederasty' which proposed the existence of a 'Sotadic zone' girdling the earth, in which 'the Vice' of sexual relations between men 'is popular and endemic, held at the worst to be a mere peccadillo' (Burton 1970: 159).[1]

Straight or gay, the myth of Third World sexual liberalism has had many modern adherents. The cultural evidence is widespread. From Gaugin's paintings of Tahitian girls to the Club Med's grass huts and sarongs and chocolate Bounty adverts' imagery of heterosexual languor; from Melville's accounts of Polynesian idylls to the gay soft-porn video *Bronze*,[2] tropical islands, conventionally warm and far from 'home', provide the perfect breeding ground for white men's dreams.

Much like imperialism itself, our present concept of masculinity is a product of the Enlightenment – the so-called 'Age of Reason' – especially in so far as masculinity is closely associated with rationality, whereas femininity (and, for that matter, effeminacy) is still supposedly chained to nature (Seidler 1994).[3] Furthermore, built into these ideas was an implicit – or often explicit – understanding that some masculinities are more rational, and therefore more masculine, than others. The black man is childish and therefore, like children, both irrational and close to nature. It is entirely within this tradition of sexual and racial distinctions that the South Sea island narrative develops.

Typically, the island refuge from tempest (or, latterly, from nuclear war) includes the following physical features: a coral reef which, once crossed without mishap, offers shelter from the direct force of the ocean and abundant fishing grounds; a calm and shallow lagoon; a curved, sandy beach (where the castaway first comes round from an exhausted sleep after fighting to survive the shipwreck, and where various useful artefacts are also washed ashore); at either end of the beach, rocks; a fringe of palm trees among which are to be found the castaway's earliest refreshments (coconuts), plus materials for a first fire and rudimentary shelter; a freshwater stream, often running down to the sea from a clear, swimmable pool, surrounded by trees, at the foot of a waterfall (scope here for aquatic erotic fantasies); perhaps a cave, to be used for storage or eventually to form the rear rooms of a house; thick jungle, well provided with fruit trees; a clearing in which primitive peoples have erected, or carved in rock, an inelegant idol or fetish to which, at certain phases of the moon, they return to sacrifice human beings (this place is usually on 'the other side' of the island, and it is on that 'other' side that the savages/cannibals beach their outriggers); a marsh or mud-patch (of which, more later); and, of course, a hill or mountain on which laboriously to build a signal bonfire which, at the crucial moment when a schooner is passing, will have been allowed to die out or will not have been lit at all.

In a very literal sense, life on the island is free. In the words of the narrator of *The Swiss Family Robinson*, 'Money is only a means of exchange in human society; but here on this solitary coast, Nature is more generous than man, and asks no payment for the benefits she bestows' (Wyss 1910: 33). However, the individual castaway appears not to be at liberty to choose how to live his or her life here: for the first priority is to Europeanise island life by domesticating nature. Some stories – *Robinson Crusoe* itself perhaps, *The Swiss Family Robinson* certainly – take this process so far that daily life becomes a systematic manifestation of incongruities, every moment of which either tempts the reader into scornful disbelief or demands a degree of credulity bordering on a dream state. There is a secure aptness to the fact that a great 'realist' film, faithful to the book, was made of *Robinson Crusoe* by the great surrealist director Luis Bunuel.

While it may be possible to 'normalise' one's life on the island by building a house and farming the land, most writers appear to agree that the most difficult initial area of adjustment is in the matter of solitude. By establishing a routine of work to ensure more than a beastly level of survival, the castaway is, in a sense, socialising the lack of society in the place. One will speak to himself; another will place a second chair at his makeshift dining table; a third will call himself Governor and invent a Constitution. It may be that these are acts of madness, or that they keep madness at bay. It is certain, though, that they are not 'natural'.

Island fiction may be taken to represent any situation of human isolation. The details of physical geography are themselves not the whole point. When writing about adolescent life on the Nebraskan prairie, for instance, Willa Cather was in the habit of invoking her own girlhood reading in references to classic island stories. In *My Antonia*, for instance, Jim Burden recalls his prairie boyhood as a comparable adventure: 'In the afternoons, when grandmother sat upstairs darning, or making husking-gloves, I read "The Swiss Family Robinson" aloud to her, and I felt that the Swiss family had no advantages over us in the way of an adventurous life' (Cather 1983a: 66). And again later: 'I got "Robinson Crusoe" and tried to read, but his life on the island seemed dull compared to ours' (100). In *O Pioneers!*, Carl has a collection of magic-lantern slides consisting of, in his words, 'Oh, hunting pictures in Germany, and Robinson Crusoe and funny pictures about cannibals' (Cather 1983b: 17). Resilience on the prairie finds an easy point of comparison in castaway tales: 'Alexandra often said that if her mother was cast upon a desert island, she would thank God for her deliverance, make a garden, and find something to preserve' (29). And the prairie family seems to look to the most famous island family as a fascinating kind of model: 'Carl and Oscar sat down to a game of checkers, while Alexandra read "The Swiss Family Robinson" aloud to her mother and Emil. It was not long before the two boys at the table neglected their game to listen. They were all big children together, and they found the adventures of the family in the tree house so absorbing that they gave them their undivided attention' (63).[4]

Coral or 'desert' islands often represent a temporal and earthly idea of paradise; but in order to do so they must provide one thing more than threatless tropical abundance. Peaceful solitude may be all very well in its way, but a life of marooned perfection must finally include good company – not the mere loyalty of a dog or tamed savage, but a sexual partner. Island stories so often turn out to be erotic stories in which warmth and nakedness combine to create the conditions for sexual freedom and free sex.[5] Such dreams of sexual well-being take two main forms: asocial, in which a pair of human beings find each other's bodies sufficient society to overcome the need for any form of human intercourse other than the sexual; and social, in which an isolated, insular environment provides the perfect conditions for the existence and survival of a genuinely 'loving' community.

Before there is a company or society, however, isolated castaways may have to project their sexual needs on to the landscape itself.[6] Muriel Spark's Robinson inhabits an island, named after himself, with a crudely human shape. Its physical features are named accordingly: the North Leg, the West Leg, the North Knee, the North Arm, the South Arm, the Headlands (Spark 1964: 6). Michel Tournier's Robinson names his island 'Speranza' after 'a hot-blooded Italian girl whom he had once known'. Studying a map of the place, he notices that 'viewed from a certain angle the island resembled a female body, headless but nevertheless a woman, seated with her legs drawn up beneath her in an attitude wherein submission, fear and simple abandonment were inextricably mingled' (Tournier 1974: 42). In the course of the story that follows, this headless, completely passive woman becomes Crusoe's mistress. At various times, he thinks of himself as her excrement (82); he retreats, naked and foetal, into a smooth, deep cavern which he thinks of as her womb ('In this deep place the feminine nature of Speranza became wholly maternal, and because the weakening of the bounds of time and space enabled Robinson to plunge as never before into the forgotten world of his childhood, he was haunted by the memory of his mother' – 89); he receives a venereal bite on his penis while penetrating a tree (99–100); and he makes love with the soil in a sexual embrace which eventually has issue in the shape of a mandrake plant (103–4, 111). Only the arrival of the man Friday breaks into the intensity of Robinson's relationship with the land, and as the two men become closer some of Robinson's capacity for love is transferred from the soil to Friday. When he discovers a cluster of mandrakes with variegated leaves and then catches Friday 'in the act of fornication with the earth of Speranza' (146), Crusoe is furious. But it is not clear whether he is more jealous of Friday than of the island herself.

One of the comforts of solitude is, of course, masturbation. But such a solipsistic act may not suit the needs of a man who has just found himself in charge of a complete world. Bernard Malamud's Calvin Cohn, searching the ship whose wreck he survived, finds a book called *A Manual of Sexual Skills for Singles*, but decides to leave it behind (Malamud 1983: 43). On the night before his ship is wrecked, Michel Tournier's Robinson is given a tarot reading by the ship's captain: 'A snake biting its tail is the symbol of that self-enclosed eroticism, in which there is no leak or flaw. It is the zenith of human perfectibility, infinitely difficult to achieve, more difficult still to sustain. It seems that you are destined to rise even to these heights' (Tournier 1974: 11).[7] However, most of the island writers appear to be in agreement that perfectibility demands company.[8]

NATURAL LOVE

The Western cultures' interest in 'going back' to the place and condition they call 'nature' dates largely from changes in perception influenced by the works of Jean-Jacques Rousseau (1712–78), changes coinciding with European explorations of the South Seas and encounters with their islanders.[9] According to Rousseau's belief, humankind was in essence virtuous and free, but had been corrupted by society's inequities and materialism. A return to nature – at least such as was practicable – could restore a modicum of happiness to human affairs.[10] Rousseau's account of the ideal ways in which to organise a harmonious society were laid out in *The Social Contract* (*Du contrat social*, 1762).

In the last few years of his life, Rousseau befriended Jacques-Henri Bernardin de Saint-Pierre, an engineer and moral philosopher whose enthusiasm for the older man's ideas eventually, ten years after Rousseau's death, bore fruit in the shape of the phenomenally popular novel *Paul and Virginia* (*Paul et Virginie*, 1762). In this sentimental romance, a boy and a girl grow up together in the isolated social innocence and natural abundance of the Ile de France (Mauritius). They learn only what is necessary to happiness within nature: 'Paul and Virginia had no clocks or almanacks, no books of chronology, history or philosophy. They regulated their lives according to the cycles of Nature'. Not only do they possess a sufficiency of learning, but what in France would show up as an academic insufficiency actually ensures their happiness: 'Their want and their ignorance only added to their felicity' (Bernardin 1989: 70).

Of course, Paul and Virginia love each other. That is the whole point of the story. Their love represents the ideal condition from which heterosexual relations have strayed under the influence of social corruption. One is never surprised to discover that Bernardin's argument inevitably has to borrow from the scriptures to make its central point about prelapsarian purity:

> They were in the morning of life and had all its freshness: so must our first parents have appeared in the garden of Eden when, coming from the hands of God, they saw each other and drew near and talked for the first time as brother and sister; Virginia, modest, trusting and mild like Eve, and Paul, another Adam, having the stature of a man and the simplicity of a child.
>
> Bernardin 1989: 70–1

While it may be that they spoke to each other, in the first instance, as 'brother and sister' (which means chastely, without flirting), nevertheless the last clause quoted informs us that the two lovers – Adam and Eve, Paul and Virginia – are mature. In effect, by stating that Paul has 'the stature of a man', Bernardin means that he has a man's genitals. By now Virginia's virginity is very much to the point. Their continuing purity in each other's company is, therefore, all the more impressive. It comes as no surprise to learn that, during an unhappy childhood, Bernardin had been consoled by readings of his two favourite books, a volume of the lives of the saints and Daniel Defoe's *Robinson Crusoe* (1719).[11]

The essentials of Bernardin's tale – boy, girl, island – were shrewdly revived by H. de Vere Stacpoole in his 1908 novel *The Blue Lagoon*. In the four subsequent decades, this book stayed in print and over one million copies of it were sold in English alone. In his preface to the twenty-first edition, of 1946, Stacpoole sketches an impression of the enthusiasm with which the book had been received:

> *The Blue Lagoon* ... took the public at once by its newness and remote charm, and almost at once began to travel the world, leaving behind it all sorts of things other than its readers: 'Blue Lagoon' swimming pools, canoe lakes, bathing beaches, inns, and crockery ware. Paris scented itself with a perfume Blue Lagoon; Como, during the late [Second World] war and the present peace, founded a Blue Lagoon Rest Club for servicemen.
>
> On the stage it left a play that ran for nine months, and on the screen a silent film as successful, in its way, as the play. The story is now in course of production as a talking film.
>
> Stacpoole 1980: 7–8

The 'talking film' was Frank Launder's version of the story, released in 1949. As well as reminding us that commercial spin-offs are by no means peculiar to the cultural circumstances of our own time, Stacpoole's preface confirms the abiding popularity – even through several of the most turbulent decades of modern times – of the dream of 'innocent' and 'natural' love, far distant from the supposedly sordid realities of urban life and the unceasing cacophony of mass communications. It is worth noting, too, that businesses appear to have capitalised on the story's popularity not only in relation to the characteristically 1930s, outdoor, mass activities of boating and bathing, but also in the more rarefied, 'chic' circles one associates with the Parisian perfumeries.[12]

The rear cover of the 1980 Futura paperback edition of the novel, timed to coincide with Randal Kleiser's film of the story, bears the heading 'THE IMMORTAL BESTSELLER – A LOVE STORY THAT GOES BEYOND TIME'. The subsequent blurb ends as follows: 'Nature was generous. They lived off the bounty of the jungle and the fruits of the blue lagoon. The boy grew tall, the girl beautiful. They swam naked, caressed and enchanted by the warm winds and the glow of the tropical moon. And they experienced a love as timeless as the passing of the seasons ...' This marketing insistence on immortality and timelessness is at best ingenuous – the castaways' insular security depends on its being set in an era of sailing vessels rather than helicopters – but it is understandable, perhaps, given the resilience of the novel's sales figures and the success of the various film versions.

The two most recent films, Randal Kleiser's *The Blue Lagoon* (1980) and William A. Graham's *Return to the Blue Lagoon* (1991) concentrate heavily on the drama of 'growing up', with particular interest, of course, in the physical and psychological effects of puberty. What had until then been a life of peaceful co-existence becomes, at puberty, a baffling interlude in which boy and girl cease to understand each other and draw apart, each keeping secrets from the other and resenting the other's moody silences. The boys spend more time out on the reef, aggressively spearing fish as if incipient manhood depended on it; and the girls spend more time in the forest, introspectively touching themselves and washing in the freshwater pool.

Whereas these girls enter womanhood via a moment of implied pain, in the actual bloodshed of menstruation, the boys enter manhood by way of pleasure: masturbation in the first film, a noctural erection during evidently pleasant dreams in the second. At least in its physical characteristics, female puberty is imagined as far more of an ordeal than the male's emergence into sexual desire and satiation. It may be that this imbalance accounts for the fact that the boys are expected to endure initiation rites further to those which their developing bodies impose on them (I shall return to these). Incipient virility is thus constructed not only biologically, but also, even here in the apparent isolation of island life, socially (needless to say, with a camera crew watching, everything is socialised, not least the taboo of showing the whole of the body: full biological development simply cannot be filmed).

Teenage girls are the main target of the marketing of the *Blue Lagoon* films. It presumably follows – given that the marketers are unlikely to acknowledge that any significant proportion of this potential market is lesbian – that these girls' interests will be aroused mainly by the promotion of the male lead (Christopher Atkins/Brian Krause). The female (Brooke Shields/Milla Jovovich) is, in effect, a stand-in for the potential spectator herself in her wish-fulfilment fantasies.

Take the example of publicity for the later film (Plate 9.1), *Return to the Blue Lagoon* (1991). Teen magazines used photographs of the stranded couple together – indeed, the same photo appeared on the front covers of *Young Americans* and *My Guy* – but they tended to concentrate on the boy. Thus, *My Guy* features an interview with Brian Krause, while *Young Americans* goes one better by featuring him, in full colour and loincloth, as their centrefold 'poster'. The *My Guy* interview, gently mocking Krause's rather goofy manner, concentrates on sexually suggestive topics. What was it like wearing a loincloth for three weeks? ('Gee, yes, at first I was really embarrassed wearing it. After a couple of days up and down the beach it felt kinda more natural.') Did he ever 'fall out' of it? ('I beg your pardon? Oh no, hey, that *woulda* been embarrassing.') Did he 'fancy' Milla Jovovich? (He avoids giving a straight answer.) (Anonymous 1991b: 16).

It appears that not only in the film but even in the filming traditional gender divisions were strictly adhered to. According to the *Young Americans* feature, 'Brian had a

IN A SECLUDED PARADISE...
SURROUNDED BY A CORAL SEA,
A BOY AND A GIRL GREW UP ALONE.
NOW THEY ARE EXPERIENCING
THE FIRST AWAKENINGS OF LOVE.
A LOVE THAT CAN ONLY
BE THREATENED...
BY DISCOVERY.

RETURN TO THE BLUE LAGOON

THE STORY OF NATURAL LOVE CONTINUES...

At Cinemas Across the Country from AUGUST 9TH

Plate 9.1 Publicity poster from *Return to the Blue Lagoon*
Photograph: Greg Woods

personal trainer and worked out for months before shooting began in order to prepare for the physical demands of the movie. He toughened up his feet by walking barefoot on coral and climbing rocks. With the help of a native islander, Brian also learned to carve and throw spears, make vine ropes, climb coconut trees and weave baskets from palm fronds'. On the other hand, 'Milla's preparations weren't quite so strenuous but she did live in a small wooden shack with no modern conveniences during filming and she had to scrub all her clothes by hand down at the river with the local women' (Anonymous 1991c: 22). In this way, the myth of the ordeal of masculinity is gently purveyed to a new generation of female readers as they prepare to enter the heterosexual marketplace.

Two differences between the two recent *Blue Lagoon* films (1980 and 1991) should be noted. Firstly, the later film does not show its protagonists naked, as the earlier film had, when they are swimming. This increased anxiety about exposure of the body is also evident over the even longer period between the British and American films of *Lord of the Flies* (1963 and 1990). Secondly, at the start of *Return to the Blue Lagoon*, the two children and the girl's mother are set adrift from a cholera-ridden ship. Their island is, therefore, an escape from disease; so too, by extension, is their love. It does not seem unreasonable to read this as the later film's crude response to the AIDS epidemic.

Posters and full-page ads for the film show the couple in a loose embrace, superimposed on a tropical sunset, complete with the all-important signifiers, palm tree and calm sea. Under the film's title runs the revealing subheading 'THE STORY OF NATURAL LOVE CONTINUES'. Since the phrase 'natural love' inevitably invokes its obverse, '*un*natural love' (which in this century means homosexuality), one could be forgiven for imagining that this was to be a film which aggressively propagandised heterosexuality. Which, of course, it does. At least to the eye of the gay spectator, the ostensible contrasting of 'natural' heterosexuality on a coral island with debased heterosexuality in the cities of the western world is a mere pretext for the more sinister intention of contrasting 'natural' heterosexuality with 'unnatural' and diseased homosexuality. While 'the story of natural love continues', that of unnatural love peters out in urban AIDS wards.

COLONIAL ROMANCE

The archipelagos on which island stories are set tend not to be completely uninhabited. White castaways are liable to meet up with, or fearfully to avoid meeting, members of the indigenous, black population. These 'natives', 'savages' and 'cannibals' are, more often than not, mere visitors to, rather than inhabitants of, the specific island on which the white protagonist is marooned. Often, they arrive only when the moon is full; their presence is signalled by ritual drumbeats; and they leave behind them the detritus of appallingly uncivilised acts: cannibalism and human sacrifice. Should their intended victim escape into the jungle, the white castaway will soon meet him, possibly to impress him into domestic service. When white islander meets black, as when Crusoe helps Friday escape from his cannibalistic captors, we have what Peter Hulme has called 'the true colonial encounter when the complex matter of the European/native relationship must be negotiated' (Hulme 1986: 201). As the father/narrator of *The Swiss Family Robinson* so succinctly puts it, 'Human creatures ... are the colonists of God' (Wyss 1910: 114). Well, some are; and others are God's chosen subjects and slaves.

From the beginning, the Friday figure has always been sent, not simply as an average savage on his best behaviour, house-trained and semi-civilised, but as the *ideal* savage. Daniel Defoe's first description of Friday clearly defines him in terms of the type of savage he might have been but is not:

> He was a comely, handsome fellow, perfectly well made, with straight strong limbs, *not* too large, tall and well shaped, and as I reckon, about twenty-six years of age. He had a very good countenance – *not* a fierce and surly aspect, but seemed to have something very manly in his face; and yet he had all the sweetness and softness of a European in his countenance too, especially when he smiled. His hair was long and black, *not* curled like wool; his forehead very high and large, and a great vivacity and sparkling sharpness in his eyes. The colour of his skin was *not* quite black, but very tawny, as the Brazilians and Virginians and other natives of America are, but of a bright kind of a dun olive colour, that had in it something very agreeable, though *not* very easy to describe. His face was round and plump; his nose small, *not* flat like the negroes; a very good mouth, thin lips, and his fine teeth well set and white as ivory.[13]
>
> Defoe 1719: 205, my emphases

The interesting thing about these resounding negatives – and others which are left implicit in the text, such as 'a very good mouth, thin lips, *not* thick like the negroes' – is that they conjure up a phantom savage at Friday's side, more impressive because more frightening than he. This phantom is not comely and handsome, but too large; not sweet and soft (albeit manly) but fierce and surly; with dull eyes, fuzzy hair, black or nauseous yellow skin, a flat nose, thick lips and bad teeth. Since Friday is 'stark naked' at this moment (206), so too is the phantom savage whom his beauty conjures up. It is their nakedness that elicits Crusoe's inventory of their parts. As it turns out in the following pages, only when he has clothed Friday can Crusoe begin to determine and evaluate his character. In effect, while still naked, the man is still a mere savage and therefore has no character at all. He is a mere appearance, like a slave on the auctioneer's block – a commodity rendered down to the sum of his parts.

The good black man, who having learnt to wear clothes proceeds to learn, never loses the phantom who stands at his side: the bad black man, or savage, or cannibal. His clothing never entirely succeeds in civilising his body. There is about him always the threat, which is also the forlorn dream of postcolonial white racists, that he will 'go back' to 'where he belongs' – which is to say, variously, back to Africa, back to the tropics, back to the jungle, back to the island; and implicit in all these locations to which a man's blackness threatens to return him, there is a more fundamental, atavistic reversion: to basics, to infancy, to primitivism, to paganism, to nakedness, to cannibalism and, ultimately, to the condition of the animals over which the Bible gives 'man' (white men) the exclusive licence to hold sway.

Both Michel Tournier's *Friday* and Marianne Wiggins' *John Dollar* were characterised by reviewers as reading as if *Robinson Crusoe* had been written by Freud. To my mind, Defoe's original needs no such revision for its rich veins of Freudianism to be revealed, particularly in the relationship between Crusoe and Friday. Peter Hulme (1986: 208) calls Defoe's novel 'a colonial romance' and emphasises 'the extent to which the true romance in *Robinson Crusoe* is between Crusoe and Friday' (212). This is not an entirely original insight. In his classic critical study of *Love and Death in the American Novel*, Leslie Fiedler points out the similarity between the Crusoe–Friday relationship and so many inter-racial 'companion heroes of the primitive epic' in American literature, and relates these bonding pairs back to more clearly homosexual archetypes such as Achilles and Patroclus (Fiedler 1960: 363).

Fiedler's key point is that, as well as racial distinction, there is a consequent class distinction between Crusoe and Friday, as between master and servant. Fleetingly, he suggests that in such European texts 'The servant may represent the protest of the unconscious against the ego ideals for which his master stands'. Thus, Crusoe – 'good bourgeois that he is', and white to boot – represents the id, Friday the repressed ego (364). This is in keeping with what we have already observed about Enlightenment 'man' (white men) and his monopoly on reason.

UNNATURAL HISTORY

Since human relationships are never static, any given situation may improve or deteriorate. Indeed, such changes form the nexus of many island narratives, asocial or social. Thus, in the course of a book or film, antagonists may become friends (as when, in *Hell in the Pacific*, an American and a Japanese serviceman learn to co-operate) or lovers (as when, in *Trouble in Paradise*, a rich widow falls in love with the bibulous stoker she used to despise); or, on the contrary, social order may gradually break down – as is most famously the case in William Golding's *Lord of the Flies*. These two possibilities correspond with the two outcomes which any dutiful post-Freudian might predict as a consequence of the relaxation of inhibitions and the overcoming of taboos. Island life, on the one hand, promises a liberation of sexualities, but on the other, threatens the release of aggression. Freed of civilising restraints, the human psyche takes its 'natural' courses. Brian Street (1975: 120) has observed that 'to some, the primitive environment is a terrestrial paradise, to others it is a dangerous jungle'. Actually, most accounts represent island life in an ambivalent manner which involves both paradise and hell.[14]

As we have begun to see, the traditional island story is deeply involved in a moral balancing act between culture and nature. As described by Joseph Bristow (1991: 107), a key aspect of the popularity of R.M. Ballantyne's *The Coral Island* is that it 'allows the boys to get as close as possible to being both pirates (defiant, daring, individualistic) and savages (survivors taming nature) but without turning into them'. But certain standards have to be maintained to prevent what happens in Golding's *Lord of the Flies*, where choirboys so easily become killers. The earliest signs of this moral descent into savagery coincide with the release from inhibition which comes as the boys' shorts are gradually discarded or fall apart. The next step is an increasingly negligent attitude to a rule about defecating in a specified location. With these two changes, it seems, the formal socialisations of early childhood go into reverse: potty training is unlearned, and 'man' reverts to going as naked as on the day he was born (nor is the deterioration wholly unconscious: in a revealing if fleeting moment, Ralph notices that he has been biting his nails to the quick – a habit he had given up, though he cannot remember exactly when he had last done it. His first reaction is articulated out loud: 'Be sucking my thumb next' – 104–5). However, nakedness is not itself the ultimate sign of discarded inhibition. For Golding, the physical sign of the boys' moral disintegration involves not only that first step of *revelation*, but a later and more sinister step of *concealment* or disguise: for savages wear war-paint or masks. Once first painted, Jack's face is described as a 'mask' behind which the boy hides, 'liberated from shame and self-consciousness' (Golding 1960: 61). Again, much later in the book, Jack is said to be 'safe from shame or self-consciousness behind the mask of his paint' (134). All of the boys who join Jack's 'tribe' feel the advantage of their new uniform: 'They understood only too well the liberation into savagery that the concealing paint brought' (164). Throughout the narrative, the boys' lengthening hair contributes to these effects.

The face-painting motif can take on meanings closely related to anxieties around sexuality. In Nicholas Roeg's film *Castaway*, based on the memoir by Lucy Irvine, Gerald (played by Oliver Reed) responds to Lucy's complaints about being constantly frightened by his anger by putting on her make-up and then lying on top of her as if to rape her; for she has, to his way of thinking, emasculated him by refusing to make love with him. His grotesquely made-up face seems both to arm him as a savage and to disarm him as a queer – a visible sign of the confusion masked by his behaviour throughout the narrative. His most sexist remarks to Lucy ('You're as good as a man to have around here') are unwittingly ambivalent, as are his constant references to his penis as if it were a male companion ('the old fellow downstairs'). After the make-up incident he disappears. When Lucy eventually finds him, he is naked in a mud-patch. He tells her he has been to the other side of the island. But now that he is back, they embrace in the mud, both naked, he apologetic, she weeping.[15]

One of the defining anxieties of twentieth-century island stories is that aroused by the spectre of homosexuality. Muriel Spark's Robinson is rumoured to be 'not a lady's man' (Spark 1964: 29, 96). Bernard Malamud's Calvin Cohn is morally confused by his relationship with his adopted son, the chimpanzee Buz, whose ostentatious puberty threatens to overwhelm Cohn's personality, if not his physique. As Buz snuggles up to him and sucks on one of his nipples, Cohn ponders a thorny issue for rabbinical law: 'If you had suckled the lad, could you marry him?' (Malamud 1983: 83). Sylvia Townsend Warner's Mr Fortune articulates to himself the nature of his intense feeling for the boy Lueli: 'I loved him . . . From the moment I set eyes on him I loved him. Not with what is accounted a criminal love, for though I set my desire on him it was a spiritual desire' (Warner 1978: 192–3). In a similar manner, writing in his journal, Michel Tournier's Robinson feels the need to defend to himself his fondness for Friday: 'As to my sexuality, I may note that at no time has Friday inspired me with any sodomite desire' (Tournier 1974).

Clearly, much has changed since Ralph, the narrator of R.M. Ballantyne's *The Coral Island*, felt at liberty to gush, in reference to his chums Jack and Peterkin, 'There was, indeed, no note of discord whatever in the symphony we played together on that sweet Coral Island; and I am now persuaded that this was owing to our having all tuned to the same key, namely, that of *love*! Yes, we loved one another with much fervency while we lived on that island; and, for the matter of that, we love each other still' (Ballantyne 1979: 124). This is not a tone one often finds in late twentieth-century boys' stories.

Still more unguarded to the contemporary eye is a mawkish children's book by Will Allen Dromgoole, *The Island of Beautiful Things* (1913), in which a man aged 32 and a boy aged 6 fall in love with each other. At first, the man does not know how to react to the little boy: 'he did not know that occasionally, just once in a lifetime perhaps, there is a child formed and fashioned, soul and body, to be adored, made much of, caressed in a wholesome sort of way, and lavished with all the love of one's heart. He did not know that this early visitor was a child of that sort; he was conscious only of a sudden warning to discretion' (Dromgoole 1913: 7). Although the phrases 'a wholesome sort of way' and 'warnings to discretion' are ringing gentle bells about the dangers of homosexual paedophilia, the man throws caution to the wind and commits himself to the boy.

The happy pair go on a fishing trip together and they land on an island in the river. There, they speak of love and friendship. The boy's contribution is a conventional tale of tragic comradeship: 'My mother's father was a soldier-mans, and he told me once of a battle and a big soldier-mans who "cried like a baby" when a little soldier-mans was killed [sic]' (115). In return, the man lectures the boy on degrees of emotion in friendship: on how to avoid emotional extremes, thereby ensuring 'that certainty of good fellowship that is as fragrant as a flower' (127). He tells the boy of an island 'made for all lovers'. He calls it the 'Island of Beautiful Things' and proposes to take the boy there (13). As it turns out, however, an adult and heterosexual relationship intervenes. The boy soon understands that not he but the fiancée will now be accompanying the man to the island. He will be left behind.

Presumably, the intention behind this novel is to instil in children the sense that one day they will have to move beyond same-sex relationships, however passionately chaste, into the maturity of wedlock. The island on the fishing trip provides the turning point, a moment of knowledge in which the man emerges from the passing phase of his homosexual paedophilia and attains a somewhat belated heterosexual puberty; and the boy is educated into exercising restraint where same-sex love is concerned. It is only on the island, away from society, that such problematic issues can be discussed and such fundamental changes can be negotiated.

'Friendship ripens quickly in the tropical sunshine' says Charles Warren Stoddard in a 1903 memoir of his 1864 tour of Hawaii; so, he continues, 'it was not many days before the young native and I were inseparable' (Stoddard 1987: 71). The native in question was a boy called Kane-Aloha. Remembering the spontaneous friendship with him leads Stoddard to the following unguarded conclusion: 'I shall not have written in vain if I, for a few moments only, have afforded interest or pleasure to the careful student of the Unnatural History of Civilization' (81). The ambiguity of these closing phrases refers both to the belief that civilisation and nature are mutually exclusive, and, daringly, to the concept of what we might call 'the natural history of Unnatural Civilisation' – meaning those societies which do not stigmatise love between men.

BACK TO BASICS

For some marooned men, island life may well be a merciful release from shipboard machismo. H.G. Wells' Mr Blettsworthy is tauntingly called 'Miss' instead of 'Mister' by the ship's engineer (Wells 1933: 55). In Caleb Deschanel's film *Crusoe* (1988), a malicious crew member urinates in Crusoe's shaving water. However, arrival and survival may also entail a degree of gendered humiliation, resulting in a reassessment of the significance of the body's manhood. This may take merely verbal forms, as in the witless American film of *Lord of the Flies* (1990), in which, inevitably, Jack calls Piggy 'Miss Piggy' and, later, 'Miss Piggy-Tits'. He calls the whole of Ralph's gang 'girls' (in return, Ralph acknowledges Jack's ostentatious, populist machismo, calling him 'Rambo').

But far more serious, physical humiliations may occur. When Mr Blettsworthy first lands on Rampole Island, he is treated with elaborate involvement – if, by Western standards, with scant respect – by the islanders: 'Old men, hairless with age, make incomprehensible gestures to me. I am moved to respond with weird gesticulations. Later I lie naked and bound in the sun while the women scald and sear my flesh' (Wells 1933:

105). Peter Benchley's Blair Maynard is likewise tied up. Men spit in his face; a woman slashes open his swimming trunks and grabs his genitals; later she gives him a medicinal enema while he lies 'gasping, his face in the dirt'; she allows him no privacy as he squats to empty himself (Benchley 1979: 121–32). Later, when he has been stung by jellyfish while trying to escape from the island, she urinates on him to relieve his pain (160).

In John Boorman's film *Hell in the Pacific*, set in the Second World War, the American castaway urinates on the Japanese from a position of security in the boughs of a tree. The Japanese captures the American and ties him up and blindfolds him. The American escapes and ties the Japanese up in the same cruciform manner. He tries to train the Japanese to run and fetch sticks, like a dog. The struggle between their two proud masculinities is also, of course, a joust between antagonistic nationalisms. Attrition can only last so long, however. In the end, inevitably, the two males bond.

Much of the sojourn of Fernando Arrabal's architect and Assyrian emperor on their island is spent in either the recounting of past humiliations or the enacting of humiliations present. The architect rides the emperor as if on horseback (Arrabal 1967: 13). Playing the role of his brother, the emperor accuses himself of drinking his own urine (65) and of raping his brother (67) when they were both still children. With evident faith in the aptness of cannibalism to their location, the emperor, condemned to death, commands the architect to eat him (87).

Humiliation is a form of exposure, a route through the carapace of socialisation. It reveals men at their least manly by laying bare the sites of infantile eroticism, and reduces the supposed nobility of combat to a spiteful childishness whose victim loses all claim to virile, adult dignity. Bodily waste becomes an instrument with which the posturings of masculinity are radically undermined: it reduces the status of the very flesh it defiles to abject waste. I have raised these examples of extreme degradation not merely for their value as oddities; but rather because they point us in a direction towards which most island stories tend. In spatial terms, we are heading for the mud-patch; chronologically, we are returning to infancy and beyond.

A valuable gloss on the solitary castaway narratives is provided in a short story called 'Crusoe' by Victor Sage (1984). In this version, Crusoe is a paranoid obsessive, so eager to make self-sufficient technological sense of his marooned existence that he ruins his island and dangerously exposes all the Freudian undercurrents of his inner life. Harnessing the gases produced by his own and animal's excrement, he is able to rig up a sophisticated electrical generator to run his lighting, fridge and so on. But to maintain his standard of living as the weather gets warmer, he finds he needs more and more fuel, to such an extent that he ends up spending all his time, day and night, both carting dung and eating in order to defecate. The island swiftly becomes a dung heap and he runs out of power.

That this transformation of the earthly paradise into a pile of faeces stems from a fundamental need to regress, is confirmed in a later section of the story, when Crusoe muses along the following lines:

> Back, back, seeking to get up the anus of my father, back into his intestines where it really matters. Back, back. Getting at my mother's red truth by unzipping her labia all the way up to the chin. Before my brothers do it first.
>
> Sage 1984: 157

Being marooned he can do nothing practical to act out his fantasy of oedipal bisexuality – whereby, as in Freud's classic accounts of the inner lives of children, penis, baby and turd become indistinguishable as they pass through the mucous passages of the parental body. In the final section of the story, therefore, Crusoe invents two characters, Pearl and Arthur, and narrates their life story in regressive order, from adulthood through puberty to birth and conception.

Sage's story leads us to the point at which we have to acknowledge that our culture's mythology has come up with an island which represents a state of mind. Sage was not the first to toy with this idea. In an explicit earlier instance, H.G. Wells' *Mr Blettsworthy on Rampole Island*, it turns out that the events on the island, not to mention the island itself, for all that they have been narrated as if with the ring of physical fact, are figments of Mr Blettsworthy's imagination. Nor should we be surprised by a popular television series like *Fantasy Island*, whose eponymous location provided fantasies not only to the people who visited it, but also to those of us who viewed them from the safety of our own homes.

The motif of accumulating excrement is not Sage's isolated whim either. It relates closely to ways in which island narratives so often turn from the pure, white sand of the open beach to a dark, filthy region of potentially engulfing mud. More often than not, mud seems to signify the indulgence of an atavistic impulse – *nostalgie de la boue*. White people who roll on mud not only revert to an infantile relationship with excremental soil – they literally *soil* themselves – but also, if only temporarily, become 'primitive', which is to say black. Mud reminds them, not only of their roots in their own polymorphously perverse infancy, but also of their Darwinian origins among primitive peoples and, looking even further back into pre-history, among the primates.

Having started his exile on the island in a state of fevered expectation of escape, Michel Tournier's Robinson is devastated when the ship he constructs proves too heavy to launch. His immediate, depressive response is to give up on his life as a human being and seek out the marshy environment in which the wild swine wallow. There, he takes up a new kind of life:

> Exiled from the mass of his fellows, who had sustained him as a part of humanity without his realizing it, he felt that he no longer had the strength to stand on his

own two feet. He lived on unmentionable foods, gnawing them with his face to the ground. He relieved himself where he lay, and rarely failed to roll in the damp warmth of his own excrement. He moved less and less, and his brief excursions always ended in his return to the mire. Here, in its warm coverlet of slime, his body lost all weight, while the toxic emanations of the stagnant water drugged his mind.

Tournier 1974: 35

Not that this is wasted time: for the 'mire' teaches him about 'his capacity for turning inward upon himself and withdrawing from the external world' (36), which is, at its simplest level, an invaluable lesson in how to live alone. This is not the last occasion on which Robinson will need to spend a period in the mud. Nor is he the only man to have to subject himself to this degree of self-abnegatory regression. As he says, 'Each man has his slippery slope' (46).

The successive reversion to and transcedence of the 'mire' constitute a rite of passage through which island life requires the white castaway, especially in films, to pass. We have seen that, in *Castaway*, Gerald ends up wallowing in guilt, self-pity and the mud-patch after having crossed to the 'other side' of the island. In the 1980 version of *The Blue Lagoon*, as in Stacpoole's novel, the two lovers and their baby escape being 'rescued' and taken back to civilisation because all three have been playing in thick mud and are mistaken for savages by the boy's father, who is searching for them on a passing ship. In *Return to the Blue Lagoon*, during the disruptions of puberty, the boy storms off to the 'other side' of the island where, to hide from a band of savages, he smears himself in mud. Although he and one savage do confront each other, a spark of humane recognition passes between them and they do not fight. When he crosses back to the safe side of the island, the boy is washed clean (and white) by the girl, they kiss and he says, 'I want us to be husband and wife'. Their adolescence is over. In Caleb Deschanel's *Crusoe*, it is only after an extensive session of mud-wrestling between Crusoe and the Friday-figure, and Friday's saving Crusoe from being sucked down into the sludge, that the two men's relationship is harmonised.

But the most terrifying instance of atavism occurs at the climax of Marianne Wiggins' *John Dollar*, when the marooned girls have to watch as their own fathers are brought ashore by pygmies and are then stripped, burnt alive and eaten. After the feast, the cannibals defecate on the beach before sailing away. The girls' traumatised, ritualistic response to what they have witnessed is to smear themselves with the excremental transubstantiation of their fathers' flesh. Black with the shit of black men and the ashes of white, they construct 'a kind of totem' with their fathers' bones (Wiggins 1989: 199–205).

THE RULE OF LAW

As well as a psychological morass, the island story is a testing ground for social ideas, whether truly new or just whimsical. In his 1914 play *The Admirable Crichton* (filmed by Lewis Gilbert in 1957, though re-entitled *Paradise Lagoon* for the American market), J.M. Barrie used the island model in order to shape a satirical discussion of whether class distinctions occur naturally. By stranding an aristocratic family and their servants on a desert island, Barrie is able to show that while the abilities and skills of the dominant, landed class are distinctly limited, those of the class which serves them are, when combined with human generosity, potentially life-saving.

Aldous Huxley's novel *Island* (1962), set in an ideal society on the island of Pala, is more concerned with the beneficial social effects of Hindu pacifism and a distinctly Californian vision of exotic sexual well-being. Crucial to the happiness of the islanders is the concept of 'Maithuna' or 'the yoga of love'. This is explained to the European protagonist as follows, in terms of 'Freud's point about the sexuality of children. What we're born with, what we experience all through infancy and childhood, is a sexuality that isn't concentrated on the genitals; it's a sexuality diffused throughout the whole organism. That's the paradise we inherit. But the paradise gets lost as the child grows up. *Maithuna* is the organized attempt to regain that paradise' (Huxley 1976: 89). Paradise, therefore, is a tropical island setting combined with guiltless polymorphous perversity. On Pala, all forms of love are revered: for this Eden is 'a place where the Fall was an exploded doctrine' (250). Huxley is no dreamer, however – or rather, his dreams are not set outside history. They are, therefore, doomed to fail. If Pala is paradise, it is no more secure than Adam and Eve's. After all, it is subject to economic reality: 'The tree in the midst of the garden was called the Tree of Consumer Goods, and to the inhabitants of every underdeveloped Eden, the tiniest taste of its fruit ... had power to bring the shameful knowledge that, industrially speaking, they were stark naked' (157). Economic imperialism takes its toll and paradise is lost.

However, it does not take the open advocation of particular social beliefs, whether satirically or earnestly propounded, for an island text to show evidence of a social agenda. Authors isolate individuals, couples or small groups of characters mainly to make points about what the rest of us are like, or what we might become. And most authors seem agreed that one of the first things men do – and I mean 'men' – when set adrift from the rule of law is to organise a new, or re-establish a version of the old, system of laws to suit the circumstances of island life. Rather than go straight back to nature, they pause in an anxious attempt to remain civilised. The laws they make are revealing.

One of the first rules any community of castaways must draw up concerns sanitation: the confinement of excretion to a particular location on the island, preferably one washed by the tides. Even while still living in solitude, Michel Tournier's Robinson formally writes down a rule stating that 'It is forbidden to perform one's natural functions except in the places reserved for the purpose' (Tournier 1974: 61). Robinson's penal code provides an appropriate punishment for infringement of the rule: 'Whosoever pollutes the island with his excrement shall fast for one day' (62). One of the first signs of social disorder in *Lord of the Flies* is the boys' growing unwillingness to defecate far from the shelters in which they sleep.

The fact is that displaced societies of boys and men require more than what Bernardin and Stacpoole envisaged as the ample rule of 'natural law'. Golding's choirboys need the conch shell to take the place of a conductor's baton; but this natural symbol (both of the authority of the older boys and of the democratic principle of meetings at which any boy may speak, as long as he is holding the shell) eventually falls into disrepute and disuse. It is replaced by the authority of intimidation and the rule of violence. In Bernard Malamud's *God's Grace*, Calvin Cohn tries to educate his community of apes into a liberal moral awareness ('Please keep in mind that others have the right to share food sources equally, as free living beings. That's saying that freedom depends on mutual obligation, which is the bottom line, I'm sure you'll agree' – Malamud 1983: 95) but his dream of responsible social co-operation ends in cannibalism. In a reversal of the biblical story of Abraham and Isaac, Calvin Cohn is sacrificed by his 'son', the chimpanzee Buz. God does not intervene; but as a remnant of moral order, George the gorilla does say Kaddish for Cohn.

Hence the perceived need for enforceable laws. Even while he is still alone on his island, Michel Tournier's Robinson institutes a written Charter and a closely related Penal Code (Tournier 1974: 60–3, 66). The descendants of the buccaneers in Peter Benchley's *The Island* are governed by a written Covenant consisting of five main 'articles', followed by various 'amendments', some of them mutually exclusive. Thus, whereas a nineteenth-century amendment permits male homosexual prostitution, a later one states that this official tolerance will be rescinded as soon as the island's female population has been sufficiently built up. An amendment of 1900 enshrines in law the principle that childhood is the ideally pure human condition, and that adolescence is inevitably a scene of corruption (Benchley 1979: 146–9).

Perhaps the most elaborate and touching of these invented legal systems is 'the Law' of the animal/humans who so painfully inhabit H.G. Wells' great novella *The Island of Doctor Moreau* (1993). Their law consists of five main prohibitions: 'Not to go on all-Fours', 'Not to suck up Drink', 'Not to eat Flesh nor Fish', 'Not to claw Bark of Trees' and – ambiguously in the context of policed masculinities – 'Not to chase other Men'. Taken together, these rules are obviously designed (in vain) to prevent reversion to the animal condition. The creatures have to chant them constantly, lest they forget their semi-human responsibility to transcend their semi-animal origins. The chanting of each rule is followed by a statement of fact and a rhetorical question: '*that* is the Law. Are we not Men?' (Wells 1993: 56–7).

Tugged in two incompatible directions, masculinity exists in a state of disequilibrium between nature and society. Frantic in the need to appear rational and not unnatural, men seek only a limited range of bodily pleasures while subjecting themselves to myriad restrictions, prohibitions and pains. Even when solitary, the life of the man is a kind of behavioural neighbourhood-watch scheme, in which 'being a man' requires the policing of one's own and others' masculinity. What the texts I have been examining tend to show is that, watchfulness notwithstanding, masculinity has a built-in tendency to fail in its key aims. For all their claims to civilised manhood, most castaways ultimately lapse into either blissful or appalled anality. To my mind, it seems that all the men in all these stories have been making that same statement and asking that same rhetorical question: '*that* is the law. Are we not Men?'

In conclusion, we can establish the following generalisations: (a) all island stories participate in a post-Enlightenment debate about the respective merits of reason and nature; (b) notions of masculinity are constructed within the parameters of this debate; (c) most twentieth-century island stories involve the testing of standards of masculinity, usually demonstrating greater or lesser individual lapses from such standards; (d) nevertheless, most stories ostensibly endorse the standards of 'natural' masculine behaviour which their central characters strain to live up to.

In addition, we may observe that marooned islanders are white; that some meet black men whose affinity with nature is both disturbing and desirable; and that the white man is therefore liable to respond to the black in an ambivalent manner, striving to make him some kind of lover as well as a slave. Although every white male aims to civilise his island space and its black islander, he must first undergo a dramatic psychological adjustment to his new situation in relation to the natural world, either (as a growing boy) by passing through the feverish ructions of puberty, or (as an adult) by regressing to a temporarily infantile/savage condition from which to grow up anew. It is by overcoming the boyhood and savagery that are implicit in this progressive/regressive moment that white manhood emerges triumphant and endorsed. Thus established in both his power bases – of whiteness and masculinity – the castaway is confirmed as the master of all that he surveys. He is the law.

ACKNOWLEDGEMENTS

I am grateful for certain additional items of information I received from Tim Franks and David Shenton.

NOTES

1 Robert Aldrich has examined this issue in specific relation to northern European gay men and the Mediterranean cultures (Aldrich 1993).

2 Marketed by Millivres, a company which both publishes glossy gay pin-up magazines and runs a good gay 'sex shop' in Camden Town, *Bronze* shows off (a) the bodies of a group of stranded young men, (b) their skimpy shorts and swimming trunks, all available in the London shop or mail order, and (c) an even more seductive island in the Seychelles. The men do not have sex with each other, presumably, both because of censorship laws and because Millivres are hoping to sell the fantasy to straight women as well as gay men.

3 Despite the best efforts of many feminist commentators, this crudely definitive division of the sexes still has its fervent – and fashionable – supporters, even among intellectuals, even among women. See, for instance, Paglia (1990).

4 The *Robinson Crusoe* mythology has even proved itself flexibly applicable to the requirements of science fiction, notably, and quite respectably, in Byron Haskin's film *Robinson Crusoe on Mars* (1964).

5 One of the greatest clichés in the cartoonist's repertoire is the tiny desert island – a mound of sand barely big enough to support one palm tree – sustaining the lives of two shipwrecked figures: a heavily bearded man in tattered trousers, and a beautiful younger woman with a small bikini and big breasts. That this situation is used as a joke does not detract from its serious attractiveness to heterosexual men.

6 I have written elsewhere about ways in which gay men's writing has often eroticised landscapes by establishing parallels between the human body and the land (Woods 1987: 37–42).

7 The prosaic sexual meaning of Orobouros, the autophagous snake, may be autofellatio, an arcane act which appears in various gay texts (Woods 1987: 20–1).

8 Company may not be much use if it is of the wrong kind. The heterosexual GIs at the beginning of Rodgers and Hammerstein's *South Pacific* belt out the famous compaint, 'What ain't we got? We ain't got dames!'

9 Louis Antoine de Bougainville arrived at Tahiti in 1767 during his circumnavigation of the globe. James Cook stayed there in 1769.

10 In the thematic/symbolic area of the return to nature, there is a good deal of overlap between island narratives and the even more popular mythology of white men left to make a life for themselves among the animals and 'savages' of the jungle. Tarzan is, of course, the best-known of these figures. (The Bo Derek vehicle *Greystoke* (1983), directed by John Derek, is as much a heteroerotic castaway fantasy as it is a conventional Tarzan movie.) More widely in favour with recent cultural theorists is the story of Mr Kurtz, missionary turned savage god, in Joseph Conrad's *Heart of Darkness* and recontextualised by Francis Coppola in *Apocalypse Now* (1979). Two other highly effective film treatments of related themes are John Boorman's *The Emerald Forest* (1985) and Hector Babenco's *At Play in the Fields of The Lord* (1991), based on the novel by Peter Matthiessen. In the former, a white man tracks

down his son, who has been brought up by Amazonian Indians. In the latter, an American (of native North American origins) parachutes into an Amazonian tribal village and lives there as an emissary from one of the tribe's more feared gods.

11 In those days, as now, commercial success demanded sequels. In the same year as the original novel, 1719, Defoe published *The Further Adventures of Robinson Crusoe*; and *The Serious Reflections of Robinson Crusoe* followed in 1720. Many of the more recent island narratives refer back to Defoe's original, retelling the story of the same characters (Michel Tournier's *Friday*) or at least using the name of the original central character (Johann Rudolf Wyss' *The Swiss Family Robinson*, Muriel Spark's *Robinson*, Victor Sage's 'Crusoe'). Later texts often include explicit references to the original Crusoe's adventures (Wyss 1910: 24, 65, 126, 233, 236; Warner 1978: 37, 41, 59, 92). Even some non-fictional texts pay such small but not insignificant debts of homage (Stoddard 1987: 37, 41, 59, 92). In a similar manner, some twentieth-century texts refer back to *The Swiss Family Robinson* (Barrie 1945: 99; Warner 1978: 88). And it is well known that William Golding's sceptical *Lord of the Flies* was closely based on R.M. Ballantyne's idealistic *The Coral Island*. As if by way of acknowledgement of this debt, the later novel contains two explicit references back to the earlier (Golding 1960: 34, 192).

12 Stacpoole's novel's popularity has a context of the development of anthropology and its eventual popular dissemination. Sir James Fraser had published *The Golden Bough* in 1890. Bronislaw Malinowski visited New Guinea in 1914 and the Trobriand Islands in 1915–16 and 1917–18. Margaret Mead's

Coming of Age in Samoa came out in 1928.

13 By contrast, Michel Tournier's version of Friday is described as being darker than the men who bring him to the island to be sacrificed, 'and somewhat negroid in feature, in general different from the other men, and it may have been this which caused him to be singled out for sacrifice' (Tournier 1974: 116).

14 A limp comedy on heterosexual men's hellish behaviour in paradise is offered in the film *Our Girl Friday* (1953). A young woman, Sadie (Joan Collins), is marooned with three straight men: an Irish stoker (Kenneth More), a journalist (George Cole) and the elderly Professor Gibble (Robertson Hare), who happens to have delivered a shipboard lecture on the need for humanity to return to nature. An instrumental version of the song 'If you were the only girl in the world' is played over the opening credits; and it is around the failure of this apocalyptic fantasy of heterosexual males (much beloved of cartoonists) that the film draws out its single joke. When they first land on the island, the three men discreetly swear a mutual pact not to lay hands on Sadie; but in due course each of them in turn tries making love with her. As the three men start deceiving and spying on each other, island life descends into an atmosphere of antagonism and competition. Male heterosexuality is jokingly shown to be uncivilised when released from social restraints. But sado-Sadie turns out to be more than a match for the most devious of male tricksters. It is she, woman, who is most at home in nature. When the four of them are rescued (prior to being wrecked again as the film ends) the Rousseauist professor has changed his mind and enthusiastically extols the benefits of civilisation after all.

15 It may be worthy of note that the very idea of being marooned with Oliver Reed was sufficiently disturbing, and widely recognisable as such, to be the subject of a joke in a British TV sketch by comedians Dawn French and Jennifer Saunders.

FILMS

Hector Babenco, *At Play in the Fields of the Lord* (USA, 1991)

John Boorman, *The Emerald Forest* (GB, 1985)

John Boorman, *Hell in the Pacific* (USA, 1969)

Peter Brook, *Lord of the Flies* (GB, 1963)

Luis Bunuel, *The Adventures of Robinson Crusoe* (Mexico, 1953)

Francis Coppola, *Apocalypse Now* (USA, 1979)

John Derek, *Greystoke* (USA, 1983)

Caleb Deschanel, *Crusoe* (USA, 1988)

Di Drew, *Trouble in Paradise* (USA, 1988)

Louis Gilbert, *The Admirable Crichton* (GB, 1957)

William A. Graham, *Return to the Blue Lagoon* (USA, 1991)

Byron Haskin, *Robinson Crusoe on Mars* (USA, 1964)

Harry Hook, *Lord of the Flies* (USA, 1990)

Max India, *Bronze: A Tropical Fantasy* (GB, 1992)

Randal Kleiser, *The Blue Lagoon* (USA, 1980)

Noel Langley, *Our Girl Friday* (GB, 1953)

Frank Launder, *The Blue Lagoon* (GB, 1949)

Joshua Logan, *South Pacific* (USA, 1958)

Nicholas Roeg, *Castaway* (GB, 1986)

10

SEXUALITY AND URBAN SPACE

a framework for analysis

•

Lawrence Knopp

Cities and sexualities both shape and are shaped by the dynamics of human social life. They reflect the ways in which social life is organised, the ways in which it is represented, perceived and understood, and the ways in which various groups cope with and react to these conditions. The gender-based spatial divisions of labour characteristic of many cities, for example, both shape and are shaped by people's sexual lives (especially in Western[1] industrial societies). For example, heterosexuality is still often promoted as nothing less than the glue holding these spatial divisions of labour (and, indeed, Western society) together. But on the other hand, these divisions of labour create single-sex environments in which homosexuality has the space, potentially, to flourish (Knopp 1992).

The density and cultural complexity of cities, meanwhile, has led to frequent portrayals of sexual diversity and freedom as peculiarly urban phenomena. As a result, minority sexual subcultures, and the communities and social movements sometimes associated with these, have tended to be more institutionally developed in cities than elsewhere.[2] On the other hand, the concentration of these movements and subcultures in urban space has made it easier to both demonise and control them (and to sanctify majority cultures and spaces). Hence the portrayal of gentrified gay neighbourhoods such as San Francisco's Castro district as centres of hedonism and self-indulgence, of other gay entertainment areas (such as San Francisco's South-of-Market) as dangerous sadomasochistic underworlds, of red-light districts as threatening to 'family values', of 'non-white' neighbourhoods as centres of rape,[3] or, alternatively, of suburbs as places of blissful monogamous (and patriarchal) heterosexuality.

These contradictions, and many others, are reflected in the spatial structures and sexual codings of cities, as well as in individual and collective experiences of urban life. Yet as David Bell and Gill Valentine point out in their introduction to this volume, there remains within the discipline of geography a certain 'squeamishness' about exploring these connections (see also McNee 1984). This persists in spite of a relative explosion of work in other disciplines which concerns itself with relationships between sexuality and space, including discussions of urbanism (Wilson 1991; Grosz 1992; Bech 1993; Duyves 1992a), nationalism (Mosse 1985; Parker *et al.* 1992), colonialism (Lake 1994); and architecture/design (Wigley 1992; Ingram 1993).

The small amount of work which *has* been done in this area has tended to reflect the particular concerns and social milieux of those doing it. This has meant a focus on urban gay male and lesbian identities and communities (Levine 1979a; Ketteringham 1979, 1983; McNee 1984; Castells and Murphy 1982; Castells 1983; Lauria and Knopp 1985; Adler and Brenner 1992; Valentine 1993c; Rothenberg and Almgren 1992; and Rothenberg in this volume). Much less attention has been paid to heterosexualities, bisexualities, sexualities organised around practices that may be only contingently related to gender (e.g. sadomasochism and certain fetishes), and (particularly problematically) radical, self-consciously fluid sexualities which reject association with such notions as 'identity' and 'community' altogether (but see Bell 1995; Binnie 1992a, 1993a). Also neglected have been connections between particular sexualities and spaces in small-town and rural environments, those between sexualities, space and other social relations (such as race – but see Rose 1993b: 125–7 and Elder in this volume), and issues surrounding sexuality and the spatial dynamics of particular social systems (e.g. feudalism, patriarchal capitalism, etc. (but see Knopp 1992)).

This chapter addresses some of these gaps. In particular, I develop and illustrate a framework for examining the relationships between certain sexualities and certain aspects of urbanisation in the contemporary West. In so doing, however, I implicitly treat 'sexualities', as well as 'the urban' and 'the West', as if they were self-evident and unproblematic empirical 'facts'. This deflects attention from the diversity within these categories, from their often constricting and oppressive effects, and from the complex social processes and power relations which produce them in the first place. However, because people often relate to such categories as if they were self-evident and unproblematic empirical facts, they have a social power which is every bit as significant as that of many more so-called 'material' concerns (e.g. jobs, families, pensions, etc.). This recognition of the problematic yet powerful nature of the categories 'sexuality' and 'urban' guides the analysis which follows.

URBANISM AND SEXUALITIES

Traditional approaches to understanding urbanism can usefully be divided into materialist, idealist and humanist (Saunders 1986). To oversimplify a bit: materialists see the dynamics of the material production and reproduction of human life as shaping cities; idealists see the interplays between great ideas as doing this (especially the philosophies and decisions of policy-makers); and humanists see cities more as a kind of subjective experience, to which people ascribe meanings. In the 1970s and 1980s, many analysts noticed that in the contemporary world few if any of the material, political or even cultural processes discussed by these three camps are peculiar to definable geographical units that could be called cities (Saunders 1986; Paris 1983). On the basis of this some concluded that 'the problem of space . . . can and must be severed from the concern with specific social processes' (Saunders 1986: 278).[4]

But at about the same time more general social theorists were reaffirming geographers' traditional claim that both space and place matter profoundly in human social life (Giddens 1979; Thrift 1983; Sayer 1989; Lefebvre 1991; Gottdiener 1985). Their arguments drew particularly strongly on a humanist insistence that the experience of *place* is socially very powerful. Now most urbanists, regardless of their philosophical perspective, tend to acknowledge this. Many materialists (including many Marxists), for example, now see the 'image' and 'experience' of the city as important material stakes in the urbanisation process (e.g. Harvey 1989, 1993; Logan and Molotch 1987; Cox and Mair 1988). Urban images and experiences are now seen as manipulated, struggled over and reformulated in ways which are every bit as important to the accumulation (or loss) of social power by different groups as more traditionally material concerns (e.g. control of the production process).

The city and the social processes constituting it are most usefully thought of, therefore, as social products in which material forces, the power of ideas and the human desire to ascribe meaning are inseparable. The same holds true for various sub-areas within the city. I will demonstrate how this approach can be applied shortly, in the context of a discussion of the evolution of contemporary Western cities. Firstly, however, I will identify some particular sexualities which tend to be associated with cities, and particular areas within them, in Western societies.

One of the more detailed general descriptions of Western cities' sexuality, developed from a humanist perspective, is Henning Bech's (1993).[5] Drawing on Lofland (1973), he describes the modern Western city as a 'world of strangers', a particular 'life-space', with 'a logic [and sexuality] of its own'. The city's sexuality is described as an eroticisation of many of the characteristic experiences of modern urban life: anonymity, voyeurism, exhibitionism, consumption, authority (and challenges to it), tactility, motion, danger, power, navigation and restlessness.[6] This kind of sexuality, Bech argues, is 'only possible within the city', because it depends upon the 'large, dense and permanent cluster of heterogeneous human beings in circulation' which is the modern city. It is modern medicine and psychoanalysis, meanwhile, that Bech credits with sexualising these particular experiences. For ironically, both have, in the process of trying to make sense of modern sexualities, actually contributed to their constitutions, particularly by sexualising objects and surfaces (especially body parts). This, in turn, has been part of modern science's more general response to the anxieties precipitated by changes in various social relations (especially gender relations) in the nineteenth and twentieth centuries. Thus the city, as a world of strangers in which people relate to each other as objects and surfaces, becomes an archetypal space of modern sexuality.

There are numerous problems with this formulation.[7] But it is nevertheless quite useful, for Bech describes in detail particular ways in which at least some parts of urban areas have been sexualised in modern Western societies. He also offers the beginnings of an explanation for these. His general description, if not his explanation, would appear in many ways to be fair (although it probably applies more to continental European than Anglo-American and other English-speaking cities).[8] There are other descriptions and explanations as well, however. Elizabeth Wilson (1991), for example, sees densely populated urban spaces as potentially liberating and empowering for women. For this reason such spaces are often associated ideologically with women's sexualities, which are in turn constructed ideologically as irrational, uncontrollable and dangerous. Thus the control of 'disorder' in the city is seen by Wilson as very much about the control of women, and particularly women's sexualities. My own work, and that of several others, has emphasised the homosexualisation of gentrified areas in cities by both dominant interests and gays (mostly white middle-class men) seeking economic and political power as well as sexual freedom (Lauria and Knopp 1985; Knopp 1987, 1990a; Castells and Murphy 1982; Castells 1983; Ketteringham 1979, 1983; Winters 1979). A few others have discussed the coding of these (and other) spaces as lesbian or heterosexually female (Rose 1984; Adler and Brenner 1992; Bondi 1992c; Rothenberg in this volume). Mattias Duyves (1992a), Jon Binnie (1992a, 1993a), David Bell (in this volume), Peter Keogh (1992) and Garry Wotherspoon (1991), meanwhile, have emphasised the alternative codings of certain public spaces by gay men for specifically sexual purposes (e.g. cottaging, cruising, etc.). And Davis (1991, 1992), Geltmaker (1992), and I (Knopp 1992) have emphasised the contested nature of predominantly heterosexually coded urban spaces, such as shopping malls, sports bars and suburbs.

The sexual codings of cities, spaces within cities and the populations associated with them, then, are varied and complex. A few generalisations do seem possible, however: (1) Many of contemporary societies' conflicts and contradictions find expression in these codings; (2) these codings emphasise both erotic and more functional conceptions of sexuality, depending upon the particular areas and populations involved; (3) areas and populations which represent failures of or challenges to aspects of the dominant order (e.g. slums; gentrified areas) tend to be coded in both dominant and alternative cultures as erotic (i.e. as both dangerous and potentially liberatory), while those seen as less problematic tend either to be desexualised or to stress more functional approaches to sexuality; (4) these codings are connected to power relations; and (5) they are (in this latter respect) fiercely contested.

Bech's sociological interpretation of the role of psychoanalysis, and Wilson's of urban design and planning, suggest one link between these sexualisations and power relations: changes in gender relations. Bech argues that modern medicine and psychoanalysis responded to anxieties associated with nineteenth- and twentieth-century revolutions in gender relations by projecting them onto infantile cognitive processes and object-relations, including those through which people develop gender and sexual identities. These then became associated with what Bech sees as a very objectified urban experience. People experience the city, he argues, as well as the other people in it, as objects and surfaces in rapid, dense and impersonal circulation, not primarily as people. In a similar vein, Wilson argues that the architects of modern cities projected anxieties about gender relations onto the maps and infrastructures of cities. Certain areas became feminised and demonised, and infrastructures designed, to facilitate the containment and control of women. These are both useful perspectives but they need to be further developed and linked to other changes in social relations (e.g. industrialisation, suburbanisation, racial segregation) going on at the same time.

Harvey's (1992) and my own recent work (Knopp 1992) suggest what some of these further links may be, but in a more contemporary context. We have both emphasised connections between culture (and in my case, sexuality) and class interests, in the sense that cultural (and sexual) codings may now be important elements of a city's or neighbourhood's image and experience. These have in turn become central to facilitating capital accumulation and the reproduction of class relations. Glen Elder's contribution to this volume highlights the importance of race-based power relations, by focusing on the sexual practices and imaginings that are and are not possible under different racialised political and economic regimes in South Africa. And it must also be emphasised that very real *sexual* interests are at stake here, in that those who benefit from certain codings are those whose particular sexual practices and preferences are privileged in those codings. But rather than developing each of these separately I wish now to develop and illustrate a more integrated approach which sees the links between these processes as all-important. For I want to stress that the various sexual codings associated with cities are sites of *multiple* struggles and contradictions, and as such are instrumental in producing, reproducing and transforming both social relations of various kinds (including sexual relations), and space itself.

CONTRADICTIONS AND STRUGGLE: THE SEXUAL AND SPATIAL DYNAMICS OF URBANISATION

In contemporary Western cities, power is still quite closely associated with the production and consumption of commodities, and with white, non-working-class, heterosexually identified men. It is appropriated and exercised, however, through mechanisms in which people who are oppressed in one respect (e.g. as working-class or 'non-white') may benefit from oppression in other ways (e.g. as men). These complex and contradictory patterns have been produced, reproduced and contested in the spatial structures of Western societies. These include importantly the built environments, spatial consciousnesses and lived experiences of cities.

To understand this process, it is useful to consider the nineteenth-century industrial context from which most contemporary Western cities evolved. In the nineteenth century, cities were typically rigidly segregated by class, race and ethnicity, characterised by very traditional gender-based spatial divisions of labour, dominantly coded as heterosexual, and imagined and experienced in terms of public and private spheres of existence.[9] The designs of neighbourhoods, homes, workplaces, commercial and leisure spaces all reflected this. They both presumed and reproduced, among other things, a heterosexualised exchange of physical, emotional and material values in the home, and a racial hierarchy in which white families and societies enjoyed most fully the benefits of a social wage paid for, in part, by transfers of value from non-whites (both inside and outside Western societies) to whites.

The contradictions in this arrangement were numerous. One very important one was a tension between the fixed nature of many aspects of the city's spatial structure (including the social and sexual structures of place-based communities) and the tendency of competition among different factions of privileged classes to produce new and more economically productive spatial structures before the investments in the old ones had been fully amortised (Harvey 1985).[10] Another, closely linked to this, was the tension between a reliance on particular class, race, gender and sexual structures and the tendency of these structures to create new, potentially disruptive collective and personal consciousnesses. Bech's psychoanalytic interpretation of modern sexualities' fetishising of surfaces, anonymity, etc., can be seen as a particular manifestation of this latter contradiction. But the collective anxiety which he attributes specifically to changes in gender relations can be seen as arising more generally from the sharp distinction between public and private experience which characterised the nineteenth- and early twentieth-century industrial city. The growing consciousness of a 'private' sphere of existence facilitated the development of a wide range of new subjectivities and rising expectations of both individual and collective fulfilment and growth (Zaretsky 1976). This meant that people could explore identities and communities based on the possibility of nonconformist and non-commodified roles and practices. But these opportunities at the same time undermined nineteenth- and early twentieth-century cities' gender-based divisions of labour. They also varied according to people's gender, race, class and sexual locations, as wealth and power continued to concentrate in fewer and fewer hands. Significant contradictions were therefore present in the urbanisation process.

The experience of 'public' life in the city was no less contradictory. Many previously non-commodified public experiences (much theatre and sport, for example) were produced and consumed in commodity form, especially by men. Ironically this was a means for these people to develop their 'personal' identities and 'individual' potentials. But, as I have said, the demand for new experiences included many that were potentially disruptive. As sexual experiences in particular became increasingly dissected, categorised and commodified (e.g. in the ways Bech describes), the possibility of new (but socially disruptive) sexual experiences being profitably produced also increased. The proliferation of commodified homosexual experience, for example, led to a homosexual *consciousness* among some people, and this was very threatening to the heterosexualised gender relations underlying the industrial city.

But these various experiences and contradictions also varied depending upon people's social and spatial locations. White middle-class women and men, for example, were in many respects most likely to experience private life as an opportunity for individual fulfilment through the consumption of experiences and commodities within and outside the home. The white, middle-class and (in the case of gay politics and identities) male biases in much twentieth-century feminism and homosexual consciousness almost certainly reflect this. Working-class white women, on the other hand, were more likely to experience private life as an unwaged world of work and consumption with limited autonomy enjoyed at those times of day when men were away working for wages. The alternative sexual possibilities in this circumstance were, therefore, somewhat more constrained (though still present, since such women often found themselves developing co-operative networks with other women). For working-class non-white women, meanwhile, private life was often experienced still differently, as a balancing act between unwaged and waged domestic and non-domestic labour. The alternative sexual possibilities here were in some ways most constrained of all, although in others they might have been quite substantial (e.g. in the spaces they occupied with other non-white women while engaging in waged labour outside the home). For men of all classes and colours, meanwhile, private life tended (though to varying degrees) to be experienced as the exercise of authority and consumption of values in the home, as well as the consumption of commodified experiences outside the home. Consequently the freedom to explore alternative sexualities was perhaps greater for most men, in general, than for most women (although virulently homophobic and heterosexist ideologies emerged in response to this freedom and penetrated the cultures of many male-coded spaces and experiences).

One result of all this was complex race, class and gender-stratified social movements and everyday struggles organised around sexuality. Waves of 'homophile' and, later, gay and lesbian activism (Plate 10.1) dot the histories of late nineteenth- and twentieth-century Western societies (Steakley 1975; Weeks 1977; Altman 1982; D'Emilio 1983; Katz 1976; Duberman *et al.* 1989). Most have been particularly well developed in cities. But these were structured by cross-cutting and complex internal struggles as well. The various cultural codings of urban space reflect *all* of these struggles, as do various waves of social and political reform and economic restructuring.

Initially, the interests and social power of capital, white people, men and heterosexuals can be seen as having converged in such a way as to combat these and other social movements and struggles by coding all non-middle-class, non-white, non-male and non-heterosexual spaces and experiences in cities as in some way sexually depraved and uncontrollable (though in different ways). The social problems associated with

Plate 10.1 Gay and lesbian activists march on Washington
Photograph: Larry Knopp

Plate 10.2 Gentrified housing in a quasi-gay neighbourhood, New Orleans
Photograph: Larry Knopp

nineteenth-century working-class communities (poverty, disease, etc.), for example, frequently were (and continue to be) blamed on the alleged sexual irresponsibility of their residents (Kearns and Withers 1991). Similarly, areas defined as 'black' in Western cities have often also been perceived as sexually dangerous (especially to white women), and this is associated with both black men's and black women's alleged uncontrollable sexualities. Women and women's spaces, meanwhile, have often been presumed by their very existence to be inviting sexual assaults. And homosexual people and spaces have been associated with all manner of depravity and disease, not the least of which, in the contemporary era, is AIDS. In a recent controversy surrounding an alleged 'gay conspiracy to pervert justice' in Scotland, for example, gay spaces such as bars were constantly portrayed as depraved and disgusting by the tabloid press (Knopp 1994).[11]

But even these codings have from the beginning been contested in ways which reflect struggles internal to these various groups, as well as changes in class relations and other political and economic conditions. In the recent Scottish case, some gays may actually have exploited cultural fears surrounding homosexuality to advance their own personal interests or to retaliate against other gays whom they saw as privileged hypocrites (Campbell 1993a, 1993b). More commonly, relatively privileged sexual non-conformists (e.g. white gay men) have forged networks and institutions which facilitate the practice of their particular sexualities *as well as* the perpetuation of other structures of oppression. The intersection of these networks and institutions with recent industrial and occupational restructurings (the expansion of mid-level managerial, other white-collar and certain service-sector jobs, whose cultural milieux are socially tolerant) have developed into the material bases of the largely urban-based, predominantly white, and male-dominated gay social and political movements (Lauria and Knopp 1985). These movements have taken their own alternative codings of space 'out of the closet' and into the public sphere, but usually within racist, sexist and pro-capitalist discourses (for an example in which these are discussed see Knopp 1990b[12]). They have influenced a wide range of predominantly heterosexually coded realms such as neighbourhoods, schools, government bureaucracies, courts, private firms, shopping areas, parks and suburbs. Their most obvious impact has been the proliferation of visible (but disproportionately white, male and middle-class) lesbian and gay commercial, residential (Plate 10.2) and leisure spaces. Vibrant gay commercial and entertainment scenes, for example, as well as the 'pink economies' of cities such as Amsterdam, London, San Francisco and Sydney, and much gay gentrification, have attracted a great deal of popular media attention over the last decade (see Jon Binnie in this volume). But these scenes have been developed primarily by and for white middle-class male markets, and have been financed by 'progressive' (often gay) capital eager to colonise new realms of experience and to undermine potential threats to its power (Knopp 1990a, 1990b).

CONCLUSION: POWER, SPACE AND DIFFERENCE

The analysis above illustrates one way in which a conception of urban spaces as social products, in which material forces, the power of ideas and the human desire to ascribe meaning are inseparable, can be applied. Along the way, it highlights the contingency, yet tremendous importance, of the connections between particular forms of race, class, gender and sexual relations in the urbanisation process. As the various contradictions within particular social systems begin to destabilise those systems, the various interests at stake scramble to form new alliances and 'new regimes of accumulation' (Harvey 1985) which enhance their power. The sexual interests of otherwise highly stratified minority sexual subcultures are no exception.

But 'power' in this context is an extremely slippery concept. It would seem fundamentally to be about the capacity to produce, reproduce and appropriate human life, and the socially-defined values associated with it, in a way consistent with one's own interests. It would also seem to be about the exercise of control over these processes. Power is realised, therefore, through social relations.

Social relations, meanwhile, would appear always to be organised around some kinds of difference. And while difference is a fundamental feature of human experience, it has no fixed form or essence. What constitutes it, ultimately, is different *experiences*. To make these mutually intelligible and socially productive (as well as destructive!), we associate our different experiences with particular markers and construct *these* as the essences of our difference. These markers may be practices, they may be objects (such as features of our bodies), or they may be abstract symbols and language. Because human beings exist in space, these differences and the social relations which they constitute (and through which they are also reconstituted) are also inherently spatial. The relations of sexuality are no exception.

But power is a strangely contradictory thing. It seems always to contain the seeds of its own subversion. As difference is constructed (spatially) to facilitate the accumulation of power, that (spatialised) difference is also empowered. This is true in even the most asymmetrical of power relations. It is manifest in the seemingly endless parade of struggles and social movements organised around difference as difference itself pro-liferates, and in their spatial manifestations as well.

In a world, then, in which spatiality and sexuality are fundamental experiences, and in which sexuality, race, class and gender have been constructed as significant axes of difference, it should come as no surprise that struggles organised around these differences feature prominently in a process like urbanisation. Their contingent interconnections, their resistance to reduction (one to the other) and their spatial dynamism are testaments to the restlessness, contingency and spatial instability of power itself. As long as human beings continue to exist in space, and as long as our bodies and experiences encompass difference as well as sameness, this contradictory situation will continue.

NOTES

1 By 'Western' I mean strongly associated, materially and ideologically, with Western economic, social, political, cultural and intellectual conditions and traditions. I acknowledge the extremely problematic nature of this term (its erasure of the roles of non-Europeans in making 'European' traditions, for example), but defend its use here as a way simply of suggesting some of the historical and geographical contingencies of my

argument. See my discussion in the second section on 'strategic essentialism'.

2 This is not always true, however. Lesbian cultures and communities in the US, for example, are sometimes more closely associated with areas not seen as particularly 'urban' (Beyer 1992; Grebinoski 1993).

3 I do not mean to suggest here that 'non-white' cultures constitute sexual subcultures, that rape is a sexuality, or that rape's association with certain 'non-white' people (i.e. black men) is anything but ideological. At the same time, I would argue that to its perpetrators rape is a sexualisation of male social dominance, and that white cultures in the West code black men in particular as potential rapists.

4 In almost the same breath, however, he acknowledges that 'all social processes occur within a spatial and temporal context' (278).

5 Actually Bech does not explicitly specify his description as 'Western'. But he does describe it as 'modern', which he in turn defines (implicitly) as Western.

6 Against the charge that what he describes is profoundly 'masculinist' (meaning male-oriented and oppressive to women), Bech invokes the argument of some feminists, including Elizabeth Wilson (1991), that such an objection desexualises women and denies them power, leaving them in need of (male) protection and control.

7 Among these is the fact that Bech attempts (albeit with appropriate caveats) to bracket off power relations from his analysis (except, interestingly, in his most gender-based sociological interpretation of the role of psychoanalysis in the production of urban sexuality). But in addition, his claim that the city as a life-space has a 'logic of its own' is at best an overstatement. Whatever the 'logic' of the urban 'life-space', it is unlikely that it is completely disconnected from the (non-city-specific) hierarchically organised social relations which constitute it, or other relations of power which emerge in the context of it. Bech's own acknowledgement that public space is 'restricted and perhaps becoming even more restricted by the interventions of commercial or political agents' (6) would seem to bear this out. Along these same lines, the claim that the sexuality he describes is 'only possible in the city' is clearly a tautology, since he defines it in terms of the city in the first place. In fact, all of the sexual experiences he describes can and do take place outside cities as well. Admittedly, many of them usually require a good deal more effort to make things happen outside cities (e.g. anonymous encounters), but this does not link them *necessarily* to such environments. Anonymity, voyeurism, tactility, motion, etc. are all human experiences that can be, and arguably have been, sexualised and desexualised in a variety of places and fashions (and for a variety of reasons), throughout history. Thus they bear no *necessary* relationship to the city. The issue is not, therefore, whether or not a particular sexuality (or sexualities) attaches *necessarily* to the city, but rather how and why urban space has been sexualised in the particular ways that it has.

8 In the American case in particular, the process of nation-building through private profit-oriented land-development (and the associated contradictory ideologies of frontier individualism and utopian communitarianism) has led to a sexualisation of the city which is (arguably) less romantic, less erotic and more

masculine than in continental Europe.

9 I wish to emphasise that this distinction between public and private is one which is profoundly ideological, but which functions as one of those powerful essentialisms (Fuss 1989) which has profound material consequences.

10 See Knopp (1992) for a fuller presentation of this aspect of my argument.

11 One headline read 'Two Judges Visited Gay Disco – But One Stormed Out in Disgust!' (*Daily Record*, Edinburgh, 1990).

12 Unfortunately, I privileged class enormously in that particular piece.

3

sexualised places
local/global

SECTION THREE

SEXUALISED PLACES: LOCAL/GLOBAL

By reading about the production of sexualised places through the use of detailed empirical material we can begin to understand how places are constituted as sexual and sexualities are constituted as located: how does lesbian presence in Park Slope create and recreate both the neighbourhood and the identities of those lesbians living there? How do rural gay men and lesbians make sense of their sexualities in what can be rigidly heteronormative environments? How are the complex microgeographies of client–prostitute interactions mapped out across the Spanish barrio? How are places like Amsterdam or Soho produced as visible sites of gay consumption? The chapters in this section work towards answering these and many other questions by exploring the constitution of these individual locations as sexualised places.

The lesbian community of Park Slope, Brooklyn, forms the focus of Tamar Rothenberg's important challenge to gentrification theories which have erased lesbian agency. Her chapter also makes an added valuable contribution to the debate over 'community' by exploring the everyday lives of Park Slope women. Angie Hart also looks closely at the lives of her respondents in an Alicante street prostitution neighbourhood. Her use of notions of social spatialisation in the context of a Spanish red-light district display an imaginative mapping of the symbolic landscape of prostitution, demolishing (as with Alison Murray's chapter) the simplistic exploitation models of sex work. By looking at both workers' and clients' spatial strategies in the barrio, Hart can observe the time–space processes structuring situated discourses of work, leisure, friendship and family. The symbolic landscape of queer consumption is the focus of Jon Binnie's chapter: his work on Soho and Amsterdam leads him to thinking about the pink economy and the constitution of gay identities in such sites of spectacular consumption. In sharp contrast, Jerry Lee Kramer uses the town of Minot, North Dakota, as a research setting for his chapter on rural gay and lesbian lives. To the people of Minot, a landscape such as Soho – or San Francisco – seems remote and strange. Being 'gay' in small-town America, Kramer shows, is like a step back in time.

11

'AND SHE TOLD TWO FRIENDS'

lesbians creating urban social space

•

Tamar Rothenberg

The Lesbian Herstory Archives takes up three floors of a beautifully renovated Victorian-era brownstone in Park Slope, Brooklyn, in New York City. Begun in 1974 out of a small Manhattan apartment, the Archives is a vital treasury of lesbian literature, from academic research to pulp fiction to obscure local periodicals to personal papers (Figure 11.1). More important for the purposes of this chapter, however, is its symbolic value. The Archives embodies two processes that have been changing the neighbourhood for more than twenty years: gentrification and the creation of lesbian space.

While literature on geographical gay communities has claimed that lesbians neither concentrate residentially nor demonstrate any connection to gentrification (Castells 1983; Lauria and Knopp 1985; Wolf 1979), I believe the example of Park Slope, Brooklyn, shows otherwise. In the process of creating a lesbian space, or perhaps more precisely a semi-lesbian or lesbian-congenial space, lesbians have been active participants in the gentrification of a neighbourhood. Despite the significant differences between the gay male space in Castells' study and the lesbian space of my study, Castells' argument for the significance of social movements in changing urban space is useful for understanding the development of an exceptionally large concentration of lesbians in a gentrifying neighbourhood.

'WE CAN HARDLY SPEAK OF LESBIAN TERRITORY'

In *The City and the Grassroots* (1983), Manuel Castells argues that the building of the Castro district as San Francisco's gay neighbourhood is inseparable from the develop-ment of the gay community as a social movement: spatial organisation was the key to the establishment of gay culture and power. Castells notes that in his discussion of 'gays', he

<u>SAL (Slope Activities for Lesbians) presents:</u>

A Lesbian Pride Benefit Party for:

The Lesbian Herstory Archives

Building Fund

Thursday, June 20th

7:30 pm at The Roost Pub
in Park Slope: 7th Ave. & 8th St.
(one block from the F train -- 7th Ave. stop)

$10 admission
(every penny goes to the Building Fund)
includes free beer & Italian dinner buffet
also: dancing, videos, raffle prizes and more

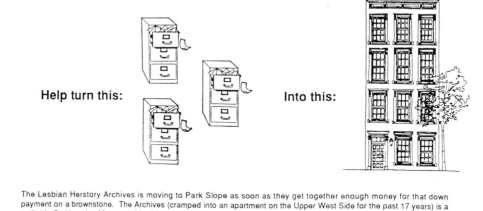

Help turn this: Into this:

The Lesbian Herstory Archives is moving to Park Slope as soon as they get together enough money for that down payment on a brownstone. The Archives (cramped into an apartment on the Upper West Side for the past 17 years) is a veritable Smithsonian Museum of Lesbian History, with extensive collections of literature, photographs and other artifacts. It is truly unique in that it serves to recognize and preserve lesbian culture.

Moving the Archives to a permanent home is an exciting and expensive undertaking. SAL finds it fitting that we pitch in during Gay Pride Month to show our financial and emotional support for this project. Demonstrate your "lesbian pride" by coming to the Benefit, or, if you can't make it, send a few bucks or kind words and we'll forward them to the Archives.

Figure 11.1 Flyer for the Lesbian Herstory Archives, Park Slope, Brooklyn, New York
Source: Tamar Rothenberg

means only gay men: 'We can hardly speak of lesbian territory in San Francisco as we can with gay men, and there is little influence by lesbians on the space of the city' (Castells 1983: 140). Castells' explanation for the lack of 'lesbian territory' lies largely in the gender-based income discrepancies between men and women. Women's wages in general are lower than men's wages; smaller incomes restrict lesbians' housing choices. In addition, a significant portion of gay men do not have families to support, giving gay men greater disposable income than heterosexual men of the same economic class.[1] As I discuss later, however, lesbians' more modest incomes are a major factor in the attraction of a neighbourhood with at least pockets of moderate rents and housing costs.

Castells also suggests cultural, even essentialist, reasons for the differing spatial patterns of gay men and women:

> Men have sought to dominate, and one expression of this domination has been spatial . . . Women have rarely had these territorial aspirations: their world attaches more importance to relationships and their networks are ones of solidarity and affection. In this gay men behave first and foremost as men and lesbians as women. So when gay men try to liberate themselves from cultural and sexual oppression, they need a physical space from which to strike out. Lesbians on the other hand tend to create their own rich, inner world and a political relationship with higher, societal levels. Thus they are 'placeless' and much more radical in their struggle. For all these reasons, lesbians tend not to acquire a geographical basis for their political organization and are less likely to achieve local power.
>
> Castells 1983: 140

Castells neglects to mention the difficulty women long have had in achieving political power, even when they have tried; women in general are less likely than men to achieve local power.

Lauria and Knopp (1985), in their study of gay men and gentrification in New Orleans, also stress the tendency of gay men, and not gay women, to concentrate in a single area and to express themselves spatially. In contrast to Castells' 'innate male territorial imperative', however, Lauria and Knopp posit the constructed differences in gay male and gay female oppression. They argue that, in this culture, males are expected to be more sure of themselves sexually, and so signs of doubt about sexuality are more threatening coming from men than from women. In addition, 'women have always been given somewhat more latitude to explore relationships of depth with one another than have men' (Lauria and Knopp 1985: 158). Therefore, gay males may feel more of a need for their own territory, a safe haven, than might lesbians.

McNee (1984) neither essentialises nor privileges one form of oppression over another, but he has little more to offer in the way of theory. McNee presents gay landscapes as generally hidden, recognisable only to those in the know. The famous gay areas in cities such as New York, San Francisco and Los Angeles are exceptions to the general rule. Lesbian subculture is more hidden than that of gay males, he says, mostly because of their lower average disposable income; lesbians have less money to run or patronise lesbian-oriented enterprises. But culturally and historically, public and semi-public social places such as taverns and clubs have been largely male domains, particularly among the middle and upper classes (Wolfe 1992).[2] Whether or not they are identifiable to outsiders as gay, lesbians already contend with harassment in public as women. Closeted gay men, those who choose to 'pass' on the street as straight, or those who do not 'appear' gay to outsiders do not face the same sorts of daily oppressions as women. Gay male bars and discos tend to be in out-of-the-way places, particularly warehouse districts. Warehouse districts tend to be (heterosexually-based) male spaces during the day and isolated and empty at night, aspects which may appeal to gay men but which many women find threatening. Therefore, a specifically gay female entertainment spot is unlikely to establish itself there unless there is already a protective gay male population in the area.

Deborah Wolf, in her study of lesbian-feminists in San Francisco in the early to mid-1970s, also follows the notion that lesbians are not territorially conscious, perhaps because the spatial agglomeration of lesbians in San Francisco is not as concentrated or as visible as that of gay men. She refers to the San Francisco lesbian community as generally ageographical, and certainly without formal geographic boundaries (Wolf 1979: 72). On the other hand, Wolf notes that

> Women do tend to live in certain ethnically mixed, older, working-class areas of the city: Bernal Heights, the Mission district, the Castro area, and the Haight-Ashbury. These areas bound each other and have in common a quality of neighborhood life, low-rent housing, and the possibility of maintaining a kind of anonymity.
>
> Wolf 1979: 98

These neighbourhoods are also accessible by public transportation to other parts of the city. Although Wolf denies geographical relevance, it appears that there is indeed an area of lesbian concentration within the city, especially if the areas are contiguous. In addition, the neighbourhoods she describes have, since her study in the mid-1970s, followed similar trends towards gentrification.

Adler and Brenner (1992) find a clear spatial concentration of lesbians in the unnamed northwestern US city that they studied. Not surprisingly, they offer a similar critique of Castells' analysis of the differences between gay men and gay women's urban settlement patterns.

BROOKLYN WOMEN TOGETHER

Park Slope, Brooklyn (Figure 11.2), has perhaps the heaviest concentration of lesbians in the US, with a population second perhaps only to that of San Francisco. 'I've gone so far as Florida, New Mexico – when I meet people and say I'm from Brooklyn, they immediately ask if I'm from Park Slope', said a graphic artist, 29. 'When I first moved here five or six years ago, I couldn't walk down the street without saying "she's gay", "she's gay".' Unlike ethnic populations, to which concentrated gay populations have been compared (Epstein 1987), there is no census or comparable data to indicate the number of gay women or men in an area. How, then, does one substantiate such claims?

As one Park Slope lesbian organiser says, 'if 25 groups say that gays and lesbians live here, then they must live here'. In 1990, Park Slope lesbians organised the first anti-gay-violence march and rally in the neighbourhood, largely in response to an attack in a Park Slope diner. Although the 1990 march called for the safety of lesbians and gay men, it was also an assertion of Park Slope as a place where lesbians *should* be able to feel safe; local lesbians were moved to action by outrage over an attack that took place in *their* neighbourhood. As part of a city-wide demonstration protesting violence against gays and lesbians in March 1991, Park Slope was chosen as the Brooklyn site for the protest (Otey 1991).

An effort by gay organisations to create 'gay-winnable' city council districts in 1991 used mailing lists and marketing data to locate the areas of New York City with the highest proportions of gay men and lesbians. According to this data, Park Slope has one of the highest proportions of gay residents in New York City outside of Greenwich Village.[3] The greatest number and proportion of people on the mid-1980s mailing list of a now-defunct organisation, Brooklyn Women Together, lived in Park Slope.

Electoral politics notwithstanding, perceptions and 'common knowledge', while lacking exactitude, may actually be more beneficial to the users of gay spaces. Word-of-mouth, not statistical information, is what lures women to a 'lesbian neighbourhood'. What matters to the people who live in a community is their experience of the place, how they feel walking down the street, the services available to them.

To build a sense of lesbian Park Slope, and to explore the development of the neighbourhood's lesbian 'community', I interviewed several residents, including gradu-ate students, occupational therapists, a performing artist, a bookstore worker, an investment banker, a word processor operator, a business owner, an assistant district attorney, a nurse and a graphic artist.[4] Most are white European-American; one woman is African-American and one is Latin American. They had lived in Park Slope anywhere from one to eighteen years, although the majority had moved to the neighbourhood in the early to mid-1980s. Most were living with their lovers as a couple.

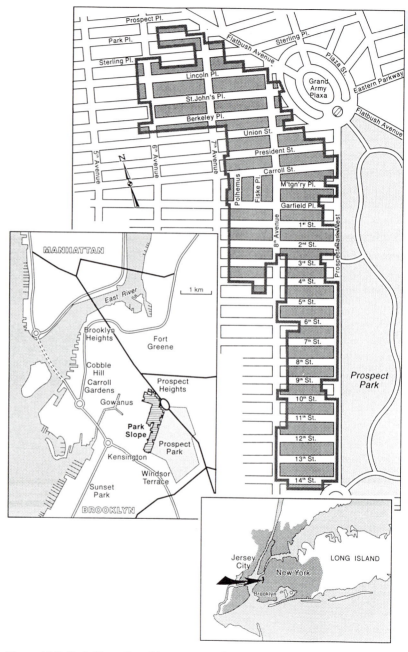

Figure 11.2 Park Slope, Brooklyn, New York
Source: Tamar Rothenberg

THE 'COMMUNITY' QUESTION

When I began my study, I did so with a preconceived notion of a lesbian community in Park Slope. 'Community' is a tricky word, however, with numerous meanings and uses. Urban geography and sociology tend to use 'community' to indicate a geographically bounded area consisting of people who share particular characteristics, such as race, ethnicity and class, and who maintain social interactions with each other (Karp *et al.* 1991). In such discussions, 'community' is often used interchangeably with 'neighbourhood' (Johnston *et al.* 1986).

But 'community' also has a sense that loosens its geographical ties. Benedict Anderson's notion of 'imagined communities', for example, rests on the idea of community as 'comradeship' and 'fraternity' (Anderson 1991: 7). It is 'imagined' in that those who perceive themselves to be members 'will never know most of their fellow-members, meet them, or even hear of them, yet in the minds of each lives the image of their communion' (Anderson 1991: 6). This particular sense of 'community' is prominent in uses of the phrase 'the gay community'. For example, the gay slogan 'we are everywhere' brings connotations of togetherness right up front in the 'we', while 'everywhere' emphasises a blanketing dispersal.

Deborah Wolf's use of 'community' focuses on the social interactions within a spatially delineated 'imagined' collectivity:

> The terms 'community', 'lesbian community', and 'women's community' are commonly used by the women themselves to refer to the continuing social networks of lesbians who are committed to the lesbian-feminist lifestyle, who participate in various activities and projects, and who congregate socially. The concept 'socio-psychological unity' is to them an important part of their sense of what a community is and who belongs to it.
>
> Wolf 1979: 73

Susan Krieger notes the different shades of meaning of 'community' in discussions about and by lesbians, concluding from social scientists' research that 'lesbian communities *define identity* for their members' (Krieger 1982: 105, quoted from Bristow and Pearn 1984: 731). The idea of community, she says,

> covers the range of social groups in which the lesbian individual may feel a sense of camaraderie with other lesbians, a sense of support, shared understanding, shared vision, shared sense of self, 'as a lesbian', vis-à-vis the outside world. Some lesbian communities are geographically specific (all the lesbians in Newark, for example); some exist within institutions (e.g. prisons);

some exist only in spirit; some are ideological (e.g. lesbian-feminist); some, primarily social.

Krieger 1982: 92

These different nuances of meaning allow lesbians to construct ideas of community to fit their particular circumstances, as Krieger elaborates in her 1983 study of a self-proclaimed lesbian community in a small midwestern US university town.

Goodman *et al.* (1983) interpret 'lesbian community' in a cultural sense, differentiating between the larger lesbian population and the visible culture with its lesbian spaces. 'If a community grows, the number and diversity of lesbian spaces does too', from lesbian bars and coffee-houses to women's counselling and information centres, lesbian businesses and lesbian political organising centres (Goodman *et al.* 1983: 71–2). Under such a formulation, it would be a stretch to comfortably label Park Slope's lesbian population as members of a particular community, as the neighbourhood has a distinct lack of designated lesbian places.

'I find every effort to define a "lesbian sensibility" or a "lesbian community" untenable', says Martha Gever (1990: 191), '[a]nd this more than fifteen years after assuming my lesbian identity'. In her discussion of lesbian and gay cinema and identities, Gever supports Krieger (1982) in noting the importance of the *idea* of community and lesbian identity in the creation of lesbian space: 'Because self-identified lesbians approach these places [lesbian and gay film festivals] with a presumption of community, no matter how fictional, these become cultural spaces' (Gever 1990: 200–1).

'AN AREA WHERE LESBIANS LIVE'

Realising the trouble with assuming a lesbian community in Park Slope, I followed Krieger (1983) and let the women I interviewed tell me if there was a Park Slope lesbian community and of what – or whom – it consisted. Most said there was indeed a lesbian community in Park Slope – and then they hesitated. 'In terms of an organised community, I don't have that sense, although many of my friends live here', said one woman. Some women decided it was more of a concentration of lesbians, or a significant lesbian population, than a community. 'Park Slope has an identity as an area where lesbians live', said a social worker. 'Within it, there are several networks of people.' A number of women said it was more like a collection of communities than a single, unified entity. Some people mentioned the lack of any lesbian cafés, bookstores, bars or other centres that could function as a community unifier. A few of the women I talked to stressed their sense of belonging to other identity groups. 'I belong to many different communities – gay, animal, gardening, homeowners', said one woman. 'I don't associate much with [the lesbian community]', said another woman; 'I'm part of the people of colour community, sometimes of the Park Slope community.'

Still, all of the women I talked to affirmed the *spatial significance* of the lesbian 'community' of Park Slope. In fact, for many the geographical aspect was *the* prime shaping force for their interpretation of the concentration as a community. Nearly everyone interviewed mentioned Seventh Avenue, the commercial strip in the heart of the Slope, as the spatial centre of both the neighbourhood and lesbian community. Several mentioned the experience of walking down Seventh Avenue on a Saturday or Sunday, seeing lots of lesbians, running into people they know and feeling comfortable.

While there is no formal lesbian-designated space in the neighbourhood, a number of local groups function as social networks as well as political bases. The Prospect Park Women's Softball League is 'about 95 per cent lesbian'. The softball league consists of twenty teams of ten players each, plus cheerleaders and hangers-on, and, according to Cindy, a coach and board member, 'plays a big role in a small part of the community'. Another group with a large lesbian affiliation is Brooklyn Women's Martial Arts (BWMA), founded in 1974. In addition to offering local self-defence and martial arts classes for women and children, BWMA also conducts self-defence, anti-racism and lesbian/gay workshops throughout the city and holds karate demonstrations at local events.

A more recently established group is SAL: originally Slope, now Social Activities for Lesbians. SAL founder Debra Jane started the organisation in 1990 after moving back to Park Slope from the suburbs. 'After a few months, I looked around and said what . . . is this? – a huge lesbian population, and nothing to do. So I figured if anyone's going to start a group, it's going to be me.' In 1991, SAL's organisation consisted of a phone number and a bulletin board in Debra Jane's apartment. Two years later, a revolving collective shares the duties, which additionally include the *SAL Quarterly*. Monthly calendars list a large number of lesbian and lesbian/gay events throughout the city, as well as local SAL events such as dinners out at local restaurants, potluck video nights at people's homes and theme parties. Many of SAL's subscribers and event-goers are single, but 'a lot are couples who like to do things and meet new people'. Two of the women I talked with had met at a SAL trip to Fire Island the previous summer and have been together ever since.[5]

While SAL has an extensive mailing list, none of the women in either of the two other social networks I worked with knew much about it. Describing the various lesbian groups in the neighbourhood, one 32-year-old woman named BWMA and the softball league. 'Then there's SAL, who nobody ever sees. They're younger women, from what I've heard, single and younger.' Another 32-year-old woman, however, said that she thought SAL 'is older than me, people in their 30s and up'. Although SAL was only a year old at the time of the interviews, such a disparity in impressions points to the difficulty in relying on hearsay for 'accurate' information and reveals the uncohesiveness of lesbian Park Slope.

MOVEMENTS: CAPITAL, SOCIAL CHANGE, POPULATIONS

Changes of a local neighbourhood need to be understood within the larger economic and social contexts (Smith 1983; Beauregard 1986; Rose 1984; Knopp 1990b). The restructuring of the industrial base that contributed to conditions of unemployment, underemployment and poverty in Brooklyn and other formerly strong industrial areas – preconditions for gentrification – is related to the overall shift to service employment. 'Service' is a split-level sector, with low-skilled and low-wage jobs on one hand, and high-wage professional, managerial and technical jobs on the other. The shift in the USA has occurred within a political context of 'new federalism' and withdrawal of government support for low-income housing. From 1978 to 1981, New York City lost 81,000 housing units, mostly low-income; 73,000 were added, but these were almost completely out of reach to low- and moderate-income people (New York Department of City Planning 1984). As Peter Marcuse notes, 'the increasing polarization of the economy is reflected in the increasing polarization of neighborhoods: at one end, abandonment, at the other, gentrification' (Marcuse 1986: 155).

In the same period of time, there have been several other changes which have profoundly transformed people's lives. By the mid-1960s, there were several loud and active social movements – including the women's movement. A good number of the women involved in the women's movement were lesbians; many 'came out' in the process. Lesbian-feminism emerged as its own branch of feminism:

> As lesbian-feminists, we are determined to struggle against patriarchy for our liberation, and the liberation of all women, because we are women who love women in terms of time, energy and commitment as well as sex – in other words, thoroughly woman-identified women.
>
> Goodman *et al.* 1983: 75

Many lesbians were influenced by, if not personally involved with, the gay liberation movement that began with the Stonewall riots in June 1969; that movement was itself greatly influenced by the ideas and the tactics of the women's movement. But many lesbians found gay men to be as sexist as heterosexual men, and preferred to congregate in feminist, anti-patriarchal, women-only or women-mostly spaces. Many lesbian-feminist spaces – ranches, communes and clustered communities – were created in rural areas and small college towns.[6] But many women were drawn to urban areas as well. As

Gill Valentine (1993c) notes, cities offer the contradictory comforts of both anonymity and community. While Kim England (1991) does not specify lesbians in her discussion of urban 'non-traditional' households, she acknowledges that 'lesbians have also been creating "safe spaces" in cities which provide a supportive environment relatively free of homophobic prejudice' (England 1991: 22). A third of the women I spoke to mentioned that the existence of a 'women's community, however loose', was what had drawn them to Park Slope in the first place.

ATTRACTED BY 'SWEAT-EQUITY' HOUSING

Like most gentrified neighbourhoods, Park Slope has old, attractive housing stock, much of which had fallen into disrepair or shabbiness. Thirty years ago, the neighbourhood was much more working class, and was targeted for federal anti-poverty programmes. It was during the 1960s that Park Slope became 'established' as an 'artsy-lefty' neighbourhood. Park Slope attracted young, educated people of the middle class drawn by the neighbourhood's racial and economic diversity as well as its affordability and cultural and aesthetic amenities. For thirty years, in-movers have slid down the New York City neighbourhood scale of affordability – the Village, the Upper West Side, Brooklyn Heights – until they got to Park Slope. According to the 1990 census, Park Slope's population is roughly 71 per cent white, 12 per cent black, 4 per cent Asian and 13 per cent 'other'; about a quarter of the population is Hispanic.[7]

The establishment of a lesbian community – loosely defined – in Park Slope appears to be related to the timing of early gentrification and the particular politically-oriented population who moved in. 'Park Slope historically has been a real active community', said one lesbian resident who has lived in the neighbourhood since 1983. 'There was an influx of political activists when you could get housing cheap. Political activists are attracted by the idea of sweat-equity housing.' When Marge, a teacher, moved to Park Slope in 1973, the neighbourhood was 'very leftover sixties, very laid back, like the Village without the [high] rent'. There was a women's bookstore, La Papaya, but the accelerated rate of gentrification that began around 1977 drove commercial rents way out of reach of the owners, and the bookstore folded after a few years' business.

The bookstore's existence may have been brief, but it designated the neighbourhood as having a 'women's community'. Another indicator was the short-lived New York Women's School, which offered courses to women, including one on how to come out, in the late 1970s. I believe that the timing of Park Slope's gentrification and the women's movement – particularly the directions of lesbian-feminism, cultural feminism and radical feminism – was essential in creating Park Slope as the centre of lesbian population in New York.

Association with particular social movements emerges in what could be seen as 'generational' differences among lesbians. Deborah Wolf (1979) discusses an 'old gay'/ 'new gay' split among the women in her study, 'old gay' having come out or lived as a lesbian prior to about 1969: 'The textures of old gay life as it emerges from subjective descriptions was filled with secrecy and fear, and was focused on coming to terms with a stigmatised identity' (Wolf 1979: 23). Trisha Franzen (1993), however, discusses a similar split, but finds distinctions based more on class than on age. The lesbian-feminists in Franzen's study of lesbian 'communities' or 'networks' in Albuquerque, New Mexico, were primarily centred at the university; drawn to the university as students or faculty, they were active in women's studies and in women's centres, shelters and clinics. Lesbians who did not identify themselves as *feminist per se*, were generally 'native' to Albuquerque or greater New Mexico; their 'safe spaces' were designated lesbian or gay bars, or the homes of a few close friends.

The women I interviewed, almost all of whom were in their late 20s to late 30s, grew up and came out after the women's movement and the gay liberation movement were well under way. Everyone had attended college. In generational terms, a different split appears to be taking place today between women who grew up and came out under lesbian-feminism and younger lesbians (in their mid-20s and younger) who have come out in the age of AIDS and gay (or queer) activism. Younger women may be leaning away from cultural lesbian-feminism towards a lesbianism that is more closely allied with gay males and with gay rather than specifically women's causes.

'Lesbians in general, and in particular Park Slope lesbians, are more politically active than most people', said Melissa. 'But lesbian groups have a hard time organising because of their PCness [political correctness] – there are no leaders, and there are a lot of things falling through the cracks, not getting done. It's taken a lot for lesbians in a group to do something as a united force [such as the 1990 march for lesbian respect and safety].' Susana, a graduate student in a completely different social network from Melissa, noted the same trend: 'A lot of the more politically active women I know got discouraged with feminist organisations and moved into gay/lesbian organisations.'

In addition to many women's frustration with the problems of action in feminist groups, younger people are likely to go directly into the more dynamic gay/lesbian organisations. 'There's so much of this queer-nation mentality lately, with a lot of young newly active people', said Debra Jane. Some of the women I talked to expressed a sense that young lesbians are not living in Park Slope, but in Manhattan and other neighbourhoods. 'I don't see that many young women around here', said Virginia, 31. 'I think they live on the Lower East Side [of Manhattan].'

'PLACES WHERE I FEEL SAFE AS A LESBIAN'

Most of the women I interviewed rented their apartments; a few owned co-op apartments and two owned houses. Half the women I talked to knew about Park Slope's lesbian reputation before they moved there; it was an important reason, although not the only one, for the move. One of the women I talked to found out about Park Slope through her real-estate agent, who was a lesbian herself. The women who *hadn't* known about Park Slope's reputation as a lesbian area moved there to live with a lover or to live in a lesbian household. And everyone I spoke with has encouraged friends (although not only lesbians) to move to Park Slope.

That the lesbian 'community' has grown as large as it has is largely a tribute to the power of lesbian social networking. Deborah Wolf described a comparable networking pattern in San Francisco. Women in Wolf's late 1970s study located where they did based on personal contacts, signs in women's bookstores, coffee-houses and other likely places, plus the San Francisco Women's Centers and Switchboard. Wolf noted that: 'since much of the socializing in the community consists of visiting friends, women without cars try to live near each other, so that gradually, within a small radius, many lesbian households may exist' (Wolf 1979: 99). Valentine (1993c) reports a similar socialising form among lesbians in a small British city. Home visits are especially appealing because there are so few public *and* 'private' spaces where lesbians can feel comfortable being themselves; having friends nearby becomes all the more important.

Wolf also noted a large turnover in living arrangements, due both to the precariousness of having a marginal income and to the tendency of lesbian lovers to live together. Women are brought into the socio-spatial community when they move in with a lover who is already living there or when a couple moves to the neighbourhood together. When a couple break up, one or both will move out, but not necessarily out of the neighbourhood. Any vacancy then gets spread through word of mouth. In a neighbourhood like Park Slope, where there is little rent stabilisation and even less rent control, each tenant-move gives a landlord the opportunity to raise the rent for the next tenant. Some of the women I talked with had moved an average of once a year for a period of seven to ten years. Such movement allows for a lot of possible rent increases.

Many of the interviewed women saw the 'Park Slope lesbian community' as larger than the boundaries of Park Slope proper, extending into contiguous neighbourhoods. Reasons for moving to these neighbouring areas were purely economic. These areas are cheaper, especially for people seeking to buy housing, and the continuous gentrification of Park Slope has meant constantly rising rents. Many of the women I spoke to who live outside 'Park Slope proper' had previously lived closer to the centre of the neighbourhood. 'I moved to [neighbouring] Windsor Terrace mostly because of the rent, and accessibility and safety, but I wanted to stay on the borders of the Slope', said a graduate student. 'There are places where I feel safe as a lesbian; walking on Seventh Avenue, I can put my arm around my lover – but not in Windsor Terrace.' Another Windsor Terrace resident, who had moved to Park Slope in 1973, suggested that there are many lesbians in that neighbourhood. 'It was a lot cheaper, but now you see old Irish ladies being thrown out, and prices are going up', she said.

On average, the women I spoke with spend about 85 to 90 per cent of their leisure time in the Slope, socialising with friends who live in the neighbourhood, eating in local restaurants, frequenting the hardware and video stores, and food shopping, particularly at the Park Slope Food Co-op and at health food stores. People spoke of moving out of the neighbourhood only if they were to leave the city – which some were ready to do.

With constantly rising rents, however, women who can't afford to keep up with the increases move into bordering neighbourhoods, extending the area of gentrification. Women who are buying homes are also likely to be forced by cost to look into the yet-ungentrified fringes. 'As Park Slope gets more gentrified, the physical community expands', said a woman who grew up working class in Queens. 'Women are being outpriced by rent increases; they're moving south, into Windsor Terrace, Prospect Heights, other places. We're the ones who lead gentrification, without even knowing it.'

What on one hand looks like the leading edge of gentrification – spreading out further from the centre of the designated neighbourhood – is also displacement. In this respect, lesbians are much like other 'named' groups that have been credited, for better or for worse, with initiating the gentrification of a neighbourhood – such as artists or moderate-income professionals: in a crude sense, victims of their own success. Again, it becomes problematic to talk about 'lesbians' as a group without attending to other particularities, such as, in this instance, class. Maxine Wolfe (1992), in one of the few mentions of lesbians and gentrification, classifies gentrification as a process wholly detrimental to lesbian life. Lesbian bars tend to be in low-rent districts, and their owners and clientele are dependent on the maintenance of such low rents; if a neighbourhood is under gentrification pressures, landlords often need little excuse to raise rents, not least in the hopes of attracting a more 'desirable' clientele.

Peter Williams' (1986) discussion of gentrification as part of the process of class constitution is useful for understanding the development of lesbian Park Slope. Class for Williams is a concept that entails both production- and reproduction-based relationships; it is a connotation as close to the non-Marxist concept of 'social group' as to any Marxist meaning of class. As Williams points out, the middle-class gentrifiers have a culture and life path that is generally different from that of working-class residents. For many of the middle-class gentrifiers, moving to a working-class neighbourhood 'appeared to mark a break from the class-segregated past and it was presumed to offer the warm supportive communal existence denied in the suburbs, discovered at university or college, and potentially to be lost again' (Williams 1986: 71). The middle class are not trying to be working class, and they are not trying to replicate suburban life; they are working out a new sense of who and what they are as a group.

I think this holds true for some people more so than for others; and I think it is particularly applicable to lesbians. For lesbians, the 'warm supportive communal existence . . . discovered at university or college', is not just an amenity but a survival tool and an affirmation of identity. Many women discover not just a supportive community at college, but their own lesbian identity. Three-quarters of the women I spoke with said that they had come out or consciously identified themselves as lesbians from age 18 to 23, and although more than a third of the women said they came from working-class backgrounds, all had college educations. The supportive circle provides a respite from an incessantly heterosexist society, and enables them to be themselves and to meet other women as friends or potential lovers. One could argue, perhaps, that what is happening in Park Slope is the creation of an open lesbian – not 'class', but collective identity. It is middle class in terms of its economic position, but it does not share the middle-class standard of reproduction (in all its meanings), which is heterosexist.[8]

'Being a dyke and living in the Slope is like being a gay man and living in the Village; it's part of the coming-out process', said the founder of SAL. The place is associated with the creation of an identity, and the collectivity of identities transforms the place. It is still a tentative thing to be gay and open about it; many of the women I talked to feel uncomfortable holding hands or being affectionate with their lovers in public, even in Park Slope.

IN CONCLUSION

'I'm really surprised there's no real gay/lesbian force here, like a bookstore or community centre', said Anne, who grew up in the neighbourhood. 'There's a generally supportive environment, an open environment for gay people. But people here tend to be very unwilling to commit to anything, unlike the progressive together dykes in Ithaca; they know they only have themselves. The Brooklyn community is unwilling to commit, probably because there's an over-abundance of things to do here.'

In many ways, Park Slope functions for lesbians as it does for many of the other people who live in the neighbourhood; it's home, friends live nearby, there's a good amount of useful services and amenities, but for some things you just go into Manhattan. Although Park Slope may possibly be home base to the largest concentration of lesbians in New York City, the cultural and political centre – and the Gay and Lesbian Community Center – is still in Manhattan's Greenwich Village, a half-hour subway ride away.

But Park Slope appears to be well on its way to establishing itself, more and more publicly, as a lesbian neighbourhood. The 1990 and 1991 marches broadcast the lesbian presence in Park Slope to their straight neighbours. SAL has continued to herald events such as its 'Lesbian Extravaganza' dance on brightly coloured flyers taped to Seventh Avenue streetlights and bulletin boards. The *Village Voice* has referred to a lesbian character on the American TV programme *L.A. Law* as having a 'Park Slope dyke haircut' (short and passably professional). And the Lesbian Herstory Archives is settling down into its new home in Park Slope.

The very concentration of lesbians has created a recognisable social space – recognisable most importantly to each other, but increasingly to the 'straight' population as well. The concentration can be attributed in large part to lesbian social networking, the success of which has contributed to the neighbourhood's continuing gentrification, and consequently, to lesbian displacement. The social networking process and its spatial ramifications remind me of an early 1980s TV shampoo commercial. A woman's talking head fills the screen, saying that she liked the product so much: 'I told two friends, and *she* told two friends . . .'. Like images of reproducing cells, squares of women's faces fill the screen, doubling in number and doubling again, until the screen is crowded with smiling women's faces. 'And *she* told two friends, and *she* told two friends', and so on, and so on, and so on.

ACKNOWLEDGEMENTS

Many thanks to the people who contributed their information, observations, opinions and experiences. I am especially grateful to Susana Fried, Elizabeth Coffey, Linda Gutterman and Debra Jane Seltzer – the roots of networks I followed. Big thanks also to Hans Almgren for his insights and support all along the way, and to Susana Fried, Ann Mussey and Susan Blickstein for later comments and suggestions.

NOTES

1 The development of the Castro district as a gay-capital-based neighbourhood predates the AIDS epidemic, which has devastated gay male populations not only physically, emotionally and socially, but financially as well.

2 There is a rich history, however, of largely working-class lesbian bar culture (Gever 1990;

Meyerowitz 1990; Kennedy and Davis 1993).

3 The main source of data used by the Empire State Pride Agenda, the gay lobbying group which led the redistricting effort, was the Strub Masterfile. While the largest gay list available in the United States, it is based on donors to lesbian and gay organisations. Consequently,

the list includes straights as well as excludes gays without the funds to donate.

4 I used a networking method to locate participants, starting with friends in two distinct social circles who suggested other friends in the neighbourhood, who suggested yet more people to interview. I also contacted a local social organisation to tap into a third network.

5 Update: in 1993, the two got 'married' under New York City's new domestic partnership registration (*SAL Quarterly*, Spring 1993, pp. 4–5).

6 This rural lesbian movement reflects a range of lesbian-separatist utopian ideals, including women's desire to extract themselves from the patriarchal society at large, and a 'cultural feminist' emphasis on women's closeness to the land and on 'natural living'; rural and small-town lesbian communities deserve greater attention from 'herstorians' and geographers.

7 The 1990 census data is for the zip code 11215, which covers a large part of Park Slope, but includes areas that are often considered outside that neighbourhood; it also omits the northern end of Park Slope. The US census regards Hispanics as being of any race. Of the 70-plus reported ancestries in the zip code area, Irish and Italian are the most prominent, followed by German and Polish, and by Russian and English.

8 This is not to imply that lesbians do not have children. Indeed, a few women spoke to me of what they perceived as a 'baby boom' among Park Slope lesbians; one couple I interviewed have since had a baby. The neighbourhood public schools were mentioned as being 'used to' lesbian mothers, and recent school district elections sent the first openly lesbian woman to the school board.

12

TRADING PLACES

consumption, sexuality and the production

of queer space

•

Jon Binnie

> Gay culture in [its] most visible mode is anything but external to advanced
> capitalism and to precisely those features of advanced capitalism that many on
> the left are most eager to disavow. Post-Stonewall urban gay men reek of the
> commodity. We give off the smell of capitalism in rut, and therefore demand of
> theory a more dialectical view of capitalism than many people have imagination
> for.
>
> Warner 1993: xxxi

In this chapter I shall offer a preliminary exploration of some of the relationships
between political economy, sexuality and urban space. I hope to make a modest
contribution towards a re-incorporation of the *material* into emerging debates on
sexuality and space. In doing so I wish to offer a more dialectical view of capitalism, the
sort which Michael Warner is calling for above. I shall therefore ground my argument
in two specific contexts – namely the marketing of Amsterdam as a gay tourist capital
of Europe and the (post-1990) development of Soho's gay village in London. Through
this discussion I wish to examine the broader relations of production of sexualised spaces
and places by looking at the role of the market in sexual cultures.

THE LIMITS OF (CULTURAL) CAPITAL, AND THE SEXUALIZING OF THE CITY

Analyses of the cultural components of urban political–economic transformations have
proliferated in recent times. A body of literature has emerged in urban sociology and
geography which has examined the relationship between cultural production and capital

formation (Bianchini and Parkinson 1993; Budd and Whimster 1992; Philo and Kearns 1994). This work has been significant in challenging previous accounts in which culture was either ignored completely or merely tacked on to political–economic approaches. However, the agency of lesbians and gay men has been largely absent.

In the introduction to their edited collection *Selling Places*, Chris Philo and Gerry Kearns (1994: 16) at least speak of gays and lesbians as being one of the 'other peoples' of the city, who as such, stand 'outside of the "normal" spaces of urban living, often because they are actively excluded but sometimes as a conscious strategy allowing them to avoid the "normalising" pressures of bourgeois expectation'. What is genuinely lacking within urban social theory is any analysis of how these 'other' people of the city shape the urban landscape. For instance, it is hard to find even a passing mention of sexuality in the work of Marxist theorists of postmodern urban culture such as David Harvey (1989) or Fredric Jameson (1984). One can only wonder why political economy and sexuality have for so long been irreconcilable within contemporary urban social theory:

> Social theory as a quasi-institution for the past century has returned continually to the question of sexuality, but almost without recognizing why it has done so, and with an endless capacity to marginalize queer sexuality in its descriptions of the social world.
>
> Warner 1993: ix

The absence of lesbian and gay agency from the literature on urban space, and the subsumption of lesbian and gay experiences under wider categories such as 'cultural capital' and the 'postmodern urban landscape', has been commonplace within urban theory. Here it is worthwhile quoting from the critic D.A. Miller:

> It is worth recalling two complementary manoeuvres through which our culture's general discourse promotes the negation of what, to respect its specific texture, one might call *gay material* when the latter threatens to migrate from the marginality where it normally makes its home: a *faux-naif* literalism to whose satisfaction gay material can never be conclusively *proven* to exist, and a prematurely sophisticated allegorization that absorbs this material under so-called larger concerns.
>
> Miller 1992: 12–13

Given the left's general squeamishness around issues concerning sex and sexuality, it is perhaps hardly surprising that lesbian and gay material spaces should remain in the realm of the unknowable. I am arguing here that sexuality should *not* be safely incorporated into studies of urban political economy, and subsumed under more neutral categories such as the cultural politics of consumption. Neither should sexuality be marginalised as the concern of a minority and studied within the subdisciplinary spaces of feminist geography, or lesbian and gay studies. The heteronormativity of social theory is critiqued by Warner:

> So much privilege lies in heterosexual culture's exclusive ability to interpret itself as society. Het culture thinks of itself as the elemental form of human association, as the very model of inter-gender relations, as the indivisible basis of all community, and as the means of reproduction without which society wouldn't exist. Materialist thinking about society has in many cases reinforced these tendencies, inherent in heterosexual ideology, toward a totalized view of the social.
>
> Warner 1993: xxi

Despite this there have been a few noteworthy attempts to study sexuality and political economy in urban space. Within urban sociology, Manuel Castells' (1983) work has mapped lesbian and gay spaces in San Francisco. His was an early study from the 'dots-on-maps' approach – it is very much the view of a distanced outsider who doesn't get under the skin of sexuality and consumption. Within Anglo-American geography, the work of Larry Knopp has been pioneering in documenting the agency of gay men in gentrifying particular neighbourhoods within US conurbations, including Minneapolis St Paul (Knopp 1987) and New Orleans (Knopp 1990b). And in his study of visual consumption and contemporary urban life, *The Conscience of the Eye*, Richard Sennett (1991) makes a passing reference to the leather bars of New York, commenting that:

> The middle Twenties play host as well to a group of bars that cater to these leather fetishists, bars in run-down townhouses with no signs and blacked-out windows. What makes the middle-Twenties distinct is that all the customers in the leather shops are served alike – rudely. Saddles and whips are sold by harassed salesmen, wrapped by clerks ostentatiously bored. Nor do the horsey matrons seem to care much where the men with careful eyes take their purchases, no curiosity about the blacked-out windows from behind which ooze the smells of beer, leather, and urine. A city of differences and of fragments of life that do not connect: in such a city the obsessed are set free.
>
> Sennett 1991: 125

It is time for geographers to try and piece together some of these fragments in a way which can be sensitive to difference and the specificity of locality while maintaining an overall political economic perspective. I acknowledge the difficulty of such a task. However, balancing the material and the sexual, the social and individual is urgent, for as David Forrest (1994: 102) notes, '[t]he eroticisation of a "consumer ethos" is more pronounced today in the so-called conservative 1990s than in the pre-AIDS "permissive" 1960s and 1970s'.

CAPITAL GAY

While it is imperative that we challenge the heteronormativity of political economy and social theory in general, it is likewise necessary that we widen debates within the emerging fields of lesbian and gay/queer studies. Recent work in the fields of literary and visual theory has provided us with a rich source of literature on the production of sexual dissident meanings in texts (Bristow 1992; Dollimore 1991; Sedgwick 1985, 1990, 1994) and images (Dyer 1992, 1993; Fraser and Boffin 1991). Yet despite the immense significance of this work it is sometimes hard to imagine that lesbians and gay men have a specific (if rather ambiguous) relationship towards the market arena and the public sphere; and as Michael Warner (1993: x) notes in his introduction to *Fear of a Queer Planet*, 'the energies of queer studies have come more from rethinking the subjective meaning of sexuality than from rethinking the social'. Queer studies has thus far tended to concern itself with the production of meanings in texts but has neglected production in terms of goods and services, markets and capital accumulation. This absence of the material is all the more puzzling given the pivotal role the market has played in the construction of queer sexualities, as Warner readily acknowledges:

> In the lesbian and gay movement, to a much greater degree than in any comparable movement, the institutions of culture-building have been market-mediated: bars, discos, special services, newspapers, magazines, phone lines, resorts, urban commercial districts ... This structural environment has meant that the institutions of queer culture have been dominated by those with capital: typically, middle-class white men.
>
> Warner 1993: xvi–xvii

There is an urgent need to link (queer) culture to wider social formations and economic structures. Writing in the collection *Lesbians Talk Queer Notions* (Smyth 1992), Philip Derbyshire argues that current lesbian and gay political concerns and activism often seem incapable of relating sexual politics to wider social–economic processes and transformations:

Queer is not about fucking your lover when you have one room with your kid; queer is not about how you get off if you spend your life in a wheelchair. Even as I say this I feel a sort of prescriptive element saying, 'That's dull and moralising'. And if that's true, then we have a problem locating queer politics in anything that looks like a broader strategy for change.

Derbyshire in Smyth 1992: 59

It is important to stress here, therefore, that not all gay men are affluent consumers. This fact can often be overlooked in some of the hype surrounding the growth and economic strength of the pink pound, and its recent courting by straight advertisers. In one recent article in *The Independent*, 'Rewards for companies that go straight to the gay market' (Jones 1993: 27), a gay marketing consultant notes that 'Gay men and women have great wardrobes, great bathroom cabinets and flats full of gadgets'. Obviously not all gay men and lesbians can afford to have great bathroom cabinets! As Barbara Smith notes, the often taken-for-granted assumption that gay means rich, white and affluent may well hinder wider political alliances around issues of sexuality:

'Gay' means gay white men with large discretionary incomes, period … perceiving gay people this way allows one to ignore that some of us are women *and* people of color *and* working class *and* disabled *and* old. Thinking narrowly of gay people as white, middle-class, and male, which is just what the establishment media want people to think, undermines consciousness of how identities and issues overlap.

B. Smith 1993: 101

In a time of economic recession, representations of gay men as uniformly affluent consumers must of course be treated with some care, particularly in a left-oriented politics where the assumption that all gay men are affluent and middle class can be used to rationalise homophobia. However, it must still be stressed that money is the major *prerequisite* and the greatest *boundary* for the construction of autonomous, independent assertive gay male subjectivities. As Edmund White wrote in his sexual geography of the USA, *States of Desire*:

Gay liberation was founded on the principle that women and gays of both sexes form an oppressed segment of the population that has much in common with other oppressed minorities (racial, ethnic and religious). The goal has always been a change in our society that would bring equality of opportunity and restored dignity and autonomy to us all. To effect such a change progressive gays have attempted to forge links between personal experience and public life, between consciousness and politics. *The only shortcoming in this approach, however, has been a steadfast denial of the real source of power in society, i.e. money.*

White 1980: 62, my emphasis

LET'S MAKE LOTS OF MONEY

In focusing on the material spaces of the pink economy, I recognise that I occupy an ambivalent position. As a consumer and participant in the scene, I am easily seduced by the (limited) sexual freedom gay pubs and clubs may facilitate; but I am also acutely aware of the level of economic inequality that pervades the gay commercial scene (as in any area of business). Whilst it is important to struggle for, and celebrate, the ever greater choices of safe spaces – venues where gay men can be ourselves, become ourselves – one must remain ever sanguine and cynical about the role played by pink businesses in gay life. Pink businesses after all do operate as *businesses*, rather than charities. However, some businesses do contribute more to their customers than others, and some bars clearly perform an important function for non-commercial social and political groups.

Discussing queer consumption practices here I feel I am treading a fine line between two positions. On the one hand, one could lapse into a generalised level of argument moralistically denigrating queer consumerism as merely one manifestation of the cultural logic of late capitalism. In addition, it is a personal observation that many of those who feel distanced from the scene (including those who comment on it, represent it, study it) are quick to castigate it for being 'exploitative'. As Edmund White (1980) notes, people are quick to despise others' pleasure or enjoyment of a scene or a place when they've exhausted its possibilities. However, others celebrate queer consumerism as an opposi-tional or subversive social practice contesting the production and definition of space as straight (Saxton 1993c).

Queer consumerism is indeed a powerful assertion of gay economic power. But it could also (to a certain extent) be interpreted as a response to the need to ameliorate the powerlessness we sometimes feel in our lives. Could one reason why many queers enjoy going shopping so much (if and when we can afford to do so) is because shopping offers us the opportunity to assert at least some kind of power? Is it an effect of our not having power in other arenas, specifically in the realm of social rights? Paraphrasing Richard Dyer's essay on camp (1992), in the 1990s one wonders whether *it is shopping so much that keeps us going*. A quick fix of panic shopping in IKEA or Sainsbury's or Marks and Spencer helps us to create tastefully-lit homes and lovingly-tended gardens, offering a private retreat from the heterosexist (and often homophobic) public sphere. Perhaps it is not so surprising that for many British lesbians and gay men the market is seen as enticing and seductive – offering greater potential as an arbiter for 'rights' for sexual dissidents than the social market and the welfare state, whose family-based institutions and social policies are undeniably heterosexist (Van Every 1992). However, there are limits to the extent to which the free market can promote the interests of lesbians and gay men. As Jon Johnson notes, there are real limits to the power of the pink spender:

> you cannot 'consume' yourself out of being sacked purely because of your sexuality, being demonised because you are a lesbian teacher or jailed for having sex at the age of 17.
>
> J. Johnson 1994: 14

In addition, by studying the international gay tourism to Amsterdam and the development of the Soho scene, I realise that I am reinforcing the trend of studying the more visible and more open spatial expressions of sexual identity. But at the same time I recognise that:

> Research has remained focused on openly expressed sexual identity and has ignored the fact that many lesbians and gay men conceal their sexualities and so 'pass' as heterosexual at different times and places.
>
> Valentine 1992b: 237

Although concentrating on two gay capitals, I acknowledge the necessity of studying sexuality in small and medium-sized towns (Valentine 1993a, 1993b, 1993c), as well as rural areas (Kramer in this volume).

CAPITALISING SEXUAL CITIZENSHIP IN THE UK AND THE NETHERLANDS

While it is not my intention here to delve into the specifics of the legislation on sexuality in the Netherlands and the UK, some indication of differences in the regulation of sexuality between the two states is necessary. By the mid-1980s the United Kingdom had one of the most restrictive sets of legislation on gay male sexuality in Europe. The Wolfenden legislation, which decriminalised gay sex in 1967, meant that gay male sex was only lawful in private, with one other partner, and for men over 21 (lowered to 18 in 1994). However, this distinction under law between public and private has considerable consequences for the policing of expressions of affection between gay men in public spaces (see David Bell in this volume). 'Public' space here includes members-only private gay men's clubs and bars, which are intermittently raided by the police to prevent sex taking place. In contrast the Netherlands has had an equal age of consent (16) for both straight and gay sex since 1971.

In the late 1980s the gap between the legal status of gay men in the UK and the Netherlands widened further. The prevailing tendency in the Netherlands has been towards greater official recognition of lesbian and gay lifestyles and the incorporation of lesbian and gay concerns within mainstream politics. New legislation was introduced in a number of areas to accommodate the demands of lesbians and gays, including anti-discrimination legislation and moves towards registered partnerships. The Netherlands has also demonstrated a continuing commitment to the citizenship rights of lesbians and gay men, on both national and local levels. Since 1991, a number of Dutch local authorities have offered recognition of lesbian and gay partnerships, and the Dutch government has given financial support to lesbian and gay organisations. In 1991 this amounted to 2.4 million guilders (Child 1993).

In Britain during the same period we have witnessed a series of legislative setbacks which have eroded the rights of lesbians and gay men in a number of areas. The most noteworthy of these was Section 28 of the British Local Government Act 1988, infamously prohibiting the 'promotion of homosexuality' as a 'pretended family relationship', which became law in May of that year. While Section 28 has had limited consequences in practice as it is a weak piece of legislation, it has served an important ideological purpose as part of the wider Conservative family values agenda in which homosexuality was seen as one root cause of the moral decay of the nation (Evans 1993). However, mobilisation of protest around Section 28 to some extent re-energised gay politics in 1980s Britain, leading to the formation of OutRage!, one of the UK's most visible activist groups.

While I acknowledge the importance of differences in legislation between Britain and the Netherlands, and the need to campaign for the rights of lesbians and gay men in both countries, I have nevertheless been fascinated by the variety of contrasting and contradictory opinions of life in Britain and the Netherlands. The relationships between the law and one's everyday life in the city are therefore complex, and the connections are not always as clear as one would imagine. While gay men in London may possess considerably fewer rights than gay men in Amsterdam, both cities are booming as international gay capitals. It is this perception of the law that I have been concerned with exploring in my research: what do laws such as those on the age of consent and on public expressions of affection mean in practice? What are the links between social rights and our rights as consumers? In Britain in the 1990s we are witnessing a generation of young British lesbians and gay men who have no respect for laws such as the Wolfenden legislation. These laws are perceived as an irrelevant intrusion into their privacy, but equally into their right to consume. For many young gay men laws such as the age of consent are not allowed to detract from the freedom which really counts, namely the freedom to consume music, dance, spectacle and sex. This freedom to consume is after all one right which the dominant political ideology has only encouraged over the past fifteen years. After listing laws highlighting Britain's comparatively poor standing in Europe when it comes to legislation concerning sexuality, gay newspaper *Boyz* asks:

> When you consider all the negatives listed here you would be forgiven for presuming that all sane, freedom-loving boyz in Europe would make their homes anywhere but in the UK. On the contrary; London is probably the gay capital of Europe. Could it be that masochism is more rife than even we suspected?
>
> *Boyz* 19 June 1993a: 25

The same piece proceeds to quote from an 18-year-old Italian recently moved from Turin to London: 'As far as I know, nobody really gives a shit about the age of consent, what matters is the people – how they live, not the law.' With the explosion of the Soho scene one could argue that in the early 1990s it is London which is increasingly challenging Amsterdam, Berlin and Paris for the title of pink capital of Europe. Before exploring this further, I want to focus on a city which is often proclaimed as Europe's pink capital, Amsterdam.

PROMOTING AMSTERDAM

> Perhaps gay men and lesbians are natural travellers, because we're more likely to be what the demographers call DINKs (Dual Income, No Kids), or because too many of us still need R and R away from the literal straight-jacket of the workaday closet . . . or simply because the outsider perspective of the traveller is a second skin for us.
>
> Van Gelder and Brandt 1992: xiii

Amsterdam has emerged as one of the gay capitals of Europe over the past twenty years (Hekma 1992; Hekma *et al.* 1992). Today Amsterdam boasts one of the most sophisticated and developed lesbian and gay communities and commercial scenes of any city its size anywhere in the world. This has stimulated the growth of lesbian and gay tourism – a tourism which crosses national boundaries.

There are particularly strong links between British lesbians and gay men and Amsterdam. This is reflected in the British gay media, where there are numerous articles extolling Amsterdam's virtues as a liberal and tolerant place in which to express one's sexuality:

Amsterdam's reputation as the drugs-and-sex capital of Europe provokes wistful smiles and condemnation in equal measure. Amsterdammers' habitual decadence and their liberal attitude towards perversions of every persuasion have made the city an essential holiday destination for thousands of lesbians and gay men.

Nye 1994: 28

Many British gay men and lesbians have migrated to Amsterdam to take advantage of the more liberal regime. Some work in the thriving Amsterdam gay scene, where a number of businesses themselves are British-owned and run. This is specifically true of the Amsterdam leather scene. While it is hard to estimate the scale of the British community, one can draw attention to a few examples of British influence within the Amsterdam scene: there is Expectations, the London-based company selling leathergear and sex toys, which has a shop in Warmoesstraat (which is the heart of the leather scene); also British-owned is The Web – a popular leather bar located in Sint Jakobstraat, which is described as 'one of the most successful gay enterprises in Amsterdam' in *The Best Gay Guide to Amsterdam* (1992: 205).

Part of Amsterdam's attraction is its more liberal regime towards sex, reflected in the wider availability of porn and the existence of backroom bars and saunas where sex is freely and openly available. These attractions, when added to Amsterdam's other more widely known tourist sites plus its proximity to Britain (facilitating short breaks and weekend visits), make the attraction of Amsterdam for British gay men an obvious one, as respondents in my research confirm:

'I always have a good time when I go to Amsterdam. I go regularly – probably like eight or ten, maybe twelve or thirteen times a year. There's one or two people over there at one time who thought I actually lived there anyway. I just nip over – it's only 45 minutes after all. So on that basis I regard Amsterdam as just another suburb of London. It just takes a bit more to get there. I'm known there, and people there know who I stay with and all the rest of it.'

Michael, aged 38, middle class

'What I like about Amsterdam is that there are people from different countries. There are a lot more variety of books and porn, basically. Apart from the leather bars, it's easy going. I like the fact that it's more apparently cosmopolitan than London.'

Dougie, aged 30, working class

While international gay tourism to Amsterdam has been well established for a number of years, it is only within the past couple of years that it has entered public policy debate. The

City of Amsterdam has recently engaged in collaborative projects and partnerships with lesbian and gay groups and businesses to promote the city as a European gay capital. From a policy angle the recent official *promotion* of the city as a gay capital by the city council and tourist authorities is an intriguing development. This has attracted international media attention (*The Times* 31 July 1992; Simpson 1992a, 1992b) and has likewise fuelled debate within the city about what kind of image Amsterdam wishes to present to the outside world. The City of Amsterdam has now become actively involved in *promoting homosexuality* (an interesting contrast to Section 28's prohibition of the same in Britain). The city council has a councillor with specific responsibility for lesbian and gay affairs, and since the mid-1980s many highly visible public manifestations of government support for lesbian and gay rights have emerged in urban Amsterdam. One of the most impressive of these is The Homomonument (Plate 12.1).

This monument, designed by Karin Daan and built in 1987, commemorates lesbian and gay victims of homophobia worldwide (Duyves 1992b; Koenders 1987). It occupies a prominent site at Westermarkt on the Keizersgracht, close to the Anne Frank House in the heart of the city. The monument consists of three pink granite triangles which together form a larger triangle. The Homomonument represents a powerful symbolic affirmation of gay rights and sexual citizenship in Amsterdam, and in The Netherlands as a whole. As such it has become a site of pilgrimage for lesbian and gay visitors from

Plate 12.1 The Homomonument, Amsterdam, designed by Karin Daan

across the world. For many, it has an immense symbolic meaning as a place of tranquillity, of rest, of freedom. Moreover, it has come to represent a site of memory and mourning for those we have lost to AIDS.

One of the more controversial policy areas for promoting gay Amsterdam has been the promotion of international lesbian and gay tourism to the city. In the introduction to the most recent edition of *The Best Gay Guide to Amsterdam* (1992: 9), the councillor responsible for lesbian and gay affairs welcomes lesbian and gay visitors to the city:

> Welcome to Amsterdam, the city sometimes known as The Gay Capital. So, what can you – a lesbian or gay man – do in Amsterdam? You can do . . . just as much as anyone else, but without having to hide your gay identity . . . If you happened to be a citizen of the Netherlands, you could also go to the city hall and ask for your gay relationship to be formally registered.

In 1992 the Amsterdam Tourist Office (together with the car rental firm Eurodollar and the charter airline Martinair) launched an advertising campaign in the American gay press to attract more American gay tourists to Amsterdam. A spokesman for the Netherlands tourist board explains: 'They are easy to define and reach. Their income is higher than the average and they travel enormously' (quoted in Simpson 1992b: 24).

The campaign emphasised the fact that in Amsterdam one's sexual preferences are respected – the city was marketed as a free city for lesbians and gay men. The city has also given financial assistance to groups such as GALA (Gay and Lesbian Amsterdam) which organised events such as the GALA Festival in September 1992 to promote international gay tourism through urban spectacle. These events are both political and commercial in nature. The anglicised name is no coincidence here, given that it is international tourism (specifically North American) which is the target market. The latest project is to build a museum charting the historical development of the city's lesbian and gay cultures, following on from an earlier successful exhibition on Amsterdam's gay past (Album Amsterdam 1992).

Official (financial) support for these projects has been sought by lesbian and gay groups and rationalised in terms of the revenue they will generate for the city from tourism. According to an estimate by the VVV (the Dutch tourist information office) from a survey undertaken in 1992, approximately 3,000 of the 25,000 jobs in the Amsterdam tourist industry are dependent on gay tourism (Album Amsterdam 1992).

This courting of the pink dollar, and the place-promotion of Amsterdam as gay capital, must be understood in the wider context of local authorities' search for new sources of income in a time of ever greater fiscal strain. Local authority promotion of culture and the arts has of course become a key component of urban economic development strategies throughout Europe (Bianchini and Parkinson 1993).

But the campaign by the City of Amsterdam to attract international gay tourism to Amsterdam has recently been withdrawn owing to wider public debate and sensitivity over how Amsterdam is represented abroad (Duyves 1993). The responsibility for marketing the city as a pink capital has now been transferred from public bodies such as the VVV to pink businesses themselves. This abandonment of the place-marketing of Amsterdam as a free city for lesbians and gay men demonstrates that even in the supposed gay mecca of Amsterdam there are limits as to how far public bodies can go in supporting pink capital. These limitations must therefore be borne in mind when one examines the liberatory potential of the growth of pink capital in Britain in the late 1980s and early 1990s.

CRUISING BOYZTOWN – THE CONSUMING GA(Y)ZE

If you laid all the boyz on Old Compton Street end to end, you'd have a very pleasant time. Over the last twelve months, it's turned into Britain's own Disney Gayworld: a sassy sea of nice haircuts, white Levis and Retro bags. It probably has the lowest shell-suit count in Europe and the highest hair-gel figures known to man.

Peacock 1993: 17

Since the start of the current recession, Old Compton Street in the heart of London's West End has emerged as a gay commercial district. The past four years have witnessed the opening of a string of new gay bars, cafés and shops along this and adjacent streets (Tranter 1994). Since the opening of Village Soho bar/café (on the junction of Old Compton Street), a number have followed – Old Compton Street has since been described as 'the gayest 100 yards in Britain' (*Boyz* 21 May 1994: 4). Until the recent expansion of venues over the past four years in the West End, there existed two main gay bars in the area – Brief Encounter and Compton's. The opening of Village Soho marked the transformation of the area. Yet while the sheer number of new bars and cafés is impressive, equally significant is their design.

A number of commentators have stressed the design of the new bars themselves, which are light, open spaces with huge plate-glass windows (Tranter 1994). This is contrasted with gay venues in the past, where the distinction between interior and the street was clearer (Weightman 1980):

With the opening of the Village bars in London and Manto's in Canal Street in Manchester, time was called for the old image of the black painted 'blinds down' gay pub.

O'Flaherty 1994: 17

In these new venues, gay men are not hidden behind closed doors. Straight passers-by can look in and observe gay men, as can an increasing number of women (as these venues tend to be mixed – reflect the changing sexual landscape). We are highly visible. As Bill Short states:

> What is really going on, is that the gay scene is adapting to meet the needs of a generation who are more 'out' than their predecessors – an increasingly confident generation of lesbians and gay men whose sense of Pride means that they want to be visible and not ignored.
>
> Short 1993: 16

ALL HYPED UP? THE GAY MEDIA AND SOHO

This development of Soho has been accompanied by, and to a certain extent moulded by, massive hype within certain components of the gay media. There has been an active promotion and marketing of Soho as a *gay village* – something which London gay men have yearned for (the choice of the name Village Soho is thus significant). The bars here are aimed at a market which is sophisticated, continental, European. Though many of the bars and cafés are not in Soho, they are lumped together under the same rubric of West End, making the dream of a gay village in London's heart sound more real.

Accompanying the current boom in Soho has been a similar spectacular growth in the gay media (Alcorn 1991). The new trashy (but wonderful) free newspaper *Boyz* has been of particular note. It has since been joined by new glossy lifestyle monthlies such as *Phase* and *Attitude*, and most recently the weekly *Bona*. *Boyz* contains endless features on and advertisements for the new Soho venues, as well as lifestyle, beauty and fashion tips, and information and advice on safer sex. *Boyz* is distributed throughout gay venues, and first appeared four years ago. It is the sister newspaper of the weekly lesbian and gay current affairs paper *The Pink Paper* which is replete with news, politics, arts and reviews. *Boyz* is widely considered to be apolitical – devoid of news, but containing only items on fashion and lifestyle. In one interview, the then editor of *Boyz*, David Briddle, described the paper in the following way: 'It's an affirmative "lifestyle" publication, an equivalent of *Just 17* [a magazine for teenage girls] for younger men' (quoted in Alcorn 1991: 25). In passing it is worthwhile noting that *Boyz* contains short articles on and adverts for books from the world of cultural studies such as those by Richard Dyer (1992) and the collection of essays on Madonna (Frank and Smith 1993). This in itself may demonstrate a degree of sophistication among *Boyz* readers (which they are not 'meant' to possess).

Above all, *Boyz* is notable (compared to other gay media) for the quantity and quality of its safer sex advice, which gets progressively more imaginative with each issue. This information is always couched in the vernacular – it is very direct and gossipy. There is an incredible all-pervasive 'joy-to-be-aliveness' about *Boyz* which is so refreshing (though obviously it could be said to be exclusionary in its focus on youth). Examples of recent articles in *Boyz* include: 'How to recycle ex-boyfriends' (how to turn ex-boyfriends into friends); 'Supermarket cruising – how to pick up your special offers'; and 'Are you a trolley dolly? Shopping for sex at the supermarket'.

Boyz is highly ambivalent in promoting a self-conscious consumerist ethic among gay men while simultaneously giving advice on how to take control over one's own bodies and make informed choices about life on the scene. Given the often highly introverted and narrow focus of the world of the lesbian and gay/queer political establishment (and lesbian and gay/queer studies) there is something strangely democratic in the exuberance and openness towards others, to people who are different, and to those who wish to define their sexuality differently from the norm. For example, in demonstrating a greater degree of openness towards bisexuality, *Boyz* may be more reflective of the changing sexual landscape than other 'gay' periodicals, which remain on the whole biphobic.

Boyz (along with the other gay commercial media) has also been highly successful in helping to promote the development of Old Compton Street as boyztown, a ghetto for young gay boyz. In one of its many features on Soho ('Soho special', *Boyz* 22 May 1993: 18), *Boyz* quotes two gay men, Damien and Philip:

> The Compton Street season starts in May and goes on till September. In the Summer, we spend at least four hours a day here. We like the relaxed atmosphere. You can hold hands and you don't have to check who's watching.

As I mentioned earlier, in Britain public displays of intimacy between gay men are liable to prosecution. The development of the Old Compton Street scene has meant a raised profile where gay men can feel confident. This safety in numbers means that you can hold hands here under the gaze of police officers patrolling the street (Tranter 1994). Old Compton Street thus enables gay men to contest the limits of their sexual rights:

> 'The nice thing about Old Compton Street is the total safety as queer space; and the large percentage of people walking up and down who are queer in one sense or other of the word. It is one of the public places where you can be queer without risk. You can kiss in the street without anyone batting an eyelid.'
>
> Dougie

The boundaries of sexual citizenship are therefore being contested every day in the street, even if it is much queerer (and more dangerous) to hold hands on the nightbus, in the supermarket or on the underground.

However, with this increased public visibility comes the ever-present threat of queerbashing (which no amount of legislation will ever be able to prevent). In one recent incident a gay man was knifed after leaving Compton's bar on Old Compton Street at one of its busiest times of the day in what was described as a clear homophobic attack (*Capital Gay* 16 April 1993: 3). Clearly the boundaries are still rigidly in place as far as the queerbashers are concerned.

QUEER SOHO?

The development of pink businesses along this street – this 'reclamation of Soho' – has coincided with the activist group OutRage!'s reclamation of the term 'queer'. OutRage! was present at the first Soho Street Carnival in February 1993, when Old Compton Street was (temporarily) renamed Queer Street. This event was one case of businesses and activists working together. Several thousand people partied in the street as the Metropolitan Police sealed off the street to traffic, thereby completing the queer reclamation of Soho (Short 1993). To a certain extent this *queerification* of Soho reflects a return to a pre-existing sexual geography, where Soho was the focus of the London scene in the 1950s (prior to the 1970s development of Earl's Court; see Leech 1993).

Given the often considerable pressures faced by lesbians and gay men going about their everyday lives in the city and the stresses of living as outsiders in heterosexual society, it would be easy to lapse into a *celebratory* mode of cultural analysis – proclaiming Old Compton Street as some kind of gay nirvana. But as with any assertion or affirmation of gay/queer identity, this queering of Soho is highly ambivalent. This claiming of space, like the claiming of fixed identities, can lead to exclusionary practices, as the opinions of two Old Compton Street boyz (interviewed in *Boyz*) show:

> But there are still one or two straight people in the street, which annoys Stuart from Brixton. 'They should block off each end and set up gay checkpoints'.
>
> *Boyz* 22 May 1993: 18

Statements like these (even if ironic) are ridiculous given the proliferation of diverse sexual dissident identities. How could it be done? It's hard to tell what straights look like any more. Would we have to have gay fashion police checking designer labels for authenticity and banning people wearing cheap aftershave? A recognition of how different we are from straights should not blind us to the differences among sexual dissidents, and should not lead to a regulation of what is and isn't a legitimate gay or queer identity.

Alliances between queer politics and pink capital may provide evidence for those who challenge the basis of queer politics for being exclusionary. But, of course, among those excluded from Soho (and therefore less visible) are people who cannot afford the prices of food and drink or are unwilling to pay the pink premium. Also, while none of the venues is exclusive, there is still a strong male dominance.

Others feel excluded from Soho if they are not young and pretty. The atmosphere can at times seem oppressive and unfriendly, lessening any 'community' feel:

'It is queer space at a price – so long as you look right . . . if you are unfashionable then it certainly is not a totally open place. The price – you pay £1.20 for a cappucino to sit out on the street – it's not a Community Centre.'

Dougie

The development of the area has been so sudden it could be fragile. Changes in ownership of the bars could jeopardise the fine balance. We have already seen the putting up for sale of Village Soho, the bar whose establishment was most significant in setting off the queerification process. If this went straight then perhaps the bubble could burst. And the contrasting experience of the much-hyped Soho Pink Weekend and the Pink Angel Day perhaps already demonstrates signs of a retreating back towards more established venues, as well as dissatisfaction with the new venues and a suspicion of being ripped off. Perhaps the novelty of the Old Compton Street scene has already started to wear off. These tensions are reflected in a letter to *Capital Gay* (titled 'No fun in Soho'), in which Steven Reagan complains about the Soho Pink Weekend:

The Soho Pink Weekend turned out to be a money-making exercise for two bars involved, The Village and Compton's . . . Nothing by way of entertainment was organized, the only organized area was the bar area which was for their benefit and not for us 'stupid queens' who bothered to support the event.

Reagan 1994: 2

This reflects suspicion that some businesses may be taking their customers for granted and therefore stretching their loyalty. As the UK economy pulls out of recession, some are also expressing concern that this will lead to the abandoning of the courting of the pink pound, something which has already begun outside London. This fear is expressed in an article by Lynne Wallis in *The Guardian*:

The prospect of the return of disposable cash in heterosexuals' pockets may mean a resurgence of the champagne bars of the eighties, where yuppie cash flowed freely, while gays may have to return to more traditional venues in scruffy pubs on the outskirts of town.

Wallis 1993: 14–15

Whether this fear will materialise is still open to debate. However, at the time of writing, there is considerable speculation as to whether the putting up for sale of Village Soho, together with its sister bar The Yard, could signal the venues turning straight again.

CONCLUSION

In this chapter I have demonstrated that in any examination of how sexualities are produced in space, it is dangerous to neglect the material. Here I share David Forrest's assertion that:

Even the most elementary understanding of the changing nature of gay male identities can only progress when we start to consider such phenomena in the broader context of a 'total society'; as a 'product' of both the underlying material and ideological (that is, 'social') practices and what constitutes human praxis – the on-going struggle between individuals, groups and classes in their widest setting.

Forrest 1994: 109

The two case studies demonstrate that while increasingly the straight media may stereotype gay men as uniformly affluent, avid consumers and taste-makers, there are clear limits to gay consumer power. The development of the Old Compton Street scene is significant in terms of increasing visibility of lesbians and gay men in the city and marking the greater self-confidence and self-assertiveness of queer culture in the 1990s, but it is off-limits to those who either can't afford it or are excluded if they don't conform to a certain conception of what a 'gay lifestyle' is. The Amsterdam case also demonstrates all too clearly that wider interests of more powerful capital formations will mean the needs of gay businesses will continue to be marginalised, something which the possible reclaiming of Soho gay bars by straight management and clientele confirms. The pink pound is apparently as volatile as any currency on the money market.

13

BACHELOR FARMERS AND SPINSTERS

gay and lesbian identities and communities

in rural North Dakota

•

Jerry Lee Kramer

The academic literature on gay men and lesbians, while ever expanding, remains essentially incomplete concerning the special circumstances of homosexuals in rural or nonmetropolitan areas, despite the fact that a considerable number of sexual outsiders are born and raised in rural locations (and, of course, there are lesbians and gay men who choose to move to the country, or who visit it for recreational use (Bell and Valentine 1994)). Correspondingly, the available research on rural geography and sociology, community psychology and related fields has almost totally ignored the existence of gay and lesbian rural residents (Philo 1992; notable exceptions are D'Augelli and Hart 1987; D'Augelli *et al.* 1987; Moses and Buckner 1980). These omissions are especially grievous in that inquiries into the lives of nonmetropolitan lesbians and gay men can further knowledge in each of these disciplines. By studying rural lesbian and gay lives, rural sociologists and psychologists, for example, might better understand the challenges faced (and overcome) by other minorities or nonconformists in rural areas. Likewise, in gay and lesbian studies, empirical research into the strategies, behaviours and motivations of nonmetropolitan gays and lesbians can provide further insights into the wide diversity of the homosexual experience.

This chapter reports the preliminary findings of a participant observational study conducted (both formally and informally) in Minot (pronounced *MY-not*), North Dakota, between the years 1979 and 1993. Besides functioning as an introduction to the study of rural or nonmetropolitan gay, lesbian and bisexual people,[1] conveying an understanding of their lives, difficulties and social needs, I primarily seek to address a more basic question, that being whether the developmental processes of gay and lesbian identities differ between American rural and urban social environments.

METHODOLOGY: TECHNIQUES AND IMPEDIMENTS

The first complication encountered in ethnographic gay and lesbian studies involves simply locating participants. While difficult enough in urban locales, where publicly identified gay and lesbian gathering places or organisations exist, this impediment is magnified considerably in nonmetropolitan areas where gays and lesbians are forced to rely much more on invisibility and anonymity in adapting to what is for most a hostile social environment. In conducting my research, however, I was able to circumvent this by choosing as a study locality my own hometown of Minot, North Dakota. By utilising my familiarity with local informants and places – by using my *insider status* – I was able to obtain greater access into local social networks and personal confidences than might otherwise have been possible (see Styles 1979).

Initial procedures involved identifying and observing the 'sexual marketplaces', which functioned well for encountering men, with their greater propensity for utilising public spaces to locate male sexual partners (see Corzine and Kirby 1977; Humphries 1970; Ponte 1974). Meeting lesbians, however, with their documented greater utilisation of private spaces (Albro and Tully 1979; D'Augelli *et al.* 1987; Ponse 1976), proved much more difficult. Here networking with local informants proved more productive, although many of the women contacted on my behalf declined to be interviewed. While understandable, this unfortunately also restricts my ability to make as many meaningful contributions in this chapter towards an understanding of rural lesbian lives.

Another impediment in conducting this type of research involved the impossibility in many cases of practising standard ethnographic techniques. For example, many of the men I met in the sexual marketplaces visited such places precisely because they could obtain sexual gratification while limiting the amount of personal information given to others (Corzine and Kirby 1977; Humphreys 1970; Ponte 1974). Therefore my observations and discussions with these men were often more fragmentary in nature.

In order to supplement these observations of and discussions with North Dakotan gay men and lesbians, I also conducted interviews with North Dakotan expatriots now living in the Minneapolis-St Paul metropolitan area, who in addition to providing me with their own experiences, referred me to other informants still residing in the state. This use of the 'Snowball' technique seems better suited to studies of such 'invisible' populations, as the researcher is able to gain confidences within a particular social circle and then use these contacts to enlist further participants (Valentine 1993b). Before I go on to describe the lives of lesbians and gay men in Minot, I want to sketch a picture of rural North Dakota as a backdrop to frame my discussion.

NORTH DAKOTA: A GEOGRAPHICAL INTRODUCTION

Poet Kathleen Norris, in her book *Dakota: A spiritual geography* (1993), describes North and South Dakota as being 'America's Empty Quarter' or 'America's Outback'. Although the state of North Dakota is nearly as large as England and Scotland combined, its population of 638,800 gives it a density of only nine people per square mile (3.5/sq. km), compared to England and Scotland's 653 people per square mile (252/sq. km). Over a third of the population of North Dakota resides in its four urban centres (Figure 13.1), Fargo-Moorhead (population 106,400), Bismarck-Mandan (64,800), Grand Forks-East Grand Forks (55,100) and Minot (34,500), in addition to the two Air Force bases (each with over 9,000 people) adjacent to the cities of Minot and Grand Forks.

For the tourists visiting the Dakotan Great Plains, especially those flocking to the region after seeing Kevin Costner in *Dances with Wolves*, the landscape of windswept

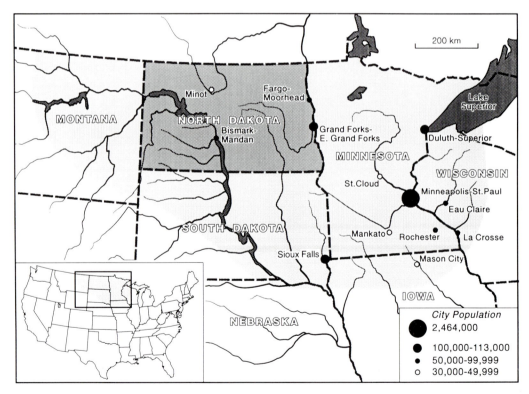

Figure 13.1 The Upper Midwest, USA
Source: Jerry Lee Kramer

treeless prairie and badlands almost invariably fills them with a sense of astonishment, both by its ascetic beauty as well as by its sense of overwhelming isolation and remoteness.

North Dakota, richer in soil than its sister but poorer in mythology (the Black Hills are located in South Dakota), nevertheless shares with its twin a reputation for climatic extremes. Lying at the exact centre of North America, its semi-arid steppes average only 17 inches (42 cm) of precipitation annually, making its climate most resemble that of Northern Kazakhstan. In addition, North Dakota possesses the absolute temperature range record for the continent, a low temperature of –60°F (–51°C) and a high of 121°F above zero (49°C) being recorded within a single town in the same year! These factors, along with its average growing season of less than one hundred days, make farming in North Dakota a much more marginal enterprise than its first Norwegian and German/Russian immigrants had been led to believe (Borchert 1987).

Oversettled under the terms of the Homestead Act from the 1880s, North Dakota began depopulating almost immediately. From a high population of 681,000 in the 1930 census, a slow and unsteady decline followed, until by 1990 the state had reached a relatively stable population of 638,800 inhabitants. Today, while the largest economic activities are still agriculturally based (small grain and cattle production and food processing), service occupations, in addition to the boom/bust cycle of oil, gas and lignite coal mining, follow closely behind. However, because each of its major industries is primary or extractive, North Dakota has always been more or less an economic colony of outside interests. Contrary to the claims of independence made by its citizens, economically, historically and culturally North Dakota functions more or less as a province within the Upper Midwest region, dominated by the Minneapolis-St Paul Metropolitan Area.

GAY AND LESBIAN LIFE IN MINOT: AN HISTORICAL OVERVIEW

Local informants paint a rather bleak picture of homosexual life in Minot before the early 1980s (Plate 13.1). The only sources of accurate and available information during this time had come from Minot's few contacts with outside migrants, mostly Air Force personnel serving at Minot Air Force Base and a few low-profile college professors teaching at Minot State College (now Minot State University). However, as both careers required their employees to remain situationally closeted, the dissemination of information through these local channels trickled rather than flowed. Likewise, in the national television media available to Minot at this time (two stations), coverage of gay and lesbian issues was usually restricted to annual (and sensationalised) coverage of the gay and lesbian pride parades in New York City or San Francisco.

Plate 13.1 The landscape is bleak for many rural lesbians and gay men
Photograph: Alissa Nesje

Women were even more isolated in Minot, as there existed no feminist organisations, and even women's softball and bowling leagues served only a small fraction of those women seeking same-sex relationships (for the importance of softball leagues as lesbian space, see Faderman 1992). Men could meet each other in Minot's sexual marketplaces: in public restrooms or under the Broadway viaduct (adjacent to the railroad depot), as well as in one of Minot's public parks. Men would also cruise through downtown Minot late at night, studying the occupants of other circling cars and pulling into parking lots to meet. While summer marked the prime 'hunting season', especially during the North Dakota State Fair in the month of July, Minot's harsh winters proved a severe deterrent to meeting potential partners during the remainder of the year. Certain heterosexual bars also functioned as clandestine meeting places for homosexuals in Minot, the most identifiable being the Clarence Parker Motor Hotel Bar downtown (recently closed), although these meeting places rotated frequently and their gay and lesbian patrons risked the everpresent possibility of discovery and exposure. But for many gay and lesbian Minoters, the only way of adapting to life in 'The Outback' was by making periodic trips to Minneapolis or other gay and lesbian 'Meccas' in the United States in order to socialise or obtain reliable information or literature.

Historically, the year 1979 proved to be a watershed for the gay and lesbian history of Minot. That year marked both the founding of Minot's first and only gay and lesbian organisation, Lutherans Concerned Missouri Valley (LCMV), as well as the opening of the first publicly identified sexual marketplace in Minot, an adult bookstore appropriately named The Last Chance.

Founded by a gay Lutheran pastor named Donald Lemke, the Lutherans Concerned group functioned for a brief time as Minot's only homophile organisation. In addition to lecturing before nursing classes at Minot State College and being interviewed in *The Minot Daily News'* first article dealing with the existence of local gays and lesbians, Pastor Lemke also provided first his parish house and then his apartment as Minot's first *de facto* gay and lesbian community centre. Besides hosting pot-luck dinners and dances for local gays and lesbians, as well as literature readings and discussion groups, he also formed an extensive lending library on gay and lesbian themes and issues.

Unfortunately, this small oasis was not to last a year. Much of Pastor Lemke's library was burned by an irate mother who blamed him for her son's homosexuality. Then Pastor Lemke's parish renounced their call to him because of his sexuality, and he could not find another local parish to welcome him. Therefore he accepted a call in New York City, where he resides today.

While this attempt at forming a visible presence of gay and lesbian community in Minot ultimately failed, the founding of an adult bookstore in the town also helped improve the quality of life for its gay and lesbian residents. The Last Chance Bookstore aided the gay and lesbian population of Minot in three important ways: first and foremost, by carrying mainstream gay and lesbian journals, publications and newspapers (one notable regional inclusion being a newsletter from Fargo's *Prairie Gay Community* organisation), the bookstore functioned as Minot's first local retail outlet of informational materials serving the gay and lesbian community. Secondly, some staff members of the Last Chance functioned as informants for both men and women, residents and visitors alike, aiding people with information about homosexuality as well as acquainting people with contacts within the (underground) gay and lesbian community of Minot.

In addition, the Last Chance Bookstore's widely reported opening and public presence, as well as its concealed parking lot, gave many gay men throughout the region their first exposure to homosexual sexuality, if not identity. Fully half the movies and magazines in 1979 were oriented towards gay men (a percentage sustained in subsequent years), and here men could meet other men easily, anonymously and in any season of the year. The store also possessed the resources to safeguard its existence through legal channels, unlike Pastor Lemke's efforts, although its possibilities were (and still are) severely limited as a locus around which a sense of lesbian or gay community could be built.

Recently, however, the pace of change for homosexuals in Minot seems to be quickening. Much of this, I believe, is a result of the increasing local availability of information regarding homosexuality. The mainstream bookstores (such as B-Daltons), being less vulnerable to local social pressures, have increased their inventories of homosexual titles within the last five years, as have both the public and university libraries (although interested patrons still must make the first daunting move of asking for many materials, as they are kept on desk reserve). To its credit, the major local mental health agency also now refers interested people to a few self-identified homosexuals who have volunteered themselves and their resources, but again this option exists only for the minuscule percentage of those people courageous (or desperate) enough to ask.

During the early 1990s feminists also began organising in Minot, but are currently struggling to secure meeting space. In addition, one interviewee admitted that while the women involved were sensitive to lesbian issues, the needs addressed were by necessity more broadly based, focusing on local women's self-awareness and self-empowerment generally. Still, while struggling and overwhelmed by the needs of the larger women's community, perhaps the mere fact that a feminist organisation exists provides hope for present and future lesbian Minoters.

MEETING MEN IN MINOT

Like a microcosm of pre-Stonewall America, the lives of rural lesbians and gay men are structured around very limited opportunities for social (and sexual) interactions.

However, for gay Minoters, ways have to be found to compensate for the lack of any coherent 'gay space'. For many men and women, their first access to transport – usually in the shape of an automobile – brings the first possibilities to act upon homoerotic desires:

> 'Even before I was of legal driving age, my mother would allow my driving to the rest area on the highway for drinking water – our water on the farm was saline … After about a year of twice-weekly trips, I began noticing the same cars frequenting the rest area, and that men seemed to just sit in their cars and watch me. Also, as I filled the water jugs inside the men's bathroom, I noticed that the toilet paper holders were almost invariably unscrewed from the walls, leaving a small hole between the stalls … It wasn't until years later that I realised that this rural rest area was a meeting place for local men seeking to meet other men travelling the highway.'
>
> Interviewee

Although there are segments of the (metropolitan) gay community who are appalled at men meeting in such locations, for those in rural areas the opportunity to meet with men, and sometimes to get information about gay lives beyond the immediate area, mark the highway rest area as an important milestone in many individuals' sexual histories.

In addition, the availability of an automobile can also facilitate what might be called 'car cruising':

> 'What you do is this: after dark, and especially on a Friday or Saturday night, you go downtown. You drive along the high street about four blocks, and stare at the people in any car that passes you, checking to see if they are looking back. Then you drive around the blocks again, to see if they are doing the same thing in reverse. If you are interested, you then pull over and flash your lights before turning them off. If they're interested, they'll come back and pull in next to you … There have been times when what occurs is a relative block party, with some five or six cars and as many as ten people in a parking lot at 2am on a Saturday night.'
>
> Interviewee

In rural areas the United States Postal Service also functions as an important, if unwitting, agency through which gays and lesbians are able to meet each other, obtain information and even discover some sense of community. Many will open post office boxes in towns far from their homes, and drive for several hours weekly (at least) just to check their mail. In this way, gay magazines and mail from contact ads can be received without risking one's anonymity (Lee 1978):

'When I first began to accept who I was, and tried to find others like me, I left a note on the bathroom wall of the University library. I just said that I was interested in meeting other men, gave a brief description of myself, and left a post office box number. About two weeks later, I received a reply from a guy who sounded nice, along with his post office box number. After about five letters back and forth, we finally felt comfortable enough to meet. We were the first gay person the other had ever met, and we kind of came out together.'

Interviewee

Such an initial, important moment of coming out usually sets the participants on a quest for further information, as they attempt to further develop their sexual identity, and to find some kind of community to belong to.

THE SEARCH FOR IDENTITY AND COMMUNITY

Differentiating between the socialisation processes of rural or nonmetropolitan homosexuals and their urban counterparts remains an extremely difficult task. Many of the psycho-social components of rural gay men and lesbian women are shared by their urban counterparts, while others differ only in degree or vary widely between individuals. It is, however, by drawing a composite portrait of rural social environmental factors and their combined effects on the general process of homosexual self-identification, that differences seem to emerge.

Perhaps the greatest difference between metropolitan and nonmetropolitan social environments lies in the availability and accuracy of locally obtainable information about homosexuals and homosexuality, and how this greater deficit affects the identity formation process of local gay men and lesbian women. The process of 'coming out' as a gay, lesbian or bisexual person almost invariably involves an abrupt and painful renegotiation of one's social and psychological contract with society (Coleman 1987; De Monteflores and Schultz 1978; McDonald 1982). Involved as well is the gathering and reintegration of available information, needed in order to produce a 'new' identity. In an environment where reliable and easily obtained information is absent, as it is in many rural environments, the entire process of coming to terms with one's new sexual identity may be compromised.

Historically, rural environments lag behind population centres in the diffusion of social change. This gap between urban and rural is especially pronounced concerning race and gender relations, as well as homosexuality (D'Augelli and Hart 1987; Molnar and Lawson 1984). In addition, nonmetropolitan media sources frequently assist rather than hinder this phenomenon. Harriet Engel Gross and Sharyne Merritt (1981), in their study of urban/rural differences in coverage and content of women's issues (including homosexuality) in newspaper 'lifestyle' pages, documented a greater amount of social 'gatekeeping' in nonmetropolitan newspaper coverage, suggesting that rural and nonmetropolitan media sources function more as social executors than educators, sustaining traditional images of reality rather than validating or explaining new information to the public.

Newspapers are certainly not unique in this regard. Television and radio broadcasts function similarly, defining through sensory images the prevailing social definitions and prescriptions (Silverstone 1994). In addition, while media social control of gender and sexual roles is not limited to nonmetropolitan rural media, in rural social environments where alternative informational sources are not readily obtainable, the effects on the identity development of men and women with homosexual proclivities seem more pronounced. In using the term *alternative informational resources*, I largely refer to informational exchanges occurring through personal contacts with self-identified homosexuals or homophile organisations or people – some actor or agency which can provide more accurate information to counter the myths and stereotypes about homosexuality predominating in American society and broadcast by 'mainstream' media.

In an urban environment such as Minneapolis-St Paul, one has a good chance of encountering homosexuals or homosexuality, even if accidentally. However, in a nonmetropolitan environment such as Minot, one doesn't come into contact with openly gay, lesbian or bisexual people readily. The gay community is invisible to all but the most diligent searcher, and even within the identifiable social or sexual marketplaces, many self-identified homosexuals limit personal informational exchanges. This reticence or invisibility results from the more justifiable fears of the consequences of exposure in nonmetropolitan environments where anonymity is a rare commodity. Likewise, there exist no gay or lesbian organisations locally, the nearest being some 250 miles away in Fargo. Of the local social networks which do exist, many are tightly knit and don't readily accept new members, again functioning to minimise the risk of exposure.

In short, while finding reliable, current and easily obtained written information about homosexuality seems more difficult in nonmetropolitan areas than in urban centres, meeting self-identified lesbians and gays themselves is even more difficult. This is especially worrisome as researchers are finding that the presence of openly gay men or lesbian women seems to play a key role in influencing others with homosexual feelings or behaviours to likewise self-identify. Thomas Weinberg (1978) found that the identity histories of his male subjects illustrated that a large percentage began to reinterpret their behaviour as 'homosexual' only after they came into contact with self-defined homosexuals. As he states:

it was found that coming into some sort of sustained contact with gay people was very important in changing the meanings of their feelings, fantasies, and behaviors and in helping them make the link between 'doing' and 'being'. Other homosexuals enabled the men to develop less negatively evaluative definitions of homosexuality and to learn what the gay community would regard as more accurate ideas about homosexuals.

Weinberg 1978: 155–6

Many of the men I came into contact with in Minot possessed such inaccurate imagery of the meanings of being gay, defining gay men as being effeminate, as being transvestites who live in large cities (an image propagated by local media coverage of gay pride events), as being pederasts or otherwise immoral or deviant. These men instead saw themselves as too 'normal' to be gay, or saw their own behaviour as a temporary phase, attributable to high libido or the effects of alcohol (cf. Doll 1992). The split between homosexual activity and homosexual identity, also documented in Glen Elder's chapter on South African mines, is a major feature of nonmetropolitan homosexualities. For women with homosexual feelings, the importance of identifiable alternative role models seems even more important, and of the women I've come into contact with in Minot, their process of forming homosexual identities frequently followed a more difficult and circuitous route, precisely because of the seeming lack of viable alternatives to the predominant gender and sexual socialisation models throughout our society.

While gains have been made in changing the socialisation process of women in American society, it nevertheless remains the case that young women are taught not to stray from the socially prescribed goal of family and companionate heterosexual marriage. This traditional socialisation script predominates even more in nonmetropolitan areas such as North Dakota, where women are known as 'gals', are socialised to begin families soon after high school, and where lesbianism (or the charge of lesbianism) remains a utilised justification for the denial of custody rights (see Miller 1989).

For those men and women able to successfully resolve the tensions between their stigmatised status and their homosexual feelings, the form their resultant identities take may also differ from their more 'liberated' urban counterparts. In 1979, when I first began my observations in Minot, North Dakota, life for the local self-identified gay men and lesbians resembled more a 'time capsule' of homosexual life in 1950s America (cf. the life stories collected in Nardi et al. 1994). Perhaps it is the case that each of these social environments, by applying more intensive pressures to assimilate, produce similar accommodative survival strategies and resultant identities in those constituents possessing homosexual behaviours or feelings. In rural or nonmetropolitan social environments, one consequence of the lack of accurate or easily obtainable information about homosexuality, coupled with the scarcity of personal contacts with openly gay or lesbian people, seems to be that for both men and women the process of making or accepting the connection between homosexual feelings or behaviours with *homosexuality* (as an identity) may be compromised. This suggests that in nonmetropolitan environments, fewer men and women with homosexual feelings or behaviours may grow to attain a gay or lesbian identity, or may do so at a later stage in their lives than do their urban counterparts. This finds support in Troiden and Goode's (1980) study on male homosexual identity development:

> We speculate that persons residing in large cities have greater access to gay opportunity structures and for this reason, perhaps, arrive at homosexual self-definitions at slightly younger mean ages than men who reside in less populous areas.
>
> Troiden and Goode 1980: 385

Because of this lack of accurate information about homosexuality in non-metropolitan areas, it may also be the case that nonmetropolitan gay men and lesbian women internalise to a greater degree the stigmatising values of the dominant culture, thereby intensifying the internal dissonance all homosexuals feel during the process of personal identity resynthesis. This dissonance usually involves a lowered sense of self-esteem, greater self-deprecation and a general feeling of hopelessness and inferiority (Rounds 1988). Frederick R. Lynch, in his study of suburban gay men, found that:

> Male homosexuals living in outlying areas ... were slightly more circumspect in revealing their identities, worried more about exposure, 'passed' (for hetero-sexuals) more often, anticipated more intolerance and discrimination, had fewer homosexual relationships, less homosexual sex, [and] less social involvement with homosexuals.
>
> Lynch 1987: 192

This lack of information or role models to counter societal stereotypes, combined with the general unpopulated nature of rural environments themselves, presents still other psycho-social difficulties for nonmetropolitan gay men and lesbians in their search for potential partners and a sense of gay or lesbian community and identity. From my observations in Minot, it seems likely that gays and lesbians in nonmetropolitan areas are more likely to form or remain in incompatible relationships, or rely on social networks with people they have little in common with but their sexual preference. Loneliness and isolation, common rural maladies in and of themselves, may also be intensified in nonmetropolitan minorities, with their smaller numbers and their identification (or self-identification) as seeming 'out of place'. As a result, while many rural communities experience elevated rates of alcoholism, drug abuse, depression and other socially deviant behaviours, the rates may be even greater in rural minority residents, including gays and lesbians (D'Augelli and Hart 1987).

SUMMARY

This chapter attempts to identify basic problems and needs for rural or nonmetropolitan gay men and lesbians, although its scope is exploratory rather than definitive. Many of the ideas presented here are untested, and therefore more studies need conducting in order to gain greater insight into the lives and needs of nonmetropolitan gay and lesbian residents. However, while such research will no doubt benefit nonmetropolitan gays, lesbians and bisexuals, perhaps more immediate actions would provide more concrete benefits towards improving the quality of nonmetropolitan gay, lesbian and bisexual lives. Because nonmetropolitan sexual outsiders must frequently depend on the diffusion of social change from America's urban centres, perhaps urban gay and lesbian individuals and organisations should expand the realm of their activism to encompass their entire regions. Because nonmetropolitan hinterlands furnish much of the growth of urban gay and lesbian communities through migration, helping such migrants adopt healthier identities earlier would enhance the overall vitality and strength of inner-city queer communities as well.

Relatively low-cost responses which urban gay and lesbian communities could undertake include sponsoring speakers' bureaus or social events in nonmetropolitan areas, or providing grants to establish local meeting places or libraries. However, perhaps the greatest impact on the lives of nonmetropolitan gay men and lesbian women could be attained by the simple dissemination of more reliable and accurate information through the hinterlands by urban homophile organisations. This would speed the diffusion of social change regionally, as well as aid directly those men and women searching for homosexual identities or social networks in nonmetropolitan libraries or bookstores. This donation of books to local libraries, or negotiating distribution sites for community gay and lesbian newspapers in nonmetropolitan areas, could greatly ease difficulties in establishing local social networks and help support budding local activists.

In conclusion, then, perhaps the picture of life I've painted so far is pretty bleak, and in some ways it is. Gay men and lesbians in Minot, as in other rural or nonmetropolitan areas, seem further hindered in their search for identity and community than their urban counterparts and lack many of the choices urban homosexuals possess in developing relationships and social networks. While the quality of life for gays and lesbians does seem to be improving in Minot, as in other nonmetropolitan or rural areas, the overall pace of change is glacial relative to that of America's urban centres.

However, despite the constraints imposed by their social environment, it remains the case that positive change seems to be occurring in the lives of lesbians, gay men and bisexual people in Minot, a fact largely attributable to the increasing availability of information obtainable locally. In recent years, I've met some gay men and lesbian women who outwardly differ very little from their urban counterparts, at least within the private spaces of their homes. These people seem to have formed relatively positive homosexual identities, as well as a sense of gay and lesbian community, often without any direct contacts with urban homosexual communities. This suggests that while a gap still exists between nonmetropolitan lesbians and gay men and their urban counterparts, this discrepancy may be narrowing, thereby providing them with expanded choices concerning the environments, identities and lifestyles they wish to adopt.

NOTE

1 While I use the terms homosexual, gay, lesbian and bisexual, I am aware that many of the men and women in rural areas who do have homoerotic feelings, experiences and behaviours would not identify as any of these. I have met more than a few men in Minot who engage in illicit homosexual behaviour while identifying as heterosexual (or, more probably, as just 'normal'). Many are (or have been) married, some have children.

14

(RE)CONSTRUCTING A SPANISH
RED-LIGHT DISTRICT
prostitution, space and power

•

Angie Hart

Over the years, *el barrio*, the street-prostitution neighbourhood in which I conducted anthropological field research during 1990 and 1991, had accumulated myriad identities. Situated in a town on the East coast of Spain with a thriving service industry and a significant tourist base, the barrio is home to many of the town's monuments and museums. Located at the heart of the town, it welcomes visitors to its pretty gypsy quarter where rhododendrons and geraniums flourish outside the authentic white-washed cottages.

Further down the hill, still in the barrio, a religious waxworks museum draws in its clientele opposite a prime prostitution and drug-dealing site. In this part of town, just two minutes down from the pretty gypsy cottages, stray cats and dogs form intimidating gangs on derelict wastelands. Few of the barely habitable houses and flats have a toilet, let alone a bathroom, and the area has earned its reputation as a centre of mugging and other violent crimes. Intrepid visitors keen on high culture gingerly pick their way through to the waxworks museum. The museum curator has been trying to get rid of the prostitutes for years. He isn't the only one. On a sunny day in August 1990, I was eating sunflower seeds with Martina who was waiting for business under someone else's balcony. A woman came out on to the balcony above us. She screamed at us to go somewhere else, she spat down at us, called us 'filthy whores' and threw water out of the window. 'Fuck off, you're not wanted in the barrio,' she yelled. This incident was a typical example of one type of reaction from residents to the presence of prostitutes.

This chapter looks at the different identities of the barrio in different contexts. It considers the various power relationships of many of the actors who have a hand in constructing the barrio's identity and are in turn themselves, to some extent, constructed *by* its identity. Above all, I consider the barrio in relation to its identity as a prostitution site.

However, I do not simply consider the barrio environment in its strictly physical sense. It is both uncontroversial and obvious (at least to geographers) that people's identities are in part constructed through the spatial locations they inhabit and frequent. Nevertheless, these spaces do not simply have physical presence. They are imbued with symbolic meaning, often at once contradictory, confusing and changing, and the barrio is a particularly pertinent example of this.

It is thus important to consider the barrio's power – as both a symbolic concept and a physical reality – to affect the construction of the identities of individuals who frequent it. This theoretical stance builds on the insights of Rob Shields. His study of *Places on the Margin* (1991) employs the concept of spatialisation:

> I use the term *social spatialisation* to designate the ongoing social construction of the spatial at the level of the social imaginary (collective mythologies, presuppositions) as well as interventions in the landscape (for example, the built environment). This term allows us to name an object of study which encompasses both the cultural logic of the spatial and its expression and elaboration in language and more concrete actions, constructions and institutional arrangements.
>
> Shields 1991: 31

Following Shields, then, it is important to consider the 'barrio' in terms of the social (re)construction of the spatial in time (the inverted commas around the word 'barrio' are used to imply the sense of barrio spatialisation – both a physical space and a more complex symbolic one).

Once one begins to reflect on the 'barrio', a logical step is to further reflect on the realm outside the 'barrio', thereby directly locating clients and prostitutes in relation to the constraints/enablements of barrio spatialisation. I also examine how these constraints/enablements affected the relationships that clients were able to have with prostitutes, and also briefly look at the ways in which prostitutes themselves were constrained/enabled through spatialisation. Hence I explore the power of spatialisation to limit people's activities and to create inequalities in life. I also consider how, through spatialisation, individuals affect and/or control the lives of others.

The barrio was (and at the time of writing still is) a confined spatial location in which many people spent considerable amounts of time. There was, on most days, considerable activity in the barrio with people constantly coming and going, but always with a locus of people there. The barrio may be described as a (confined?) space for a number of the town's 'marginal' people, many of whom were not employed in the formal economy. Hence, on a number of levels the 'barrio' is located in terms of peripherality – marginal individuals often *felt* confined to it, and the 'barrio' was constructed locally for people who, often readily identified as 'marginal', were not accepted (or felt to be accepted) in other areas. Hence certain 'marginal' individuals were, on a day-to-day basis, relatively confined to the barrio. They had to live there because they could not obtain accommodation in other locations, and many were reliant on having constant access to barrio 'amenities' such as drugs and the Catholic social welfare agency, Cáritas.

This sense of spatialisation-as-confinement seems to me to be usefully compared to Foucault's notion of the disciplining powers of certain institutions such as prisons and 'mental asylums' (1977). Whilst the 'barrio' was certainly not an *institution* in Foucault's sense (see Foucault 1977: 178), some of his ideas about confined disciplined bodies may be related to it. However, I depart from Foucault by conceptualising barrio individuals as agents rather than simply as bodies (see Giddens 1984: 154). Questions about the disciplining abilities of the 'barrio' and of leisure discourses to structure informants' actions are certainly postulated. However, I also attempt to examine how informants interacted with these disciplining abilities, exploring how they themselves coped with and used the very (disciplining) discourses of centrality and periphery through which aspects of their identities were constructed.

CLIENTS IN AND OUT OF THE 'BARRIO': A LIFE OF LEISURE?

Whilst, as I explained above, some individuals felt and/or were more or less permanently confined to the barrio, many individuals in the 'barrio' had much more complex identities. Many were 'confined' to the barrio space only for parts of their time. A number of people who spent many hours outside the barrio went there at various times of the day. These included some prostitutes, clients and *mirones* (voyeurs). Many of the clients in the barrio were either unemployed or retired. Many of these men spent considerable amounts of time in the barrio, drinking or simply standing in the bars, or wandering the streets chatting to people occasionally or just walking around.

Although some of the clients could be described as unambiguously marginal persons who were 'at home' in this context – for example those who were unemployed, alcoholic, unwashed and who lived in the barrio – this was not generally the case. More accurately, most clients were ambiguous marginals, in a number of different senses. Hegemonic values in Spanish society do not condone the action of men buying sex openly, so most clients went to the barrio in secret. Whilst they were there, they were engaging in what was considered to be a stigmatised and marginal activity. But when they were back in the 'mainstream', clients were 'non-clients' – at least as long as they were not discovered. Hence, they were in a more powerful position than many barrio residents, as they were generally able to move between spheres. This ability to move between spheres allowed clients to be (re)constructed through different spatialisations.

Many of the clients were ambiguous marginals in another sense. A number of them had been employed in formal wage labour for most of their lives. Once retired, they found themselves with considerable 'free' time, with few responsibilities restricting how they spent that time. With no obvious position in the formal economy (except for the weak role of pensioner), they became, in another sense, marginal persons.

The 'barrio' was clearly of considerable importance for many clients and could even be described as a kind of leisure spatialisation. However, their individual experiences of this were affected by their personal histories. For example, retired or unemployed clients had a different experience of the 'barrio' than did those who were employed.

One genre in sociology conceptualises social time through the division of time into *work* and time off from work, or *leisure* time (see, for example, Haywood *et al.* 1989; Rojek 1985; Seabrook 1988). There are numerous debates on how to differentiate between the two spheres, discussions of the definitions of work and leisure, and so forth (for example, Haywood *et al.* 1989). There are also some studies that call into question any such divide, arguing that leisure is simply another part of work within a capitalist economy (see Jones 1986; Rojek 1985; Seabrook 1988).

However, any discussion of 'time off' is complex in that one must always define what an individual is having time 'off' from. Few contemporary authors see this as a simple issue. With unemployment a reality for many people at some point in their lives, and retirement an experience for most, time off from work is beginning to lose its privileged status as a hegemonic structuring of leisure.

Increasingly, authors writing about leisure have begun to explore issues of power in describing leisure as time off from work. They point to the vast numbers of people who are unemployed, retired and/or on low incomes who have 'forced leisure' (for example, Haywood *et al.* 1989; Seabrook 1988). Whilst many authors define leisure as free-choice activities, these authors point out that if individuals have little money (and/or education), they are severely limited in the number of choices available. Some authors have also explored the negative psychological and symbolic aspects of 'leisure time' in relation to ageing (Haywood *et al.* 1989) and gender (for example, Seabrook 1988).

Discourses articulated during my period of 'participant observation' in Alicante point to how fluid the meanings of these categories of leisure and work may be. Many clients were retired or unemployed or worked in casual employment. Consequently they had a lot of 'free' time, or rather time when they were not engaged in paid employment. The barrio provided a social arena in which they were able to *do time*, thereby avoiding many of the negative aspects of life for the unemployed or retired as discussed above. However, this notion of 'doing time' may also be related to the manner in which people often had little more than a physical presence in the barrio at certain times; the prostitutes 'did time' in this confined space, the clients 'did time' (and indeed I certainly 'did time'). Was this then (again almost in a Foucauldian disciplined sense) 'forced leisure' for the clients and 'forced work' for the prostitutes?

I am certain that most clients and most prostitutes in the barrio would not think so. Individuals were often engaged in more than simply 'doing time'. Numerous animated conversations went on in the bar and on the pavement area outside it. It was possible to watch television in the bar, have a conversation, or simply 'be'. On numerous occasions people spent time in the bar without buying a drink or at the very most buying only one. These are all ways of spending time. For some prostitutes the presence of clients was important. They provided a welcome relief from silence, from boredom and from chatting to each other.[1]

The realm of prostitution as leisure is generally excluded from all discussions in academic texts of leisure. Academic texts on prostitution largely exclude leisure in association with clients. Clients are thought to go to prostitutes for all kinds of reasons, but leisure is the least of them. Authors of leisure texts prefer to stick to more wholesome subjects despite the fact that, as Rojek points out (1985: 21), sex is one of the five major 'leisure' pursuits in contemporary Britain. Perhaps more surprisingly, authors who write about the work/leisure divide also exclude the categories of 'love' and 'romance' from their texts. This may be because such entities are not thought of as being readily quantifiable (most of the research on leisure is quantitative). It may be that because of the popular privileging of discourses of 'love' and 'romance', they are seen as being somehow beyond time and space, with transcendent status – perhaps not a part of the 'real world' worthy of documentation.

One sociologist of leisure studies who does not entirely essentialise a definition of leisure is Chris Rojek (1985). He points out that in much of the literature, leisure is associated with the experience of personal authenticity. Following Frith, he asserts:

> There is clearly a bridge of consensus regarding the medium and the content of leisure experience which spans the divisions of multi-paradigmatic rivalry. The medium is self, and the content is pleasure . . . Leisure experience is not an essence in human societies, but an effect of systems of legitimation.
>
> Rojek 1985: 173–5

Thus for Rojek there can be no fundamental definition of leisure. Such a definition depends on particular contexts. However, he says, this hegemonic discourse of leisure as personal fulfilment/authenticity is a powerful 'moralising discourse'. This discourse is behind the absence of discussions of sex-as-leisure. In contemporary Western society, although sexual activity may be legitimate in some senses, it is illegitimate in the sense of not being generally accepted as something to talk about and, in a less clear sense, as something *to do*. It is not a leisure pursuit that most people boast about. Its meaning is quite different to that of, say, gardening – a popular, acceptable leisure activity.

Within the barrio, sex as leisure was not an unambiguously illicit or illegitimate pursuit. Many clients did voice misgivings about being there (on a periphery), and often went so far as to describe their presence in the barrio as a 'vice'. However, they were able to enjoy this 'vice' in an atmosphere in which this was accepted as a leisure pursuit, albeit one that was considered to be rather different to others.

As clients moved outside the barrio, the concept of the 'barrio' as leisure spatialisation became increasingly less acceptable. In many circles they were unable to acknowledge that they frequented this area, even if they did not have sex with prostitutes there. This led to the prevalence of 'barrio' as *closet leisure spatialisation* beyond the immediate vicinity, and sometimes within it. Hence Venturo, a well-known barrio client who died during the latter part of my fieldwork, commented to me about his unease at going to the barrio because he was afraid that his family would find out. This particular conversation took place in the barrio; nevertheless Venturo was informed by the 'barrio' as an illegitimate, peripheral or closet spatialisation.

FRIENDSHIP, LOVE AND ROMANCE

For the prostitutes, time spent in the barrio was indeed generally considered to be work time. It was certainly the case that most of them worked long hours and had little definable leisure time. However, this was sometimes through choice, and was not even necessarily directly related to economic need. Some of them told me that they would rather spend their time chatting in the barrio, even during periods when business was quiet, than sitting at home watching television. Consequently, many of the relationships that they had with clients were not simply work relationships. Friendships enabled both clients and prostitutes to experience some kind of leisure time.

The clients went there, sometimes for sex. However, these transactions often lasted only twenty minutes. Further time was generally spent in the bars or chatting on the streets. Often men who went to the barrio were not looking for sex, but rather for a few hours' conversation. Others were looking for sex but also for conversation and company.

Friendships in the 'barrio' took many different forms, although they were often adversely affected by external forces. It often seemed as though friendships between clients and prostitutes survived remarkably well, considering the hegemonic impediments of the 'negative-illegitimate-peripheral-barrio' that militated against them.

Many of the friendships between the prostitutes and clients drew on discourses of love and romance. Recurring language used by clients and prostitutes was often equated with the language of love and romance. Both parties frequently talked in hushed, seductive tones, and played the dramatic roles of romantic woo. Of course, this romantic language was often an act put on for economic reasons. Rita told me that 'the best way to try and get a client is to do lots of sweet talk and coaxing. With regulars you don't have to, they just followed you up to your room.' However, this was not always the case. Much of the 'romantic sweet talk' was carried out between regular clients and the prostitutes with whom they had sex. Rita informed me: 'Most women aren't interested in talking to their clients, they just want a quick fuck. I'm very rare for a prostitute.'

Despite her insistence that she was unique and different, it was not my impression that this was the case in the barrio. Many of the prostitutes spent time talking and drinking with clients in the bar. During these conversations they often showed physical affection to each other, and used romantic language. Prostitutes who did not engage in such conversations were generally quieter than the others in all situations. They might not have talked at length to clients, but then nor did they talk much to other people in the barrio.

Thus a number of clients and prostitutes, in their relationships with each other, imported ideals of love and romantic courtship prevalent beyond the hegemonically perceived context of their mutual relationships. For example, a number of clients were keen to establish their fidelity (or hold onto a romantic illusion of fidelity), even if this was not strictly true. This illustrates the manner in which individuals were able to mobilise behaviours considered appropriate in the mainstream, and (re)construct them in the marginal context of the 'barrio'.

Many of the friendships that evolved in the barrio were less dramatic than the ones discussed above, and their constructions show characteristics of mechanisms for passing time. People normally think of a friend as somebody who can be relied on to help another person, somebody to confide in, and so forth (see Allan 1979; Porter and Tomaselli 1989). This was certainly one major ideal in the barrio, but few people had these kinds of friendships. This was the case with regard to most of the relationships that the clients had: between them and other clients; between them and prostitutes; and between them and other barrio persons. This was also the case with regard to acquaintances between prostitutes, and seemed to apply to a group of men who played cards together. They referred to each other as friends, although they did not spend any other time in each others' company.

TIME AND SPACE FOR THE 'FAMILY'?

Clients spoke a great deal to me about their families, who appeared to be important in their lives. This was not simply an abstract emotional and symbolic importance; families also provided a place for them, in Heidegger's terms (1972), to be in time (Da-sein). Many grandfathers enjoyed being with their grandchildren, and this was in many cases an important weekend activity for them. Such men rarely frequented the barrio at weekends. However, during the week, even if they lived with members of their families, they were generally left to their own devices as the pressures of work and school took over the time of others.

Clients who were engaged in work, and for whom their family was important, often saw the passing of time in the barrio in a different light to retired clients. This time was more rationed. They often had to lie to partners and family in order to go there. It was not a place for casual chatting. Their time in the barrio had an importance quite different to that of clients who spent considerable parts of their day there.

It was, for them, even more of an illicit way to spend time than for retired or unemployed clients. Certain clients of this type crept around the barrio, extremely concerned that they would be seen by somebody who they knew. When talking to them, I had the sense that they were always aware of the time, constantly worrying about being late, about having to find some excuse. I think that this state gave many of them an added charge of excitement – the power of the illicit. They enjoyed the thrill and the challenge of being somewhere that they should not, and lying about it. These men did not relax in the barrio; however, they had the power to enjoy the 'barrio' as an illegitimate, peripheral space since, unless they were discovered, there was always a space for them in the legitimate centre.

Hence, even when 'barrio' leisure did not necessarily take the form of a direct sexual encounter with a prostitute, it held a contradictory position for many clients. On the one hand the 'barrio-spatialisation' was one of 'free-time' activity, and was often discussed in those terms. For many it was a space for positive non-work experiences that they could enjoy. But on the other hand, many clients felt the need actively to justify to me their reasons for going to the barrio. This was exacerbated for clients who had wives and families who were concerned about what clients did with their (spare) time. For none of them was it an unambiguously acceptable way in which to spend time.

Clients' families certainly seemed to have some kind of controlling effect on the way in which clients experienced the 'barrio'. Some clients who worked and who had important family ties legitimated their presence in the barrio by doing work deals there. A mechanic who went for a drink after work every day often did small jobs for people, although he told me that most of the time he did not seek payment. He was happy with his work salary. One of the local plumbers did considerable work in the barrio and thus was often organising deals in the bar. Employed clients who had no family were not as constrained. However, it was my impression that for them, the 'barrio' took on considerable importance, similar to the way that it did for many pensioners. This reflected their peripheral identities in relation to central, monied, nuclear 'families'.

DIVIDING SPACE, DIVIDING FRIENDSHIPS?

Most of the clients and *mirones* (voyeurs) went to the barrio without their families' express knowledge. Many thought that other family members might guess that they went there, whilst others felt very strongly that they should not know. Hence the 'barrio' took on the connotation of an illegitimate space and time location in many contexts. On the other hand, it was also a safe space for activities that were not approved of in wider society. However, even if men did not engage in these disapproved-of activities, the barrio-as-space became associated with them. In addition, men had to frequent them at the appropriate times – generally when families were unavailable to them.

Despite this hegemonic notion of illegitimacy of time and space, the meanings of the 'barrio' as time and space manifestation could be manipulated. Hence some clients insisted that they went to the barrio only because there was nothing else for them to do (time), or nowhere else to go (space). It was, then, suggested that it was not as though they actively sought out the barrio for illicit activities, but rather that they had nowhere else to go. Other clients stated that they went to the barrio for reasons unconnected with prostitution (for example to arrange work deals, or to visit 'friends').

Of course it was not only the clients who generally had an interest in keeping a distance between home and their life in the barrio. The prostitutes in the barrio generally tried to keep some separation between work and home or non-work space. Although some of this applied to keeping things (especially home addresses) secret from the other workers, it was mainly aimed at the clients. Many men attempted to cross this barrier, but in doing so they often encountered resistance.

One day, Rita and Antonia were showing me Rita's photograph album. We were looking at pictures of Rita's home in Granada province and of a day trip to Murcia province that they had been on together. A regular client appeared and looked over our shoulders, trying to join in the conversation. Rita immediately shut the album and moved away, saying that it was private. Even though Rita was showing the album to Antonia in the barrio (a work space), the album belonged to a conceptual space beyond work. Hence she did not want the client to see it.

This tension between the meanings of work space and non-work space is further illustrated by Rita's reaction to Antonia (a prostitute) and Venturo's frequent siestas. Although Venturo was (theoretically) a client, Rita (and in some respects Antonia and Venturo) interpreted their siestas as mixing two separate domains. Although Rita was happy at other times to accept Venturo as a friend, in this instance she wanted the power to define how her apartment was used during what she might have termed 'non-working hours'. She chose to position Venturo firmly in the 'work' category in this instance. Hence Antonia, resisting Rita's interference, invited him up for siestas mainly when Rita was away and therefore would not realise.

Despite these kinds of intimate friendships between some clients and some prostitutes, few clients had ever visited a prostitute's home, especially on a strictly social visit. Those who did appeared to have particular types of friendships with the respective prostitutes. Such friendships were less easy to define in terms of more obvious ideas about client–prostitute relationships.

Prostitutes often went to clients' homes in relation to work, but only when other household members were not present. For example, although Antonia was extremely good friends with Venturo, she was permitted to visit his house only when his family were out. To my knowledge, Venturo never visited Antonia's house. This ambiguous state of affairs was the case with many other client–prostitute relationships. Antonia had two other clients who allowed her (and often requested her) to go to their homes when nobody else was there. These home visits did not generally take place unless sexual activity was involved. Some men preferred Antonia to go to their house for sex. One client, Joaquín, informed me that he preferred it this way because he found the act more 'friendly'. In addition, he did not like people talking about him in the barrio. If he went with Antonia in the barrio, people would talk. It was much better for her to go to his house. Sometimes he went to her house, although he could go only when her partner was not at home. Needless to say, clients who went to their own homes with prostitutes were never married men. They were widowed, divorced or single. Most of them lived alone, although some shared their homes with relatives. In such cases they would invite a prostitute home only when they knew that nobody was in.

Prostitutes rarely went to clients' homes for social visits. One prostitute explained how a particularly good friend of hers took her to his home town. However, she was unable to actually go and meet his family. Apart from the issue of home visits, it is also important to consider whether clients and prostitutes spent time together outside the barrio in any other spheres. Certain clients with whom prostitutes were most friendly were often invited on 'barrio trips'. However, clients who went on such trips were expected to respect the taboo of talking about prostitution. Prostitutes generally saw these trips as a chance to escape from their daily routine, and did not like to talk about their work. When one prostitute, Antonia, began to discuss men's penises – a work-related topic – during a coach trip to Lourdes, she was quickly silenced by other prostitutes who were on the trip. I never witnessed an incident on a trip in which a client talked about prostitution – although other people on the trips made a number of negative comments about the prostitutes.

Some clients and prostitutes met on a regular basis outside the barrio. Those who were particularly friendly with each other often went for walks, met in parks, or went to cafés together. Nevertheless, friendships between clients and prostitutes were often significantly limited because of the self-censorship that they imposed when they were outside the barrio. Many were scared of being seen together; they were afraid of reprisal from family or friends. Hence individuals could manipulate others through particular conduct beyond the 'barrio'.

There were, of course, negative sides to extra-barrio meetings between clients and prostitutes. Some clients appeared to want to wound prostitutes. Maria, for example, complained that one of her regulars, an alcoholic who also happened to be drunk at the time, saw her walking in the street with her daughter and called her a whore. Other women related similar stories. I also experienced this on many occasions myself. I would be out walking around town (either in a group or alone) and a man would remark (often loudly so that others could hear) that I was a *puta* (whore) who worked in the barrio. Young children also made such remarks.

Whether or not clients or prostitutes initiated discussion outside the barrio had a lot to do with whether or not they were alone. Of course the person with more power in this regard was the one who was either alone or in company who knew of her/his 'status'. The loser was the person who was shown up. However, I heard very few anecdotes from clients who told me that they were shown up by prostitutes. My impression is that it was more often the other way around. Besides, the prostitute invariably had more interest in not showing up her client, as she wished to keep him. She might, however, in certain circumstances, enjoy showing up the client of another prostitute.

Events immediately preceding Venturo's death illustrate the importance of extra- and intra-barrio controlled spatialisation in his relationship with Antonia. During the two weeks that Venturo was in hospital, Antonia and Rita visited him every day. I was unable to see him before he died. However, I had a number of conversations with Rita and Antonia about him. Rita said:

'We went to see him every day in hospital. His family were there too, but I told Antonia to keep quiet, and to just say that she was a friend. Of course big gob Antonia went and told them who she was. She told them the whole story, and so they threw her out. Well, Venturo's son did. Antonia wanted a widow's pension. She wanted Venturo to marry her on his deathbed. But Venturo's son wanted his money. Poor old Venturo was torn. That's probably why he died, all the hassle . . .'

When I saw Antonia for the first time after Venturo's death (some months later), it was one of the first things that she talked to me about. She seemed genuinely sad:

'Me and Rita went to see him loads in hospital. I sat by his bed and cried and cried. Poor bugger, he liked us being there and never wanted us to go. He said "girls, girls, stay a bit longer". I loved the way he was. We used to go to bed together and I even used to go to his house and we'd have baths together. He made love like a young man, not like an old one. I was very sad that his son wouldn't let me see him. So was Venturo. I'm sure that this was what killed him in the end. I didn't want anything from him. We were friends.'

Antonia went on to tell me that she was obviously not permitted to go to Venturo's funeral. Venturo and Antonia's friendship illustrates one of the ways in which intra- and extra-'barrio' spatialisation was related to power. Their experiences of extra-'barrio' spatialisation were not regulated by them. They had (according to a powerful third party) overstepped the peripheral location of their friendship. This last anecdote serves as an illustration of the way in which Venturo, when time and space in the world were running out for him, was no longer in control of the small amount that he had left. He was not free to spend his time and space with Antonia.

Despite the fact that the relationships of clients and prostitutes often could not unproblematically exceed the physical limits of the barrio, they often exceeded the emotional limits of their 'basic' relationship. Whilst numerous texts point to the inhumane, direct commercial exchange in prostitution, in my research site there was an abundance of friendly relationships between clients and prostitutes that defied the conceptual limits (in both a symbolic and physical sense) imposed on their fundamental relationship.

PASTS, PRESENTS AND FUTURES

It might be useful to borrow from Heidegger's theory of biographical time (1972) to illustrate how pasts, presents and futures can be shown to have interplayed constantly in constructing the spatialisation of the 'barrio' present. I have shown how clients (many of whom were elderly) and prostitutes constantly brought into conversations (an experience of the present) their notions of the past. Hence the barrio was, for many, a spatialisation that had a golden past.[2]

When I asked clients or prostitutes about what historical period they were referring to when talking about the past, I was rarely given dates. Many referred to the period 'under Franco', a thirty-six-year span which saw many internal changes. However, for most clients and prostitutes, the Franco years were a 'block past', perhaps vaguely informed by memories of times before that. Most people could not remember much about their lives before Franco, although, significantly, memories of childhood in particular were sometimes articulated. Despite all of the complex internal changes in the regime, people were remarkably uniform in their conceptions of the past under Franco. For them it was a time of repression, a time of tight control, which, understandably, they related mainly to their own experiences of prostitution. Individuals differed in seeing this in a negative or positive light; however, the vision was generally uniform. Hence 'the past' was set up in stark contrast to the 'present', intensifying individuals' experiences of the present as liberal.

Whilst I was in the barrio, 'the present' was also affected by people's concepts of futures, which in turn were connected to notions of pasts. During my first research trip, many people had grim conceptions of the 'barrio's' futures, but a positive notion of its pasts. Hence their experiences of 'the present' were informed by these other perspectives, becoming a kind of (dying) limbo period. Prostitution was seen to be on the decline; there were fewer prostitutes, fewer clients, and more disease. It was simply a matter of *time* before prostitution died out. Few people saw the chance for the positive rebirth of the 'barrio' that had been planned by the local authorities. This is most probably because they imagined (correctly as far as I can gather) that these plans excluded them.

The lived-out preoccupations of barrio people in relation to their futures became more apparent to me when I returned to the town after nine months' absence. This trip was, for me, an experience of stepping into my perception of the futures that barrio people had created nine months previously. On the one hand the 'barrio' was buzzing with life. Bulldozers tore down buildings and new ones had already been built. One could see that the town authorities' plans were being successfully implemented.

However, the 'barrio' as a spatialisation for marginals and ambiguous marginals was significantly debilitated. El Atlántico bar, a central location from which clients and prostitutes operated, had closed down. This meant that many clients no longer frequented the area. Those who still did wandered from bar to bar, no longer quite knowing where to place themselves, or else they stood chatting in the streets much more than previously. Some of the prostitutes found it difficult to work now that the bar had closed down. They had to stand for longer in the streets, or sit up in their dark, dingy flats waiting for regulars to call.

Further changes added to this sense of decay. A number of people had died. Others were dying. Apart from these more obvious indications of the changing nature of the 'barrio's' spatialisation, there were other less tangible signs. The streets had fewer people in them; some of the prostitutes had moved to work in other areas, and had not been replaced by many new workers. Some of them who were alcoholics or injecting drug-users appeared to be even more physically affected by their addictions. I perceived little enthusiasm for the work of the nuns in the local Catholic Welfare Centre. (Marginal) people seemed disillusioned; there was a sense of the barrio decaying, and them decaying as a result, with little control over that process. All this struck me as a stark contrast to the expensive new face lift that the 'barrio' was being given by mainstream actors in mainstream institutions.

BEYOND LEISURE – SYNCHRONIC AND DIACHRONIC SPATIALISATION

By exploring the dialectical relationships between centralities and peripheralities, I have shown that the relations between clients and prostitutes in the barrio do not fit neatly into the rubric of leisure. It would seem to me that they fit into a much more fluid conception of spatialisation. However, such an analysis should not be limited to looking at 'present' spatialisations. As I have shown above, the 'present' is created by other times, and the meanings of a particular space – and relationships within that space – change through time.

People compartmentalise their lives and also have them compartmentalised for them, but simultaneously, such compartments are often fluid, overlapping and changing. Certain clients experienced the 'barrio' as a kind of leisure spatialisation; it had many of the qualities of leisure as commonly theorised. But it had other meanings beyond those conventionally understood within the term 'leisure'.

One of the most important organising discourses of 'barrio' spatialisation was that of friendship. The discourse of friendship inter-relates to discourses of love and romance. For individuals who had poor social relations outside the 'barrio', it was a space in which they privileged discourses of friendship, love and romance. For others who, under certain conditions, took their friendships beyond the spatial boundaries of the barrio, the non-barrio space took on significant connotations: those of the non-barrio space often signified more than those of the barrio. In relation to friendships, some informants saw those that broke the bounds of the 'barrio' as more important (or even more central?) friendships than those contained within the 'barrio'. Discourses of time and space for friendship were also related to power – how people chose to include or exclude others through organising discourses in relation to spatialisation. 'Barrio' client–prostitute spatialisation cannot be explained in either strictly Marxist or strictly Foucauldian terms. Clients were not simply exploitative users of leisure facilities, and prostitutes were not simply exploited leisure providers. Neither were passive, peripheral, 'disciplined' bodies. Only by opening up the 'barrio' as a symbolic site can we begin fully to unravel the complex situated microgeographies of these client–prostitute relationships.

ACKNOWLEDGEMENTS

I am grateful to the Economic and Social Research Council (ESRC) for financing this research.

NOTES

1 This kind of relationship between clients and prostitutes, between the consumers of services and the providers of services, is not acknowledged in the negative Marxist view of leisure (as portrayed by Seabrook (1988), one of the few authors to consider the servicing of the leisure industry).

2 Research conducted by Polak (1973) suggests that these nostalgic constructions of 'the past' that constantly inform a person's experience of the present are particularly common amongst elderly persons.

4

sites of resistance

SECTION FOUR

SITES OF RESISTANCE

The performance of sexual identities in space is rarely an unconstrained pleasure. All too often, forces of regulation and discipline – from the panoptical gaze of homophobia to the physical threat of an individual queerbasher, from the state and law to the lyrics of a song – are in place to constrain (and to punish) nonconforming sexualities. At the same time, resistance is mobilised against such constraining forces: queers can (and do) bash back, on the pages of books and the streets of cities – here theory and activism truly come together, since both have the potential to function as sites of resistance, and each has its own power. By reading resistance as spatial practice – defying the propriety of place which keeps certain people out of place, without a home, or lost in space – we can see how contested and embattled terrains can be reinscribed, redefined, remapped.

AIDS activism continues to be a vital site of resistance in the face of pitiful government responses. Both David Woodhead and Michael Brown engage with AIDS work, looking respectively at campaigns and groups in London and Vancouver, and showing how safer-sex programmes have to be sensitive to issues of scale and of sexual identity. Woodhead's work on the cottage as material site for men to have sex, from which he views exclusive notions of community, is particularly resonant on these issues.

The work by Tim Davis, Tracey Skelton and David Bell all builds around a particular event, then pans out from that to witness the wider implications of gay and Irish contestation over Boston's St Patrick's Day parade, the events surrounding the release of Buju Banton's ragga song 'Boom, Bye, Bye', and the arrest of a group of British men for engaging in consenting same-sex sadomasochism. Both Skelton and Davis address the conflict between interest groups often constructed as oppositional, and Bell's chapter surveys the interplay of discourses of public and private as they are used both to regulate transgression and to mobilise it.

15

'SURVEILLANT GAYS'

HIV, space and the constitution

of identities

•

David Woodhead

The earth quaked, and the shock waves of AIDS awakened monsters from the depths of our collective imagination, monsters of a species we had long thought extinct. This plague has attracted the inevitable swarm of AIDS researchers, officials, businessmen, and journalists, and they are the ones who have monopolized the media. We people with AIDS, who devote every waking moment to our own survival, have been unable to prevent those loquacious experts from stealing our thunder and robbing us of the only thing we have left: our illness.

Dreuilhe 1987: 1

SITING RESISTANCES, SITING KNOWLEDGES[1]

This chapter is primarily about spatiality and the multi-faceted, complex and complicit notions of power and resistance. It is not just an attempt to consider political and cultural sites of resistance that have been found, and analysed, in the light of HIV and AIDS. Neither is it intended solely to consider the imagined and material geographies of resistance and power. This chapter itself is an attempt to *constitute* a site of resistance. It attempts to resist through, and within, the site of text. These attempts to resist are manifold and have several focuses. Primarily, this resistance hopes to challenge the seldom questioned 'radical' analyses of gay male resistances. Simultaneously, it attempts to resist prevailing vocabularies of cultural geography and to valorise reflexivity as a serious political and ethical praxis. A further point of resistance is found in the attempt to disrupt the cultural assumptions often made in the name of expediency during the enactment of health promotion practices relating to HIV.

When writing about power and resistance, it is important to be clear about how those notions are being conceptualised.[2] Here I want to propose a reading of power in the broad Foucauldian sense: power is not a coherent or coercive force only exercised through class relations, but 'an all pervasive, normative and positive presence, internalised by, and thus creating, the subject' (Evans 1993: 11). So power is conceived of as being realised in much finer capillary scales, through 'circuits' (Clegg 1989), in an attempt to recognise 'the mechanisms and effects of power which do not pass directly via the State apparatus, yet often sustain the State more effectively than its own institutions, enlarging and maximising its effectiveness' (Foucault 1977c: 73). Of interest to me in this chapter is how power is a consequence of expertise which supports and legitimates those who exercise it. Power is realised through the disciplining surveillant gaze, which constitutes individuals as subjects through knowledge (Fox 1993). Resistance to power comes through counter-discourses. However, as Foucault points out, 'resistance is never in a position of exteriority to power' (Foucault 1982: 209, cited in E. Martin 1992: 409). The sites of resistance, therefore, are the very sites of discipline. Whilst Foucault's position on this is apparently pessimistic, all hope is not lost, if resistance (as points, sites and moments of counter-power) is conceived of as being found in equally fragmented circuits. This recodes resistance as being fluid, momentary and local, a 'third space' of opposition and excess (Bhabha 1990).

I want to commence by positioning myself firmly within the (notoriously fluid) post-structuralist camp. Central to my self-location here is a commitment to rigorous reflexivity in my work. It is this position that I want to explore more deeply throughout this chapter, to (hopefully) create a space, a leeway that allows us to begin to unpack the stuff that constitutes research and its associated practices and power. It is hoped that by disclosing my position I might prompt the reader to be critical of what and how I write, and also to be critical of what and how he/she reads. Never should the responsibility and importance of the active audience be underplayed: it would be a further bonus if such a position were not restricted to this text, but extended to others.

In 'AIDS, keywords, and cultural work', Jan Zita Grover (1992: 231) critically reflects that '[l]ike other activists, I have found that AIDS is a 360 degree sense-surround, and there is no door out of it leading back to a faculty office for me'. This single, simple sentence has changed the way I view my work and left me with no option but to reappraise my motives and my role in the execution of my research. My interest in HIV and gay men grows from my previous work as a professional health promotion worker (implicated in the bureaucratised practices of health promotion), as a researcher who aspires to find himself located around the 'critical margin' (yet equally complicit in the middle-class pursuits of academic work complete with its inherent language games, exclusionary nuances and hierarchical structures), and as a gay man unaware of his antibody status. These are all issues I will consider in this chapter. This does not, however, give me a broad perspective but in fact closes up my position, as it means that the professionalised perspectives I have as a result of my various practices have often led me to bypass the realities that AIDS has created, leaving me firmly bracketed with Dreuilhe's 'loquacious experts' that have almost opportunistically appropriated AIDS for their own. In this context, AIDS is *not* everybody's problem. This is not to say, however, that I have no business in even writing this text,[3] but, whilst I can momentarily position myself within the broad picture I wish to interpret, I hope to show how this is seldom achieved comfortably, rarely achieved ethically,[4] and never achieved wholly.

There are several comments I wish to make before moving on with the chapter. Firstly, I wish to reflect on the metrocentrism of my writing and declare myself as having only known what it is to be 'out' in London. This has the consequence that the thoughts, aspirations and worries I have largely relate to my experience there. Secondly, I want to call into question any notion of a coherent 'gay' or 'lesbian' self. I recognise that how we strategically label ourselves, and, indeed, how we are labelled, is not just related to spatial concerns. I recognise, partly through experience, and partly through listening to the histories of colleagues and friends, how being gay and lesbian is not a unitary experience, but has to be contextualised into a wide range of other factors.

It is at this point that I want to turn to the empirical focus of this piece – the many practices that constitute HIV-related health promotion – placing them in broad cultural, ethical and spatial contexts.

TOWARDS A RADICAL CRITIQUE OF HEALTH PROMOTION[5]

It is of cultural interest that the expression *health education* is steadily being replaced by *health promotion*. This is symptomatic of a broad paradigmatic shift being felt in health services away from a central philosophy of *prescription* to one of *promotion*, *partnership* and *autonomy*. It has been suggested however, that despite the new rhetoric as a bid for legitimation, the nuts and bolts of health promotion/education remain similar, and that despite moves towards targeting efforts, some sectors of the population remain neglected (Hinvest 1993). However, to overstate this would be something of an oversimplification in a time where a whole range of radical and innovative approaches is being pioneered. Whilst it would be expedient to singularise the many practices that constitute health promotion, it would also be misleading to do so. Techniques that practitioners use fall theoretically into several categories. Examples of such theoretical models include the medical model, the behaviour change model, the educational model, the client-centred model and the societal change model (Ewles and Simmett 1992), and in practice draw on a range of available options. However, in broad terms the objective of health promotion work to furnish individuals with the *knowledge* of how to be 'healthy' remains constant. Practitioners would assert that the major exceptions to this are empowerment models of work.[6] The objective of these is somewhat different: it entails creating a climate in which the individual can comfortably, and with confidence, make choices between 'healthy' and 'unhealthy' practices. One such empowerment model would be that of *community development*, where discrete categories of populations define their own health 'needs', determine their own strategies for change and turn to the health promoter whose role is now one of facilitator, advocate and adviser. However, it would be inappropriate to think that such a process of self-empowerment takes place in a cultural vacuum. What is more, the *power* central to empowerment can never be 'pure'. Ultimately the facilitator/advocate/adviser is never able to become a vessel, ridding him/ herself of his/her subjectivities and grounded status as expert. It is not my aim here to undermine health promotion definitively, but to suggest how the increasingly advanced and sophisticated disciplines of health promotion serve (though do not necessarily *seek*) to discipline individuals, through their surveillant discursive practices.

The practice of health promotion relies on a number of basic components. Firstly, there has to be an information-giver, a mediator of knowledge, a self-positioned expert who has access to the 'truth' and sees it as their role to mediate the process of truth-giving. The health promoter is thus expected to assume the role of 'black box', where pieces of *objective*, complex, scientific information are taken in, and equally objective (yet simplified) scientific information is produced. This knowledge passed on to the client saves lives, betters people's lots and enriches experiences. Despite professional pretensions to the opposite, the discourses mobilised in the realisation of these practices are not, at core, dissimilar to those found elsewhere in health services (Fox 1993). What is more, the validity of the practices of health promotion is maintained, and indeed enhanced, by the legitimation of science, by the use of the scientific warrant. This is reinforced by the specialist 'health promotion speak' that has developed concurrently with the discipline (*sic*) itself. The perceived relationship of oracle (he/she who knows best) and the client (the individual in *need* of information, that is, *lacking* the knowledge), creates the dichotomy of the expert and the ignorant awaiting enlightenment. Bourdieu (1989: 24) comments: 'specialists agree at least in laying claim to a monopoly of legitimate competence which defines them as such in reminding people of the frontier which separates professionals from the profane'. Health promotion, despite its stated intention of choice and facilitation of choice, is a prescriptive practice. It has been noted that 'the development of bodies of knowledge surrounding health promotion should be seen within . . . the tendency towards the systematizing of professional knowledge in general' (Bunton and MacDonald 1992: 13). Indeed, the institutionalising of technological knowledge and professional expertise has become a key social issue. It is of further interest to set health promotion in the broader context of health services and the potential effects they have (Turner 1992). Arguably, the very term *Health Promotion* and its association with institutionalised medical practice is a device to legitimate a practice that is still seen as new and marginal, not least within the traditional hierarchies of knowledge within the academy and the hospital. As Bourdieu (1991: 241) notes: 'The professional or academic title is a sort of social perception, a being-perceived that is guaranteed as a right. It is symbolic capital in an institutionalised, legal (and no longer merely legitimate) form.' Just as doctors mediate between the deviance of illness and the desire to maintain the *status quo*, health promotion trades on the same revered position of medical science, and exploits the discursive association of medical science being the key to liberation (now recoded as *empowerment* in a more contemporary voice). The discipline of health promotion and the individuals who work within it are not simply caught up in these knowledges, but are complicit in their production, coding and recoding that which is 'healthy' , 'clean' and, by way of association, 'good'.[7] The reinforcing consequences of this are clear: 'knowledgeability legitimates actions, which in turn legitimate the knowledge base' (Fox 1993: 62).

CONCEPTUALISING SPACE (PART ONE)[8]

Let us now move on to consider the spatiality of health promotion. I do not wish to dwell unnecessarily on an empirical classification of where the sites of health promotion might appear on a map. The spatiality I want to consider is that which may not find its definition in mainstream geographical work, but one which tries to capture a notion of space that is more cultural and political. The hierarchical way in which geography has pieced spaces together, as one being part of another in order of empirical and/or quantifiable significance (for example, Body-Home-Community-City-Region-Nation-Global (N. Smith 1993)) not only serves to maintain the masculinist logic of geographical discourse (Rose 1993a, 1993b) but also frames arguments about identity and place in overly simplistic terms (Keith and Pile 1993). It is one of many intentions in this chapter to disrupt the now-standard geographical notions that there exist imagined spaces that are somehow less significant, less *real*, than material, physically bounded spaces. Similarly, I also want to examine the notion that somehow one of the two categories fits neatly into or onto the other. I intend to illustrate how, in complex and ever-changing ways, the material and the imagined are not discrete classes of spaces, that they are implicated *within* each other; they are *complicit*. That is to say, the spaces are not merely or necessarily oppositional. Neither are they certainly harmonious or coterminous. I hope to examine the potentially complicated theoretical situations that arise in relation to an individual's position in relation to spaces. The intention is to illustrate how presence in one of the spaces (the imagined or the material) does not necessitate absence in the other, how presence in both can be achieved at the same time, how presence in neither is common (and indeed strategic), and how absence in neither is equally as possible. This spatial matrix of multiple presence/absence possibilities is complicated when we consider that despite constant material presence, the individual's relationship to both material and imagined spaces can change over time. Conversely, material absence may mark imagined presence. Never do the material delineations correspond wholly to the imagined.

Above all, as may already be apparent, it is intended that this chapter serve to stir-up the dichotomies often relied upon to conceptualise the spatial, in order to argue that the suggested opposition between two established elements in any pairing may be inappropriate. Whilst this device (of 'the binary') has its uses when thinking about space, not least in political and oppositional terms, the distinctly modernist frame of the thesis/antithesis split may have lost its currency in this period that we may (or indeed, may not) want to refer to as postmodernity. Whilst I intend to work through a critique of some of the troubles I have with such conceptualisations of space (in relation to constituted lesbian and gay spaces), I think it would be useful to bear in mind the direction I would advocate in order to conceptualise a meaningful alternative. By trying to call into question the dualism of the imagined and the material I do not wish to simplify the work of cultural geographers in the name of expediency. I would not argue that geographers have suggested that material space ever lacks imagined status, and neither would I wish to deny materiality.[9] However, I would like to assert that space does not stand awaiting us to give meanings to it, but that space *becomes*, that space is *constituted*, through meaning. That is to say, I would like to think about how imagined spaces become *materialised*.

In order to work within the confines of an organisational logic I seem unable to rid myself of, I will have to lay out my critique of the suggested binary relationship of the imagined to the material in an ordered fashion. Once this is realised I will then attempt to illustrate how the monolithic representation I have created can then be refracted through prisms of spatiality, in an attempt to show the depth, breadth and multi-faceted characteristics of the spaces. The imagined space *par excellence* in relation to the liberationist aspirations of much gay and lesbian political mobilisation is that of the *community*,[10] and it is this space that will introduce my conceptualisation of gay male identities.

IMAGINED GEOGRAPHIES OF LESBIAN AND GAY COMMUNITIES

The liberationist rhetoric of the 'Lesbian and Gay Community' is easy to locate: published in the gay press, broadcast by other gay media, espoused in the language of reclamation politics and histories, promoted in the rhetoric of Pride, called upon in gay male-targeted HIV health promotion literature, and, of course, talked about by many lesbians and gay men themselves. It is also coined by the mainstream media whenever HIV is deemed as newsworthy, 'gay' murderers strike, or lesbians and gay men march. However, the limitations and critiques of community are now well-rehearsed (e.g. Young 1990b). On the one hand, community is a device that homogenises, suppresses internal differences, creates exclusionary boundaries and functions as a dynamo of separatism. On the other, community is a site of resistances, of strategic essentialism and strategic difference. What is more, community is a shelter, a site of shared injustice, a symbolic representation (Cohen 1985). Whichever way you look at it, 'community' – and the 'unity' that it espouses – is far from stable, and indeed we might question the function of unity in the constitution of communities:

> Is 'unity' necessary for effective political action? Is the premature insistence on the goal of unity precisely the cause of an ever more bitter fragmentation among the ranks? Certain forms of acknowledged fragmentation might facilitate coalitional action precisely because the 'unity' of the category ... is neither presupposed nor desired. Does 'unity' set up an exclusionary norm of solidarity at the level of identity that rules out the possibility of a set of actions which disrupt the very borders of identity concepts, or which seek to accomplish precisely that disruption as an explicit political arm.
>
> Butler 1991: 15

This critical position, accompanied by the notion of *strategic essentialism* (Spivak 1988; Fuss 1989), has created a theoretically challenging yet politically satisfying position to hold in relation to gay identity politics.

I am aware that this only constitutes the briefest of comments on the political uses of community as a notion (but see also Jon Binnie and Tamar Rothenberg in this volume), and does not reflect upon the disciplining techniques that such a notion draws upon. It is on this level that health promotion plays. Fundamentally, I am concerned with how the notion of community is worked and reworked through the discursive interventions that health promotion workers enact. Still working within the confines of a logic I want to disrupt, I want to move on to show how this uncertain imagined space of community is not unproblematically related to the material, as the binary might lead one to believe.

INVISIBLISED GEOGRAPHIES OF MATERIAL 'GAY' SPACES

To conceptualise 'gay' material spaces is not difficult. Examples include bars, clubs, cafés, community centres, the bedroom, parks, heaths, car parks, beaches, public toilets, saunas and sex shops. These are the constituted 'gay' spaces that somehow, through their inscription as being impenetrable to heterosexuals, pretend perfect accessibility to gay men and lesbians (the issue of bisexuality is seldom considered in this binary opposition). I want to disrupt any notion of these spaces as being exclusive, and suggest that we can *never be sure*. What is more, all gay men and lesbians do not view these spaces with equanimity, but in fact, there might exist as much disdain for certain practices being realised in certain spaces within the lesbian and gay community as there is elsewhere. It is for this reason that I have decided to write about the cottage (a public toilet used for sex) as an example of a material 'gay' space, as it is arguably the most contentious, the most criminalised and indeed, the most leaky of all 'gay' spaces. Appropriating the language of public/private spatiality, I would like to show how the cottage can be talked about in such a way as to frame the arguments around similar tenets, whilst also exploring points of difference/departure.

The public toilet is literally a constructed space. However, its function could be read as being discursively confused. The public toilet is a device to protect men from public embarrassment (read: female gaze), yet it retains a decidedly public feel. It is a private shelter for public use. It seems to blur the point of definition at which the public becomes the private and vice versa. Suddenly, the private becomes the sphere that halts women's intrusion during the realisation of basic physiological practices. In this context, we see men retiring from the public but finding or constituting a male-only space that is quite deliberately void of the feminine. The toilet becomes a public *and* a private space. It is a non-inhabited yet colonised space for male-only sporadic activity. Its material characteristics (cold hard floors and symmetrical urinals, cubicles and wash basins) reflect in the crudest of terms the properties of the masculine. It is a space authored for very particular reasons, with an intended and officially fixed meaning.

A reinterpretation of this space is apt when its intended purpose, its essential meaning, changes. The change of purpose in this case, however, is not a completely disconnected one. The public toilet still remains a public toilet and is used as such. It is still a male-only space, a world reserved for the masculine gaze. What is more, men using the toilet as a cottage may actually use it as a toilet as well. The 'new' space is interesting when, and if, it is contextualised into the original frame of meaning. It is because the toilet is a toilet and has a series of associated meanings (dirt, faeces, urine, basic physiological functions, privacy and public-ness) that has, in part, promoted and provoked the now well-established social and political responses to cottaging.

A cottage's function depends largely on its reputation. The effort to find and frequent a cottage is arguably greater than that to find a public lavatory. The *real* geographical referent – the contained space – is crucial to the existence of the cottage. However, the spatial specificity of the cottage and its public status are two major problems for those who use it. The contained space is also a containing space which leaves those men using the cottage in a vulnerable situation. That is to say, once the space becomes a forum for sexual expression and freedom it potentially becomes the forum for restraint and control. Involvement in a process of bringing the unspoken world of homosexual activity into public entails the risk of criminality, the risk of further regulation (in its crudest sense). Having an encounter in a cubicle, for example, not only entails a locking-out of the outside world, but also signifies an imprisonment. It is as if the criminalised couple or group is putting itself in an easy position for arrest. Despite the exclusionary promise of the cubicle door, the world is never wholly shut away. The open top allows the public and reminders of the public world (vehicle noise, people talking) into the cubicle. More importantly, it allows easy access for those whose interest in what is happening therein is less concerned with personal sexual activity and more with disrupting those alleged practices. Anonymous sex may be represented as liberatory, yet that liberation would soon be marred if your partner(s) turned out to be either violent or 'official', or both. The cottage may pretend safe-space status, yet it encloses the very same men it serves to protect, leaving them open to attack just at the moment they enjoy their safety. So the cottage is a sexually transgressive space, just as it becomes reputed as a meeting place for 'sexual deviants'. Cottaging affirms identity, just as it constitutes it. The cottage becomes a space for the unclothed and the plain-clothed. The anonymity of its procedures promises uncomplicated, fantastic (*sic*) sex, but it simultaneously provides the conditions for control. The cottage, therefore, not only disrupts the exclusionary pretensions of the binary, but suggests that the moment of resistance may actually be the very moment of government – a formula that is sure to sadden even the most ardent liberationist.

In rhetorical terms, cottaging becomes a device to control gay men as it is cited as an example of their immorality. However, this again relies on a crude division. It is not only gay men who use cottages – bisexual and (often married) heterosexual men do so too, to have sexual encounters with other men. And whilst some gay men are open and proud of their cottaging and are keen to assert its political and cultural importance, others are as critical of it as the 'establishment' is, and indeed collude in its regulation.

(RE)CONCEPTUALISING SPACE (PART TWO)

The standard geography alluded to earlier might conceive of the tightly-bounded material space, and what goes on in it, as being a component part of a broader community (ideally place-bound, or at the very least spatially-specific). I have hoped to highlight how inhabiting the imagined space of community is not an all-embracing, commonly-understood experience shared by all lesbians and gay men, and how presence in the many material spaces that might be labelled 'gay' (bars, clubs, cafés, parks, car parks, toilets) does not necessarily mean that those men or women present actually feel part of the lesbian and gay community, or even believe it exists, let alone desire to be a part of it if it does. In the same way, individuals might happily consider themselves members of the gay community, yet dislike the idea of visiting a constituted gay space. There are, of course, multiple possibilities for the individual who feels happy in a physical space, content with their presence in, and belonging to, an imagined space, but both experiences come to an end when a material space is vacated. Others might remain present in the imagined even when they find themselves back out in the allegedly straight world. These are examples of the spatial matrix outlined above.

Having demonstrated, hopefully, how imagined and material spaces are complicit, inseparable and discursively produced, I want to move on to reflect critically on how the boundaries of both could be seen to be excluding – how the whole discursive product of the lesbian and gay community and lesbian and gay spaces could actually constitute a bracketed, separating space that excludes a whole body of men and women because of their inability to structure their lives around the clearly defined and limited tenets of what is, in fact, constituted as a gay *lifestyle*; for example, socialising, politicising and accessorising, not to mention the pursuit of companionship and sex. The singularising effects of that which is often labelled 'gay literature', 'gay film', 'gay culture' and 'gay space' cannot be overstated. The constitutive effects of such notions isolate many and seem to alienate, marginalise and intimidate them. Being *bona fide* gay becomes a situated practice, one reserved for cognisant, compliant and moneyed 'gay' individuals. Assertions that claims to citizenship are becoming increasingly inseparable from demands for economic rights may be impressive (Evans 1993; see also Jon Binnie in this volume), however, these should not be used to singularise identity into some kind of spatialised economic game. Let us not forget Bauman's reflections on the commodification of identity where postmodernity is 'a shopping mall overflowing with goods whose major use is the joy of purchasing them; and [an] existence that feels like a life-long confinement to the shopping mall' (Bauman 1992: vii). When seen in these terms the promise of 'liberation' seems to lose its appeal.

CONCLUSIONS: GMFA, GAY SPACE AND SURVEILLANCE

In this section I will attempt to bring together the potentially disparate areas I have touched upon in this piece: reflexivity, power, space, health promotion, gay male identities. Primarily, I want to ponder the ruse of liberation: the liberation we are promised through science, through community, through the increasing commodification of our identities. There is no clear delineation between them, neither is there a clear path that joins them. The three intersect, compete, concur. They are tied together in unfathomable ways, held together in an eternally flummoxing Gordian knot (Fuss 1991). It is the very same science that seeks to explain (and naturalise) difference(s) through genetic mastering, hypothalamas measuring and thyroid monitoring that disciplines our 'deviance' through pathologisation. It is the liberatory promise of medicine that has created a host of unclear, contradictory and changing explanations for HIV and/or AIDS (Patton 1990). The logics and promises of science have been criticised, yet the scientific warrant still remains a powerful credential. It is science that fuels the existence and unprecedented growth of health promotion, which in turn relies on science to validate the promises of safer sex. Whilst I have no intention of calling into question the value of safer sex as the only way of effectively combating HIV infection for gay men, I want to underscore the irony of a political movement that embraces the very same science that has sought to discipline our sexualities (Mort 1987).

The most radical HIV organisation in the UK at the present time is Gay Men Fighting AIDS (GMFA). London based, and metrocentrist *par excellence*, GMFA is the lovechild of some of the gay community's most press-covered notables. The approach of the organisation is innovative, dynamic and unique in Britain, clearly inspired by the community mobilisation model of health promotion as developed by Cindy Patton in the USA. GMFA's premise that it is the first organisation in Britain to be singularly concerned with the health needs of gay men despite more than ten years of the AIDS epidemic is as admirable as it is disgusting. It seems incredible that such a statement stands, despite the fact that 60 per cent of those infected with HIV are gay men, or have become infected through unprotected sexual intercourse between men (King 1993: 241). GMFA is not without its critics,[11] and whilst some of their methods might be questioned, I think it only fair to note that the ease afforded to those keen to criticise is not met by the difficulties encountered by dedicated practitioners engaged in heading-up and managing such projects. GMFA is seen as an organisation working at the radical end of health promotion, seeking to bypass the bureaucratised practices implicated in much health promotion work, working in a hostile climate, committing itself to the recognition of the range of cultural, social, sexual and spatial differences of gay male identities and communities. The health promotion literature used seeks to convey safer-sex messages using the parlance of gay men on the scene. This device has the consequence of minimising the rhetoric of science. The images used seek to eroticise safer sex, rather than sanitise (or indeed invisiblise) homosexual practices. The spatial differences of being gay are recognised and seen as the principal sites of work, including those contentious sites (like the cottage) that, because of their criminalised nature, are often erased by many HIV health promotion projects. The political agenda is thus unhidden. The messages conveyed are clear and proud, and above all uncompromising. GMFA attempts to resist the medicalising discourses of science. As a totally volunteer-led organisation, it enacts the objectives of community development while exposing the political edge to HIV and AIDS often ignored by institutionalised bodies such as the Health Education Authority (McGrath 1990). Importantly, it has legitimated sex between men in the forum of public health.

This (less than empirically adequate) evaluation of the work of GMFA is made as an attempt to spell out the value of the organisation and to recognise its strengths. It is also my desire to air some misgivings about GMFA. These misgivings are not necessarily about the organisation *per se*, but about the approaches it adopts. GMFA is not the only organisation to use such techniques, it is simply the most successful. I seek not to undermine GMFA definitively but to highlight potential unintended consequences of some of the many practices enacted, roles played, claims made.

I intend to relate my concerns about the approach currently favoured by GMFA to the issues of space and spatiality outlined earlier in the chapter. In the realisation of a community-based programme, initiated at community level, the problem of defining the boundaries of the community, of considering ways in which the breadth of interests found within it can be represented, and how best to work in it, are key issues. Whilst a volunteer-led organisation is undoubtedly more democratic than a professional-led one, the volunteers themselves might not present from a wide range of sources, but from a narrow, already-aware sector that potentially has little or no contact with other men who might identify HIV as an issue in their lives. Not only is this excluding in logistical terms, but it contributes to the constitution of community as divisive. Establishing a bounded collective that becomes coded as the core, this community becomes institutionally recognised by those bodies that fund the projects of GMFA as those in need, and more dangerously, those whose needs *are being met*. The radical promise of such projects ultimately rests on the singularising moves that assume that *out* gay men can somehow reflect *en masse* the worries of *all* 'gay' men, or at least those at risk. Whilst the overt medicalising discourses of traditional health education may be resisted, there remains the problem of the power/knowledge conundrum. By training-up an army of cognisant gay men who have the brief to educate others, a situation is arguably being pioneered where a knowledgeable elite is interrupting the practices of those constituted as profane, assuming what is best for them, assuming where they can be found, and assuming they will comply. The volunteer imbued with the knowledge (and the power) associated with the GMFA warrant becomes a sophisticated and effective disciplining force. Trading on his status as gay, he is able to dispense unthreateningly the regulatory aspirations of medico-moral discourses in constituted material gay space. This, like any other space, functions in particular ways, and the appearance and intervention of the GMFA volunteer (whether the profane are aware or not) threatens to disrupt the space, marking the subjectivities of those present, and in turn reconstituting both. In broad terms, whilst the objective of promoting safer sex may be effectively met, when seen spatially the interventions raise ethical and political issues. In public sex environments, for example, the intrusions themselves not only threaten to halt sexual practices, but also to undermine individual resistances: echoes abound of Foucault's pronouncement that power realised through mechanisms seen to be extra- the state apparatus serves to support and enhance the state more effectively than its own institutions.

In broad conclusion, I wish to reiterate that it is not my intention here to senselessly criticise the work of any one organisation. The political and economic climate in which this epidemic is faced is one that creates more obstacles than it knocks down (King 1993), and the innovative approaches being developed in this time of crisis are remarkable. However, with the benefit of experience and time to critically reflect, it can be seen how even the most oppositional of bodies (coded as the most resistant of hegemonic discourses) are, in measure, complicit with the dominant. To recognise the value of spatiality, to formally assert its importance in the constitution and expression of identities and to then expediently appropriate those spaces as sites to dispense disciplining knowledges, is to initiate and enact a series of practices that can result in disciplining the actions of those individuals whose resistances are often beyond the reach of the state. Gay male practitioners (HIV prevention workers, political activists and researchers), armed with the good intentions of empowering, are complicit in sophisticated and subtle modes of self-surveillance. New ways of working will face the challenges arising from this, respecting (and indeed promoting) new resistances, taking notice of the imperative to radically recognise difference and the relationship of space, power and identity.

ACKNOWLEDGEMENTS

I would particularly like to thank Gillian Rose, Ann Taket, Jessica Allen and Carol Hinvest for offering a range of critical input and support whilst I was writing this chapter. I would also like to thank Hilary Woodhead, Kate Woodhead, Mark McPherson, Jackie Curtis and Toni Miller for their encouragement, support and calming words. Ultimately, the piece, as it stands, is my own responsibility.

NOTES

1 The complexities of these debates were indirectly made apparent to me by Ian Hodges through his admirable enthusiasm for, and commitment to, the work of Michel Foucault.

2 I wish to reflect briefly upon the ways in

which the concept of power as a notion will be employed in this chapter. Whilst such a scant comment can in no way reflect the complexities of post-structuralist debates on power, I feel that a thumbnail sketch of my position will give a flavour of the theories I hold with in relation to its conception. This seems infinitely preferable to a piece of writing that fails to recognise the importance of power relations.

3 For a clearer and more sophisticated account of this type of argument see Keith (1992).

4 I hope that this does not imply that somehow there is an absolute ethical benchmark against which any work can be measured, but simply that I worry about my ethical position in relation to this work.

5 Special thanks to Ann Taket for discussing the issues surrounding health promotion.

6 An understanding of community development models of work has grown out of several interesting (and heated) discussions with Mark McPherson, Gay Men's Outreach

Worker, Aled Richards Trust, Bristol; Becky Woodiwiss and Christina Wennell, Environmental Health Service, London Borough of Newham; and Paul Vallance, East London and City Health Promotion Service.

7 For a detailed discussion of the discursive relationship between 'clean' and 'good' see Douglas (1978). For a discussion of the association of sexual 'deviance' with dirt and 'germs' see Patton (1985).

8 Special thanks to Gillian Rose for patiently listening to my musings on space.

9 Thanks to Steve Pile and Doreen Massey for raising the issue of materiality.

10 For almost comic unquestioning of the status of 'the community' in relation to lesbian and gay politics see Seidman (1992) and Cruickshank (1992).

11 Anecdotally, I have heard GMFA referred to as Gay Men *Fleecing* AIDS and Gay **MaFiA**.

16

SEX, SCALE AND THE 'NEW URBAN POLITICS'

HIV-prevention strategies from

Yaletown, Vancouver

•

Michael Brown

WHICH 'NEW URBAN POLITICS'?

Wedged between Vancouver's downtown core and its gay West End, Yaletown presents a study in contrasts for urban political inquiry. It is a rapidly gentrifying warehouse district, whose proximity to the central city has made it a prime target for redevelopment over the past five years. On its southern edge, the former Expo '86 grounds are currently being transformed into condominium towers, staging the largest development project in the city's history.[1] Warehouses now stand next to gleaming luxury towers. Trendy cafés and art galleries are springing up in renovated 'character' buildings. Yaletown's landscape changes signify what Cox (1991, 1993) has recently labelled as 'the new urban politics' visible in North American cities. These are the politics of local economic development (often specifically large-scale construction projects). Over the past decade, they have claimed great priority and attention amongst scholars. Intertwined with an empirical focus on central-city development, there has been a mounting theoretical interest in situating city politics in the context of broader-scale forces of globalisation, more specifically the mobility of capital in affecting development in particular places (Cox and Mair 1988; Cox 1993). Indeed, much of Vancouver's development capital has been financed by offshore interests, from Asia and Hong Kong especially (Gutstein 1990; Barnes *et al.* 1992). Given this theoretical concern with global capital mobility, and an empirical focus on local development, urban political inquiry would readily draw our attention to the new urban politics shaping this portion of the central city.

Yet a simultaneous geography, which also problematises scale issues, may go unnoticed in Yaletown for students of city politics, for the area is home to most of the city's largest AIDS service organisations (ASOs). AIDS politics also highlight the issue of global–local scale processes, but do so beyond the state and market poles typical in local economic development debates (e.g. Elkin 1987). In other words, these voluntary sector organisations in Yaletown further the theoretical agenda of the new urban politics, but impel a broader definition beyond economic development. Here, the globalisation of the AIDS pandemic is paramount, and strategies to block the spread of HIV are at stake. Issues like sexuality, identity and health promotion occupy the same area where students of urban politics might more readily see a city 'regime' affecting the political geography of Yaletown (Stone and Sanders 1987).

My purpose in this chapter, then, is twofold. Firstly, I want to extend geographers' interest in the global–local relations of urban politics, by showing how the AIDS politics in Yaletown also take place at a variety of spatial scales *simultaneously*. I use the example of HIV-education and prevention work to make that point. Secondly, I warn against taking the enormous amount of work done on urban economic development as the *only* new urban politics in cities over the last fifteen years. HIV prevention politics show that there can be local and global relations that are not captured by an analysis of capital alone. Three different examples are employed to stake these points. I discuss the micro-geographies of public sex areas, prevention projects aimed globally from Yaletown, and the cyberspace of an ASO's telephone Help Line. My goal is not to detract from work on the new urban politics of development; it is to extend debates around city politics and spatiality in both empirical and theoretical directions.

FOCI IN THE 'NEW URBAN POLITICS'

Economic revitalisation has become a mainstay in urban political inquiry over the past fifteen years (e.g. Peterson 1981; Stone and Sanders 1987).[2] Undergraduate textbooks, for instance, now devote entire sections to the topic (Goldsmith and Wolman 1992; Judd and Swanstrom 1994). As Cox (1993: 433) has declared most recently, 'Quite clearly, urban development, for many scholars, is now what the study of urban politics is about.' To understand this focus, it is helpful to recognise its empirical and theoretical roots. The conservative attack on the welfare state coupled with the decline of manufacturing led to a partial demise in attention towards consumption – and production-centred local politics in North American central cities (Gottdiener 1987). The increasing globalisation of capital, and the need for cities to attract and retain investment, sparked attention on downtown (re)development, public–private partnerships, gentrification and the like (Peterson 1981; Logan and Molotch 1987). Theoretically, these broader trends produced an interest in more sophisticated conceptualisations of power, a debate over whether local politics still mattered, and a research agenda specifying the myriad ways in which they do affect both local development and international capital flows (Stone and Sanders

1987; Cox and Mair 1988, Clarke and Kirby 1989).

Attention has shifted to global–local scale relations. Cox (1993) has argued that simplistic global–local dualities are employed in the new urban politics, and more sophisticated understanding of their theoretical relationship is required. Moreover, he claims that global has been connotatively equated with economic relations only, while political relations are narrowly characterised as local and static. Such simplified, dichotomised linkages have been coming under increasing scrutiny and critique by both geographers and political scientists. For instance Alger (1990) and Magnusson (1992) have ardently insisted that a focus on the (local) state must be 'de-centred' in light of the broader links contemporary social movements have made between different places. Geographers have also worried about facile conceptualisations of space current in social theory as fixed, limited, discrete and unchanging (Massey 1993; N. Smith 1993; Smith and Katz 1993). Most recently Jonas (1994) has insisted on a 'scale politics of spatiality'. Like N. Smith (1993) he insists that spatial scales, while often opposed to one another, can just as often be nested, relational and simultaneous. Geographers, then, are voicing their frustrations with representations of political engagement where the local is a fixed set of limited processes, which are washed over by broader forces, typically the ebb and flow of capital. What follows is an attempt to uncover not only a new urban politics of HIV prevention, but their multiple spatialities.

YALETOWN

As in the US, interest in the local politics of (re)development in Canadian central cities has grown over the past decade (Stelter and Artibise 1986; Ley 1994; Perks and Jamieson 1991; Gutstein 1990). Certainly in Vancouver, a pro-growth regime has facilitated an enormous amount of construction across the urban landscape. In the city the leading municipal party, the Nonpartisan Association (NPA) has strong ties with the development community, and the numerous residential and commercial developments on the downtown peninsula have been well documented (Gutstein 1990; Ley et al. 1992). Yaletown certainly has not escaped the effects of Vancouver's pro-growth regime. Roughly bounded by Granville Street (Vancouver's main north–south artery) on the west, Robson Street to the north and False Creek to the south and east, Yaletown's proximity to the CBD has made it ripe for redevelopment (Figure 16.1). It was originally home to Canadian Pacific Rail (CPR) workers, after the company moved its headquarters to Vancouver from Yale, BC in 1886. By the 1920s, Yaletown had become a warehouse district between the CPR rail yards along False Creek, and the central business district. Through the 1980s, city planners advocated historic preservation of many buildings in the area, while also allowing for high-rise apartment and condominium towers to be built

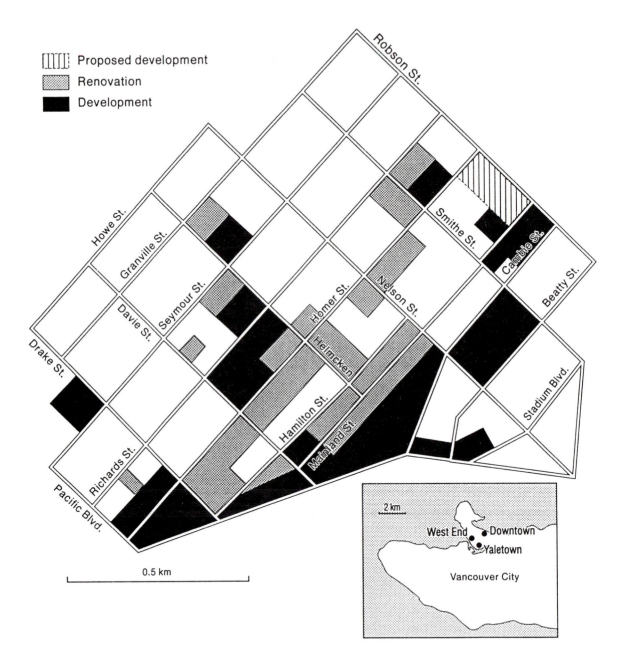

Figure 16.1 Development in Yaletown, Vancouver
Source: Michael Brown

(City of Vancouver 1988). Figure 16.1 maps the extent of development and renovation across Yaletown. Census data are only just beginning to show the effects of redevelopment in the area. Rough measures of social class show a slight change towards a more elite residential area, but are paralleled by similar shifts in the city as a whole.[3] Figure 16.1 readily depicts the ongoing 'new urban politics' of redevelopment in Yaletown. From this brief survey, it is clear that Yaletown evinces the new urban politics of the past decade, and a case study of development politics and investment capital in the global city of Vancouver would no doubt readily augment the debate over spatiality and scale in global–local relations. An alternate political geography around Yaletown is also apparent, which is slowly being reproduced and displaced by the consequences of redevelopment. It is here that we can begin to see a different urban politics.

Vancouver is a centre of the AIDS crisis in Canada, exhibiting the highest per capita rate of infection in the country (Brown 1995). The city's gay community has been at the fore of the response in the city, creating a comprehensive shadow-state set of organisations to provide services and support (Brown 1994). The response to AIDS locally has come from the West End and Yaletown, which are heavily gay areas. Moreover, Yaletown has a particular sexual geography, as shown by Figure 16.2. The area is a landscape of many desires. Adjacent to the residential West End, Yaletown conveniently houses several of the city's gay bars, including the oldest one in the city. There are also many popular straight bars in former warehouses. Two of the Lower Mainland's four gay bath houses are located in the area as well. As Figure 16.2 also points out, Yaletown is a leading area for sex-trade workers in the city. Female prostitutes tend to work along Nelson and Helmcken Streets, while male prostitutes tend to cluster along Homer and Drake Streets. Street youth also populate the area (many of whom are in the trade and are IV drug users).

Pacing this alternative social geography, Yaletown is at the centre of the local response to AIDS in Vancouver. Several of the city's leading ASOs have settled there, due to the area's proximity to the West End and St Paul's Hospital, and its cheaper rents (compared to the West End). Since 1992 the Pacific AIDS Resource Centre (PARC) at the corner of Helmcken and Seymour Streets has housed AIDS Vancouver, the Vancouver Persons With AIDS Society, the Positive Women's Network, and the Women and AIDS Project have set up here. From 1987 to 1992, AIDS Vancouver (the city's leading AIDS eduction, prevention and support organisation) was located on Richards Street in the area, and the PWA Society was located three blocks north between Yaletown and the West End. Helmcken House (an apartment complex for people living with AIDS) stands diagonally placed to PARC on Granville Street. A drop-in centre for children living on the street, called Street Youth Services, is around the corner on Richards Street. Also a mobile needle-exchange van drives through the area to prevent the spread of HIV through IV needle-sharing.

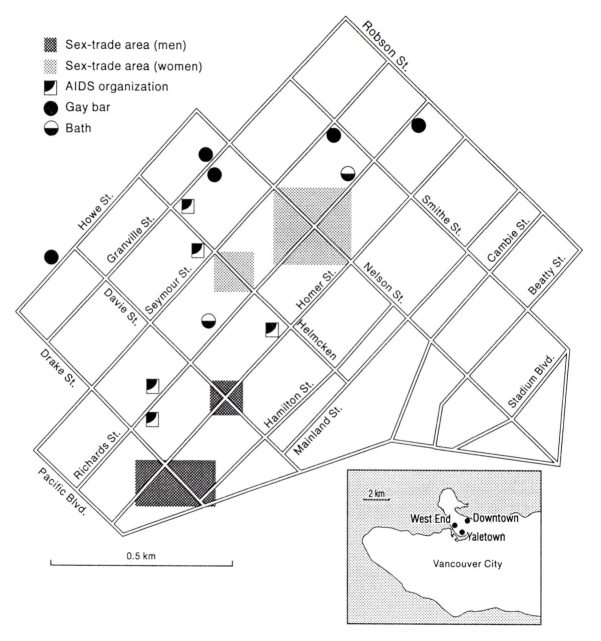

Figure 16.2 The other 'new urban politics' in Yaletown, Vancouver
Source: Michael Brown

SPATIALITY IN THE OTHER
'NEW URBAN POLITICS'

While a great deal has been written on the overlap of identity politics with HIV-education and prevention curricula (e.g. King 1993), less attention has been given to the spatiality of the strategies employed by ASOs to abate the spread of HIV. The tactics employed often work precisely on spatial bases. What is most fascinating, however, is the multiple simultaneous scales on which HIV prevention and education operate, which is precisely the point geographers interested in scale-politics have made. It is not only HIV-education/ prevention programmes that transgress the simplistic neighbourhood scale of consumption-sector politics; other 'local' activities vary from the small-scale micro-geographies of public sex areas to global efforts emanating from Yaletown to promote safe sex in Third World countries. What follows, then, is an account of the alternative sexual urban politics from this inner-city Vancouver neighbourhood.

Micro-geographies: Man-to-Man and youth services

Public sex environments have long been associated with male–male sexual practices (Humphries 1970). These sites present excellent opportunities for HIV-prevention volunteers because they can spatially link reducing risk practices with behaviour changes in those locations. Man-to-Man is the AIDS Vancouver team that is dedicated to HIV prevention among gay men specifically. Since 1990 it has operated 'Operation Latex Shield', where volunteers enter public sex environments and distribute condoms and explicit information on safer-sex practices (Plates 16.1–16.3). In the city, Operation Latex Shield has been particularly active in several of the city's gay bars, two of its bath houses, and cruising areas in parks adjacent to the West End. One volunteer described these micro-geographies in the following way:

> 'At the baths on an average evening, there'll be about five people who come into that room. People will have specific questions, "I'm confused about oral sex." "What's safe and what isn't?" But I think mainly what the function is is to make sure that condoms are distributed. We don't talk a lot. We just make it a point to walk around and hand people condoms. We don't talk to them or acknowledge them, though we usually get a "thank you" in return.'

AIDS Vancouver has also worked with a provincial outreach nurse in its Operation Latex Shield campaign. In addition to distributing safer-sex education materials, he can draw blood for an HIV antibody test in the bath house. Like the Man-to-Man volunteer quoted

Plate 16.1 'Take pride in yourself' package containing a condom, lubricant and information distributed by AIDS Vancouver's Man-to-Man programme
Photograph: Michael Brown

above, the nurse stressed the importance of doing safer-sex education work in the very location where sex takes place:

'In the bath houses we're in there and we announce our presence and what room we're in. I actually do testing there. They get the same information and the same counselling they would get here [at the Gay and Lesbian Centre], except it's done in that setting. The reason I wanted to do it – and it's proven to be true – is that, I'd say over 65 per cent of the men have come forward for the first time for the test. And they've never been tested before. The idea is strike while the iron is hot. It's *there*. It's *anonymous*. So it's *acceptable*.'

Safe-sex education and prevention outreach also targets street-entrenched youth and prostitutes in Yaletown. One counsellor noted the success of the needle-exchange programme in Vancouver while describing her work in keeping her clients HIV negative. She, too, noted the importance of micro-geographies in prevention work more precisely than the neighbourhood scale which was her clients' turf. She emphasised the relative success of the Needle Exchange's mobile van, which literally drives to the IV drug users throughout Yaletown. The Needle Exchange office (located just east of the CBD), is a social world that is too foreign for Yaletown's street kids to enter:

'I'm not sure what the latest stats say, but a few months ago the HIV rate in IV drug users was lower than that of the so-called 'general population'. That's just phenomenal when you look at user groups in most places, it's running 40–80 per cent. It's hovering around 2–3, might have crept up to 4 per cent by now, but I doubt it. But the kids here won't go that far [to the Downtown Eastside], and the Needle Van comes up here every night, but that's very different – it's here between 8 and 9 – from being able to walk into the Needle Exchange at a fixed site from 8am. to 5pm. So you've got an eight- or nine-hour period to get it together and get yourself in there and get your business done, and also the van will land back at the Balmoral Hotel and be there on Hastings Street for another couple of hours in the evening. So in that community there's way, way more access. Like, most people from the Downtown Eastside will not come up here to access a service. They just won't. This is a very alien environment for them. It's very white, and it's kind of empty

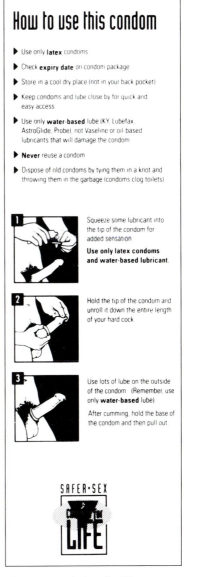

How to use this condom

▶ Use only **latex** condoms

▶ Check **expiry date** on condom package

▶ Store in a cool dry place (not in your back pocket)

▶ Keep condoms and lube close by for quick and easy access

▶ Use only **water-based** lube (KY, Lubefax, AstroGlide, Probe), not Vaseline or oil based lubricants that will damage the condom

▶ **Never** reuse a condom

▶ Dispose of old condoms by tying them in a knot and throwing them in the garbage (condoms clog toilets)

1 Squeeze some lubricant into the tip of the condom for added sensation
Use only latex condoms and water-based lubricant.

2 Hold the tip of the condom and unroll it down the entire length of your hard cock

3 Use lots of lube on the outside of the condom (Remember, use only **water-based** lube)
After cumming, hold the base of the condom and then pull out

SAFER·SEX
LIFE

Plate 16.2 'Choices for life' condom package instructions. These packages are handed out in public sex areas by AIDS Vancouver's Man-to-Man programme
Photograph: Michael Brown

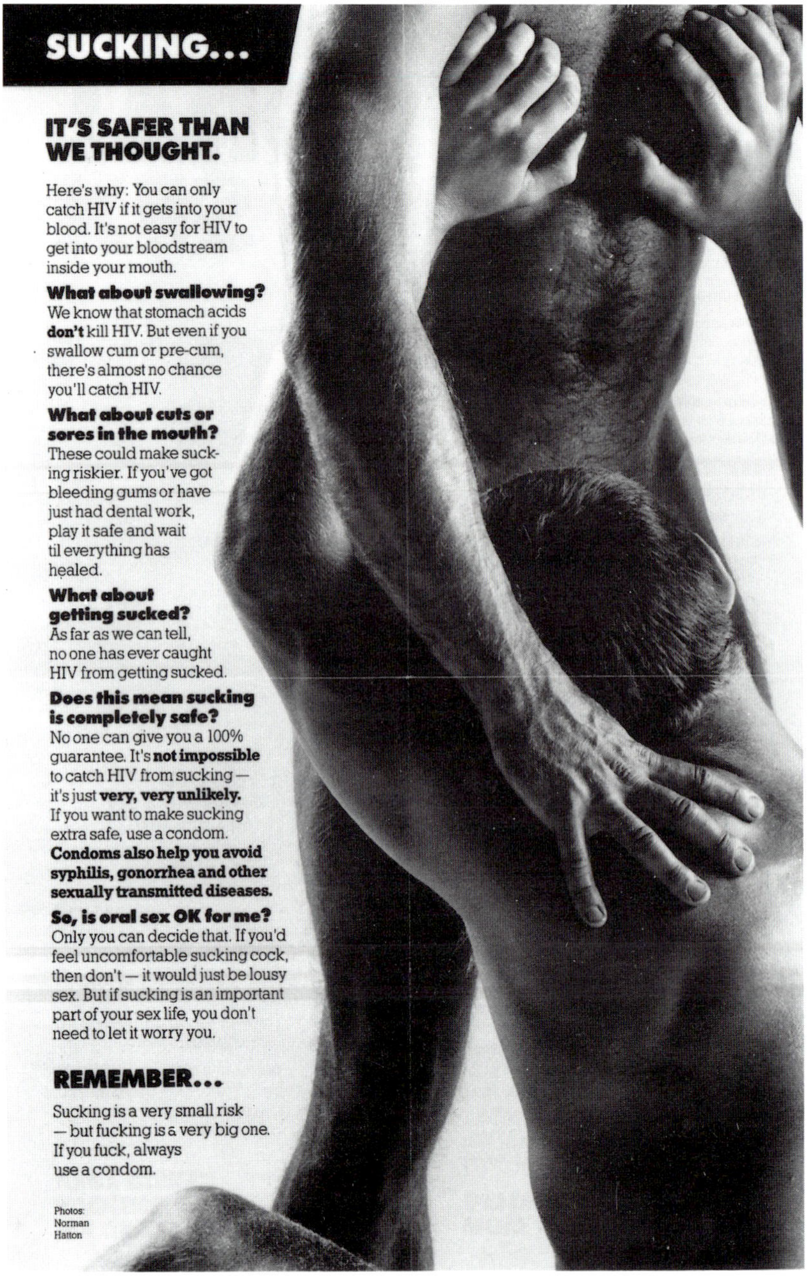

SUCKING...

IT'S SAFER THAN WE THOUGHT.

Here's why: You can only catch HIV if it gets into your blood. It's not easy for HIV to get into your bloodstream inside your mouth.

What about swallowing?
We know that stomach acids **don't** kill HIV. But even if you swallow cum or pre-cum, there's almost no chance you'll catch HIV.

What about cuts or sores in the mouth?
These could make sucking riskier. If you've got bleeding gums or have just had dental work, play it safe and wait til everything has healed.

What about getting sucked?
As far as we can tell, no one has ever caught HIV from getting sucked.

Does this mean sucking is completely safe?
No one can give you a 100% guarantee. It's **not impossible** to catch HIV from sucking — it's just **very, very unlikely.** If you want to make sucking extra safe, use a condom. **Condoms also help you avoid syphilis, gonorrhea and other sexually transmitted diseases.**

So, is oral sex OK for me?
Only you can decide that. If you'd feel uncomfortable sucking cock, then don't — it would just be lousy sex. But if sucking is an important part of your sex life, you don't need to let it worry you.

REMEMBER...

Sucking is a very small risk — but fucking is a very big one. If you fuck, always use a condom.

Photos:
Norman
Hatton

Plate 16.3 Information brochure on oral sex from AIDS Vancouver's Man-to-Man programme. Photograph: Norman Hatton

up here. It's not like the busy strip along Hastings. And it's very gay up here. All those things are problems. Now the same applies for the kids here to go down and use the Exchange. It's a very alien environment for them. It's very nonwhite. People are older. Straight. It's a community. The police are present on the street. People talk to them. They won't go. Put them in a taxi and say, "Here, go get your needle", and they won't go.'

This woman's presence in Yaletown has made her familiar to the sex-trade workers in the vicinity. That familiarity and her ardent efforts to instil safe practices among prostitutes has meant she has insisted that they see condoms as integral to their work. Consequently, she is often besieged by workers grabbing condoms before they go out on the Yaletown streets to work. She went on to note how she must always carry around a supply of condoms, so she can be a condom distributor anytime, anywhere:

'People come in here consistently and they pick up condoms, and then they go to work. We can't go have a cup of tea in the restaurant down the street without fifteen people in fifteen minutes coming in to get the condoms because our door is closed because we're having tea. They're very conscientious in terms of their work.'

In the bars and bath houses, along the streets, safer-sex education and prevention materials are commonly distributed in Yaletown at sites where public sex actually takes place. The strategy works, so it is argued, because prevention is spatially, immediately linked to high-risk behaviours, to the micro-geography of the body. Moreover, these education efforts take place on an ongoing basis through these micro-geographies. One might argue that these politics are fixed in terms of scale, at the level of the body; they have nothing to do with global politics. That conclusion can be countered by noting the enormous exchange of information and resources between Man-to-Man, the Needle Exchange and similar ASO programmes elsewhere. Many of Man-to-Man's educational brochures, for instance, are shared with the AIDS Committee of Toronto, and other groups in North America. The success of the Vancouver needle exchange has been touted in other cities as well. Finally, bear in mind that by preventing HIV transmission locally, these 'bodies' are also countering the global diffusion of AIDS. The global and local scales are quite simultaneous here.

Global links: World AIDS Day

Certainly geographers have been adept at describing the global nature of the AIDS crisis (Gould 1993). Less considered, however, have been the global links that prevent the spread of AIDS and HIV (Brown 1995). World AIDS Day in Vancouver has been named as 1 December. It is marked annually in this city, and others, with programmes designed to increase awareness of AIDS issues *both* locally and globally based around a theme set by the World Health Organisation. AIDS organisations in Vancouver have co-ordinated World AIDS Day events in the city throughout the 1990s. In 1991, the theme for World AIDS Day in Vancouver was 'AIDS: A Human Rights Challenge'. The events were co-ordinated through the city's development agencies and the AIDS shadow state in Yaletown. The overarching theme was to demonstrate the similarities and differences between responses to AIDS in Vancouver and other places (e.g. Dube and Smailes 1992). Events included a poster contest, a public forum, receptions and the annual 'Day Without Art', and concluded on 10 December, which was International Human Rights Day. And in a most fascinating display of global–local politics, local ASOs held an exchange between HIV prevention workers in Vancouver and those in Nicaragua and Mexico. An AIDS Vancouver volunteer who helped organise the exchange recalled the World AIDS Day's global–local politics:

> 'It was called *AIDS: Sharing the Challenge*, and it lasted a week. Mostly the issues that came out was how different countries are dealing, and how different communities in countries are dealing with the challenge of AIDS. In Canada, of course, it's fairly sophisticated. But it was empowering to see how much people can do with how little they do have. The problem, for example, the main problem was that they never had enough condoms. And like, for me, that's unthinkable, you know! It's like, "I want a gross of condoms, whenever I want it I have it!" ... They see people with AIDS here and we think they don't have much but the [workers from Latin America] see them and say, "My God! You have so much!" I'm sure they also learned a few strategies. And we made sure they left with lots of condoms.'

Here, safer-sex materials and information were taken elsewhere, linking the local politics of AIDS prevention in a direct, straightforward manner.

Interestingly, a theme that arose out of World AIDS Day efforts was the need to look at the AIDS crisis in British Columbia, but *outside* of Vancouver or the Lower Mainland. Another World AIDS Day volunteer spoke of the role events played in linking separate local politics together, while reminding Vancouver AIDS activists that their efforts did not operate in a vacuum:

'One of the things that is lost at AIDS Vancouver, at PWA, at so many organisations that exist and work here in the city is that it's a very different experience being a PWA in Vancouver than being a PWA who lives forty miles away from Prince George [in the British Columbia interior]. He might as well live in a Third World country, almost. And the issue there was the global consequences and the wellness issues highlighted around this epidemic are things that apply to them as well, and that there is some brotherhood involved when we talk about services and when we talk about acceptance from the community at large and just simply knowledge. I mean, here in Vancouver people may recognise now that it's no big deal to eat off the same plate as someone who's HIV, but that's not necessarily the way it is five miles out of Vancouver.'

Another World AIDS Day volunteer reiterated his argument, but took a more theoretical reflection. She insisted on the success of the event in linking immediate local struggles in Vancouver with politics elsewhere.

'But people have realised that it's important now to look at the global issue, and it's important for them to discuss that with the people that they actually work with and provide services to, because we're all part of this larger community and the community is in more of a crisis than it ever was before. I think people are starting to realise who's in power and who's in control and alternate ways to change that. And I don't know what the turning point has been, but anyways it's happened in our group. And I think that's been the most profound thing. To come that far in a short period of time, because I think everybody [now] has the same analysis. And I don't think we started there at all. And it's been a hard struggle. It's not a local thing. We have to look at what's the best for all of us in the community that we live in and not to negate – I mean, one of the things that we've actually built into our mission statement is to ensure that the voices of people living with AIDS are part of this process. So it's important not to look at the global picture in isolation from that. So the things that we're trying to do are more, I think, inclusive of the reality of people here, too. Yet it's still important to always look at the broader picture.'

In the case of World AIDS Day, the global nature of AIDS is not dichotomised with a fixed set of local responses in Vancouver. Instead, local responses in Vancouver are compared and contrasted to those in other places like Mexico and Nicaragua, as well as rural British Columbia. In this way, the local politics of AIDS in the city are simultaneously – and self-consciously – global politics as well. Moreover, the week-long string of events allowed thousands of Vancouver citizens to reflect on their own relationship to HIV risk, while constantly situating themselves in the global context of the AIDS crisis. The education efforts stemming from Yaletown ASOs and the city's development agencies were both local and global simultaneously. Moreover, their work resisted simplistic dichotomies (like AIDS = global/responses = local); World AIDS Day 1991 illustrated that responses can be both local *and* global, as is AIDS.

Cyberspace, or nowhere in particular: the Help Line

'I love the Help Line. I enjoy the fact that I am instructing straight people, straight men for instance, how to put a condom on. And I think that if they knew that this 73-year-old woman was telling them how to put a condom on they'd have a canary! But I enjoy it. I like the contact, because I feel people on a telephone will tell you things that if they met you face-to-face they wouldn't know how to put these things into words. I feel I've helped a lot of people on the Help Line. And sometimes you get really interesting calls that go on, and on, and on, and on. And you really feel you've done your intake counselling on the Help Line. That's why it's a good thing there are two of us on.'

The two examples given so far of an alternate new urban politics around HIV prevention are explicitly, and self-consciously, spatial. A third example of HIV prevention work from AIDS Vancouver's Yaletown office works precisely through *cyberspace*, that is, the space constructed by technology that mitigates distance through technological compression of time and space (Heim 1992). More familiarly, I am talking about a simple phone line. Even before AIDS Vancouver had an actual office it operated a Help Line (687-AIDS), whereby callers could anonymously get up-to-date, accurate information about AIDS and HIV. Its potential reach is enormous. For instance in 1992 close to 100 volunteers logged 7,000 hours on the phone with over 16,000 callers (AIDS Vancouver 1992). Three topics dominate: testing, transmission and safer-sex practices. As the woman quoted directly above hints, the anonymity and confidentiality of the Help Line were underscored as the reason for its success. Indeed, they are stressed in the line's advertising (Plate 16.4). Spatially, then, callers can literally *be* anywhere, while Help Line volunteers take their shifts in PARC's Yaletown office. The freedom that this cyberspace enables facilitates contact, as another volunteer explained, and thereby augments HIV education and prevention work:

'People find it very safe to call because of the anonymity. Individuals call from telephone booths on very busy streets. And sometimes you can't hear them. Or sometimes people will call and they'll say, "Well, I have to go because somebody's here." If we have a sense initially in these conversations that there might be a little bit of anxiety, we try to reassure the caller that this is anonymous. We don't have any call-tracing or any of those devices. I think the telephone is enough of a barrier to make them feel comfortable and anonymous. I mean, it takes a lot of courage to call in the first place. Certain people procrastinate for a long time, fear or anxiety or whatever it is. So we try to acknowledge that early. "We know it takes a lot of courage. Take your time." '

Another theme that volunteers struck while discussing the significance of the Help Line's anonymity and confidentiality is the context of guilt or shame that can motivate a caller. Several volunteers noted the importance of callers' anonymity and the line's confidentiality by discussing the recurrent confessional tone many callers took while discussing their sexual behaviour:

'Depending on what's happening in the media, then you'll get the calls in the morning. Or you get calls from people at night who've been sort of thinking about it all day but don't have the courage to call. Monday mornings are typically fairly "good" in the sense that people have been out on the weekend and perhaps something happened and they need to talk about it. Evenings for the last while have been actually sort of quiet, and the calls are coming in during the day. And I don't know if it's because people will call from their offices rather than their homes.'

'People want to know where to get tested. That's pretty much a standard question. Basically people are looking for information.

Plate 16.4 Bookmark
advertisement for AIDS
Vancouver's Help Line
Photograph: Michael Brown

That's one side of it. The other side is they're panicking and they're looking for someone to talk to. Quite often that was the case. They just want to talk for fifteen minutes or a half hour and get relieved of their anxiety and guilt. Like, "I went to a convention in such-and-such a place and this prostitute came in and we all got blow jobs. And could I give my wife and my kids AIDS?" Anxiety, you know? Like I can't relieve the guilt that you fooled around on your wife, but if you want to know the facts about AIDS, this is it.'

Still another reason the cyberspace of the Help Line facilitates anonymity was stressed through discussions of popular conceptions of AIDS as a 'gay disease'. Because of this linkage, many straight people would never enter PARC to get information out of fear of being seen *there* and therefore labelled as gay. Even within the gay community, AIDS stigma can dissuade people from walking into the PARC complex and obtaining safer-sex information or materials. In that case, the fear is that being seen *in* PARC might lead some to suspect that a person is HIV-positive. Volunteers discussed these points at length:

'I have a great conviction in the Help Line because I think in terms of the organisation, it's a very under-utilised programme. Because it has the most potential – probably more than any other programme we have – to reach the larger community: the people we truly need to be reaching. I think we cover quite well, you know, the gay community with Man-to-Man, the workplace with the AIDS in the Workplace programme, and now the Women and AIDS Project is on board – but as far as the larger community – whether it's gay people who are not aligned with the gay community, or whether it's the heterosexual community in the suburbs or whatever, the Help Line really is the most obvious possibility for getting them all information on AIDS and HIV. And also because it's anonymous and it's confidential, and it's eleven hours a day, six days a week. So it's a tremendous venue for people out there.'

'When you think about it – I hate the term "general population" – but we're sort of limited here, because people to some degree, even after all these years, still have this idea that HIV is still a gay disease, and so they equate that with this organisation. I mean, you're not going to get straight people walking into this place. So it's a lot safer to actually call. This way they can call the Help Line. And if they happen to be gay or lesbian but don't want to be seen in the building or have any information come through the mail to connect them to AIDS Vancouver, well the Help Line provides that kind of anonymity too. We've certainly sent stuff in, you know, brown unmarked paper envelopes. It's really safe.'

In a recent essay on the ontology of cyberspace Heim (1992), while ironically celebrating the freedom from alienation technology offers, echoes more theoretically the justification for technologies like the Help Line to facilitate HIV education. As he puts it:

> Cyberspace supplants physical space. We see this happening already in the familiar cyberspace of on-line communication – telephone, e-mail, newsgroups, etc. When on line, we break free, like the monads, from bodily existence. Telecommunication offers an unrestricted freedom of expression and personal contact, with far less hierarchy and formality than is found in the primary social world.
>
> Heim 1991: 73

Calls to the Help Line can come from anywhere, and precise locational data is impossible to collect (calls are not traced). Consequently, urban politics through the Help Line are at once local to AIDS Vancouver's Yaletown office while simultaneously global, in the sense that they open up that service to any place with a phone. With this example, we can see the importance of Smith and Katz's (1993) warning against using only absolute space in theorising scale. Here, the point is not so much the minimisation of actual distance in providing AIDS education, but rather how the phone line's cyberspace can overcome social distances that have accompanied the AIDS crisis.

CONCLUDING REMARKS

Informed by this alternative sexual geography of Yaletown, we can begin to see a very different 'new urban politics' in that area of Vancouver than those that have occupied urban political inquiry over the past fifteen years. The politics of responding to the AIDS crisis is now a significant feature of urban (and rural) geographies. This chapter has sought to acknowledge timely concerns over theorising scale while it has sought to expand the definition of what passes for 'the new urban politics'. Yaletown is at once an intense site of the local–global politics of economic development, to be sure. However, it is also the site of the new urban politics of HIV prevention and education, amidst an alternate social geography of sexuality. Like their regime-centred counterparts these AIDS politics operate at – and between – numerous spatial scales while remaining city politics. Operation Latex Shield uses micro-geographies of public sex environments to instigate safer-sex practices. World AIDS Day allows Vancouver AIDS activists, as well as 'the general population' to situate the local dimensions of the AIDS crisis in a more global perspective, while also helping to prevent the spread of HIV quite directly in other places. More obtusely, the cyberspace of the Help Line allows safer-sex information to be disseminated in a highly anonymous and confidential way, recognising while circumventing the various stigmas that are still attached to HIV infection. The Help Line allows a space for HIV prevention to exist where it would be unlikely that a service user would actually walk into the gay-coded PARC. In none of these examples were politics *solely* local or global, but some (often a highly self-conscious) mixture of both.

It might be inferred that I am arguing for a shift in urban political enquiry away from development issues. Hardly! These politics have had significant effects on the geography of North American cities, and the increasing attention on urban regimes is well warranted. The enormous development in Yaletown, and the HIV prevention work emanating from there, can both be considered 'new urban politics'. As I have tried to hint (however briefly) these politics are certainly affecting the spatial structure of Vancouver in quite visible and powerful ways. Nor should it be concluded from this essay that the politics of HIV prevention and economic development are in any way disconnected in Yaletown. The same rent surface that made Yaletown an affordable location for AIDS service organisations (which would have congregated otherwise in the West End) also facilitates the development of land parcels there.[4] Gay men who can be both deliverers and clients of AIDS organisations also comprise the gentrifiers who are occupying Yaletown's new residential towers. Tensions have also been brewing between the area's gentrifiers and its sex-trade workers. Increasingly, the male workers (who often congregate along Homer Street just outside the front doors of these new residences) are being pressured by residents and police to 'move along', towards more remote, deserted and dangerous blocks in the area. The two political geographies of Yaletown are certainly bound up with each other. My concern, however, is that the tremendous attention paid to local economic development matters blinds students of city politics to the other, often highly concealed, often very painful AIDS politics in cities like Vancouver, and neighbourhoods like Yaletown. In a single part of the city there can be multiple spatialities, as well as multiple politics.

ACKNOWLEDGEMENTS

Thanks to David Bell, Anne-Marie Bouthiette, Robyn Dowling, David Ley, Geraldine Pratt, Heather Smith and Bruce Willems-Braun for their helpful comments on previous drafts of this chapter, and to AIDS Vancouver for permission to reproduce its materials.

NOTES

1 For background on the 1986 World Exposition in Vancouver see Ley and Olds (1988). The Expo lands and adjacent parcels that are being redeveloped, it should be noted, actually stretch far beyond Yaletown, skirting the entire north shore of False Creek.

2 Even where the alternate politics of HIV-prevention loom *so* large in place the new urban politics of development are favoured. See for instance DeLeon's (1992) history of progressive politics in San Francisco from 1975 91.

3 Census data do not show a starker contrast in Yaletown's social upgrading perhaps because of the recency of development since 1991. For instance, just under half (49 per cent) of 1991 households in the area live in buildings constructed since 1986.

4 While both Yaletown and the West End are culturally recognised as gay space in the city, the West End has historically been seen as the centre of Vancouver's gay community. Hence ASOs wanted to remain in, or as close as possible to, the West End when PARC was being conceived.

17

'BOOM, BYE, BYE'

Jamaican ragga and gay resistance

•

Tracey Skelton

A PERSONAL GEOGRAPHY

By way of an introduction, I want to map the geographical experiences which led to my writing this chapter, which will chart resistance to the homophobia[1] of particular performances of Jamaican ragga[2] and statements made by ragga stars.[3] This personal geography will, I hope, illustrate some facets of the geographies involved and also position myself as author.

I have visited the Caribbean several times for research and for attendance at conferences organised by the Caribbean Studies Association, at which I presented two papers in Cuba and Grenada.[4] My first visit in 1986 allowed me to live in a small village on the island of Montserrat for almost a year.[5] My most recent visit in December 1992 took me to Jamaica for my first Caribbean holiday (Plate 17.1). It was my first, and probably last, experience of the long-haul package tour. It was on this holiday that the contours of thought around the subject of homophobia and Jamaican culture began to be laid down for me.

In Jamaica shortly before Christmas I was invited by two waiters of a local hotel and restaurant to be their guest at the staff Christmas party. It was an idyllic setting – candles and coloured lights around the garden, the beach and the sea metres away, excellent Jamaican food. As the staff and their guests, both tourists and Jamaicans, sat at long tables to eat, the music was turned up. There was reggae that I recognised from my other Caribbean trips, the timeless classics of Bob Marley and the Wailers, and newer music which I knew to be ragga. Here I was in the home of Caribbean reggae and I wanted to find out more about the recent musical trends. Jamaican reggae/ragga is held in great esteem by the youth of other Caribbean islands, who see it as the true voice of resistance and struggle both for the Caribbean against the outside world and for those who fall

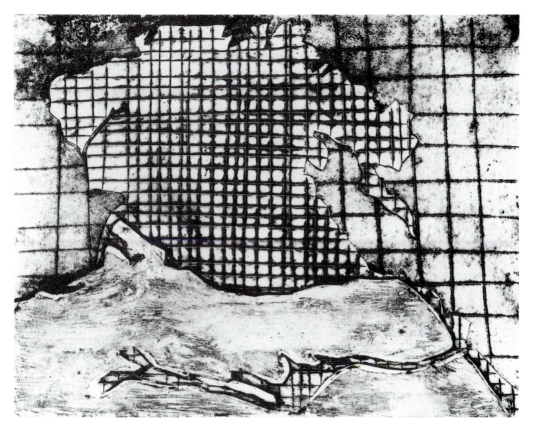

Plate 17.1 Kingston
Artist and ©: Michelle Keegan

outside the mainstream of Caribbean society: the youth, the unemployed, the politically disenfranchised. Young Jamaican men around our table were quick with their musical commentaries. Musical performance and success stories of poor Jamaican youths 'risin' up' through music became the focus of discussion. Someone mentioned the prowess of Buju Banton and then, as just another part of the debate, came vitriolic homophobic comments from two of the men.[6] They argued that the 'batty-bwoys' (homosexuals)[7] who were trying to get the 'yout' and his music banned should be shot, 'boom, boom!'; that in Jamaica they knew what to do with such men, 'stone them until dem a bawl and haffi dead'. Jamaica did not have such men and if they were there in Jamaica, real Jamaican men knew what to do with such 'anti-men', 'kill them, bwap, stone them dead'. They talked about Shabba (Shabba Ranks) and how he was a shining example to the youth of Jamaica, how he made them feel proud to be Jamaican men. When I tried to present an alternative view and show that I would not collude with homophobia, their anger was such that I thought it both diplomatic and safer to let the debate lie.

A second incident occurred in a taxi taking us into town for an evening meal. As we drove past a bakery called 'Patty Place', my friend misread the sign and asked, 'Is that Batty Place bakery a good one?' The driver had the same immediate reaction to the word 'batty'. He told us that it was not called that, that nothing in Jamaica could ever bear that name, and that 'batty' was the name used for the 'dirty sinful' men who were homosexuals. He talked of the 'white man's disease'. He continued: 'Because all of the white men up in those places, New York, London and so, are homosexuals, you women have to come down to Jamaica to find some real men, to find some real sex.' The conversation continued along similar lines of violence to the one at the party, describing

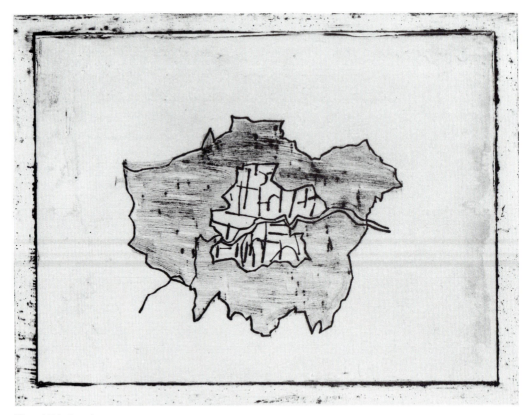

Plate 17.2 London
Artist and ©: Michelle Keegan

the 'best' way to deal with the 'nasty faggots them'. Again there was a weak attempt not to collude as we both said we had several men friends who were gay and that tolerance was important. That was met with the reply that we should keep away from such men, otherwise we would catch the disease and even AIDS too; and that to be with such men was to risk the same kind of attack he advocated, death by stoning or hanging. We were nervous about sitting in the back of a taxi with a man articulating such violence, so we asked to be dropped a fair way from our destination.

Negative attitudes towards homosexuality in the Caribbean were not new to me. I had tried to investigate attitudes towards homosexuality in my PhD research and had had several conversations with people in the islands where I had stayed over the past years (Antigua, Barbados, Grenada). Nothing I had encountered in these islands was as violent in its verbal reaction to the issue of homosexuality as what I heard in Jamaica. What was central to the Jamaican men's rhetoric was that Jamaicans knew what to do about such problems, that the penalty for such practices had to be death, and that the Jamaican ragga performers had every right to spread the Jamaican message against homosexuality.

In a new geographical setting – London (Plate 17.2) – I talked with academic colleagues about my experience in Jamaica. They told me about Buju Banton's song 'Boom, Bye, Bye' and also an interview on Channel 4's TV Programme *The Word* (4 December) with Shabba Ranks and white rap star Marky Mark where both men had endorsed Banton's homophobic view. I decided that I wanted to investigate the controversy further and to develop some understanding of this phenomenon of Jamaican culture. In view of my own ineffectual stance against the homophobia, I was also keen to record the processes and geographies of resistance that appeared to be developing.

THEORETICAL POSITIONS

This chapter attempts to locate the particular controversy of Jamaican ragga and its homophobic elements within wider discourses that debate race, sexuality and masculinity, and argues that there needs to be a conceptualisation which can allow a conjunction of these three.[8] This nexus has been developed within recent writing by black feminists, particularly those writing from African-American experience (Collins 1991; Goldsby 1993; hooks 1991, 1992; Wallace 1990). There has also been writing which has described and analysed masculinity and race in the British context, most notably that by Kobena Mercer and Isaac Julien (1988) and by Julien (1992). African-American masculinity, some of it heterosexual and some of it homosexual, has been investigated by several commentators (Beam 1986; Harper 1993; Majors and Billson 1992; Nelson 1991; Riggs 1991). Connections have been made between the Black Power nationalism of the 1960s in the US and sexism and homophobia in the African-American population

(Harper 1993; hooks 1992; Nelson 1991; Ransby and Matthews 1993). While such debates have a geography and context related to the United States, much of Afro-Caribbean history echoes at some level that of African-American due to the experience of slavery and the plantation economy. In recent times the US-based Black Power movement has had an important impact on Jamaican nationalism and Black Power movements (Thomas 1992; Levi 1992). Jamaica is the Caribbean island most influenced by political and cultural debate within the United States. Discourses around race and sexual politics have been few within the Caribbean context, although with some exceptions (e.g. Cooper 1992). My research has found very little material which has examined the issues surrounding homosexuality in the Caribbean apart from two articles on lesbian identities (Silvera 1992; Wekker 1993) and short mentions in Dann (1987) and my own previous work. Hence, although Jamaican ragga is a Caribbean cultural product I argue that there is enough similarity between the socio-cultural, sexual–political contexts of Jamaican and African-American society that the aforementioned intellectual debates can be utilised to examine the phenomenon of homophobia in Jamaican ragga and gay resistance to it.

MAPPING THE RESISTANCE

Trying to chart the resistance to the homophobia of ragga is, as with any process of resistance, complex. It is fairly easy to identify dates of specific actions taken but it is much more difficult to map the development of a culture of resistance, particularly when such resistance is the practice of a group of people who are invariably marginalised and rendered invisible. When such resistance appears at first sight to be white gay resistance against black Jamaican culture then the complexities of race and sexuality further complicate the issue. Such a dualism is inevitably going to be the one which informs mainstream media representations which are constructed by, and construct, racist societies while following traditions of Western thought which elevates binary divisions. Their analysis will suggest that all the victims are white gays and all the antagonists are black; and that all black people are homophobic. If tensions and divisions within and between the marginalised groups are created and actively maintained then it is much easier to maintain a cultural and geographical hegemony: white, Western, male and heterosexual.

Closer inspection of the resistance shows that it does not follow lines of race but rather of sexuality. In all presentations of the debate within the gay press the emphasis has been that this is a straight/gay problem and not a black/white problem. This particular form of homophobia is targeted at all gays, both black and white. It is the mainstream press, in particular *The Independent* and *The Guardian*, which, entering the debate almost a year after it was begun in the gay press, attempt to establish a binary divide based on race.[9] I will return to such reportage below.

Now though I want to pinpoint key events within the resistance. What becomes clear is that from the outset (August 1992 with the release of 'Boom, Bye, Bye') various gay organisations have been agentic, taking the lead in resistance through both democratic processes and imaginative direct action. Such action has taken place on both sides of the Atlantic and has developed out of coalitions between groups such as Black Lesbian and Gays Against Media Homophobia (BLGAMH), OutRage!, The Anti-Racist Alliance and the Gay and Lesbian Alliance Against Defamation (GLAAD). What is also evident is that while the initial resistance was against homophobic lyrics and statements, the resistance which has begun to emerge in Britain within the particular locality of Brixton and South London, is now against increasing levels of homophobic abuse and violence largely, but not exclusively, perpetrated by straight black youths. In September 1993 *Capital Gay* established a telephone Homophobia Hotline for people to report homophobic abuse. On 17 September the paper reported:

> Most of those who reported attacks were white gay men under 40. The majority of perpetrators of incidents were reported to be young black straight men (61 per cent), with 29 per cent of incidents involving both white and black attackers, and 10 per cent young white straight men.

Let us traject back to August 1992 and consider both the means of resistance and its geography. Buju Banton's hit ragga song 'Boom, Bye, Bye' was released. It contained lyrics which tell the audience that if any homosexual makes an advance towards them then 'is like, boom, bye, bye, inna batty bwoy 'ead' because 'rudebwoy nuh promote no nasty man, dem haffi de'd'; and that if 'batty-bwoys' want to escape their deaths then they have to 'get up an' run'. BLGAMH launched a campaign to have the record banned and joined forces with OutRage!, a (predominantly white) gay direct action group. In September OutRage! reported Banton to the Director of Public Prosecutions (DPP) for his song's incitement to violence. The song was banned some time later but still played on pirate radio stations and in clubs. In October the US-based GLAAD condemned New York (Plate 17.3) radio stations playing the record. The Mayor of New York publicly damned the record for its homophobia. Mercury, the record label Banton signed to in September 1992, disaffiliated itself from the record and offered to sponsor US Public Service Announcements on anti-gay violence in conjunction with GLAAD. Most New York radio stations stopped playing the record. Banton published a statement on 26 October in which he disavows violence but refuses to condone homosexuality. His statement read:

Plate 17.3 New York
Artist and ©: Michelle Keegan

I do not advocate violence against anyone and it was never my intention to incite violent acts with 'Boom, Bye, Bye'. However I must state that I do not condone homosexuality as this lifestyle runs contrary to my religious beliefs ... The anti-gay sentiment expressed in the lyrics is very much part of Caribbean culture ... As a product of the Caribbean, my commitment is to be a voice for my community, but not to advocate violence. In no way should the views of the song be construed as being condoned by Mercury Records.

Banton, 26 October 1992

That last sentence of the statement clearly shows that the pressure to make such a pronouncement came from Banton's record company, presumably anxious that the controversy should be contained with limited financial damage.[10]

In November *Gay Times* printed a letter from Paul Miles criticising WOMAD (World of Music Arts and Dance). He stated that WOMAD planned to hold a winter festival in Brighton and that one of the 'highlights' was the appearance of Buju Banton. He reported comments made by Basil Anderson, bookings director of WOMAD:

> You've got to take into account that this guy's the Number One reggae star in Jamaica . . . Jamaicans have always felt strongly against homosexuals so he is just singing what he believes . . . We had a debate here at WOMAD and some people felt very strongly that he should be cancelled, but I personally feel he's just expressing what is his culture.
>
> Miles 1992: 31

Clearly Anderson, and some of WOMAD's managers, felt that the expression of culture had to take precedence over the issue of a homophobic song.[11]

A key triangulation point on the map of resistance was 4 December when Shabba Ranks, and white rap star Marky Mark, appeared on the live television show *The Word*. The programme showed a pre-recorded feature on the 'Boom, Bye, Bye' song, and presenter Mark Lamarr asked Shabba Ranks for his opinion. Ranks responded with:

> 'Most definitely right now I'm on . . . the supporting side of all Jamaicans, be concerned for people in progress. But within this world, people be living the way they feel they want to live. If you don't have free will to move about, you got freedom of speech, freedom of opinion. [Lamarr said that surely freedom of speech to say 'shoot gay people' was wrong.] Well, most definitely, for him who forfeit the law of God Almighty, you deserve crucifixion. Most definitely. The Bible, I live by the Bible, which is the righteousness of every human being, and the Bible stated that man should multiply and the multiplication is done by a male and a female.'
>
> *The Word*, 4 December 1993

In the live showing of the programme on Friday 4 December Lamarr countered Ranks' comments and an argument ensued; the audience also hissed and booed Ranks' words and applauded Lamarr's responses. Presenter Dani Behr intervened, asking Ranks and Mark to perform a song together. In fact Ranks performed 'Ting a Ling' alone. The programme broadcast on Saturday 5 December edited out much of Lamarr's critical responses and the audience was not shown booing. It showed the free-style song which Ranks and Mark performed side by side. Ranks sang one of the lines from 'Boom, Bye, Bye', and Marky Mark shouted: 'Shabba Ranks! Speaks his mind, speaks his opinion and if you all can't deal with it, step the fuck off!'

In January 1993, BLGAMH decided to build a closer coalition with US-based GLAAD by sending them copies of *The Word* programmes. GLAAD contacted Budweiser who were sponsoring a Bobby Brown tour on which Ranks was the support. Also in January a lengthy article entitled: 'Batty Boys in Babylon: West Indian gay culture comes out in Brooklyn, and so does violence', by Peter Noel was published in *The Village Voice* (Noel 1993: 29–36). It was the front-page headline and eight pages long. In it Noel records homophobic comments, beliefs and actions of Afro- and Indo-Caribbean men. The pattern of language and ideas of violent punishment are almost identical to those I had heard in Jamaica. The connection between what the men said and the Buju Banton song was explicit; they mimicked shooting, talked of shooting the 'bwoys' in the head, claimed Banton as the dancehall Don. What this article also demonstrated is that the homophobia of Buju Banton's song is not isolated. In the dance halls of Brooklyn many DJs toast lyrics on the mike that are just as if not more violent.[12]

Returning to Britain, despite the fairly constant reporting of the controversy in the gay press, the transmission of *The Word* and the formal banning of 'Boom, Bye, Bye', *The Voice* newspaper ('Britain's best black newspaper'), in its review on the previous year's interviews commented on both Buju Banton and Shabba Ranks:

> Buju Banton might be a homophobic big-head, but apart from that, he's great . . . Shabba Ranks, the ragga star who touts machismo in tunes such as 'Trailer Load a Girls' and 'Gun Pon Me', the man who turned slackness into a philosophy of life, is a real cherub.
>
> *The Voice* 1993: 28

Clearly there is no attempt to distance themselves from any of the homophobia of the two stars but rather to endorse both homophobia and sexism.

February in the US saw Ranks withdraw from the Bobby Brown tour citing exhaustion. The *Soul Train Awards* dropped Ranks and he hired the most successful black PR agency in the United States (who represent Eddie Murphy and Oprah Winfrey). GLAAD sent a copy of *The Word* video to *The Tonight Show* on which Ranks was due to appear. In Britain the Anti-Racist Alliance issued an open letter on homophobia signed by ten prominent members of the black community (but no black MPs). This is another example of the 'resistance coalition' maintaining a position which emphasises the straight/gay antagonism: an attempt to ensure that the discourse revolves around issues of sexuality and not race. The ten wrote:

> As black people opposed to homophobia we are shocked and totally opposed to the anti-gay and sexist sentiments expressed in recent material produced by some

black reggae artists. These songs divide and weaken our communities ... Black people and lesbians and gay men all stand to lose from the rise of fascism ...

quoted in Saxton 1993b: 4[13]

What is now apparent is that GLAAD is able to mobilise much more resistance within New York. It uses the threat of consumer boycott to make promoters and the media act upon demands for decision-making in relation to the song and the three key protagonists (Banton, Mark and Ranks). In London, BLGAMH and other groups work within the democratic processes available: a report to the DPP; peaceful picketing of Banton's concerts; radio, television and newspaper interviews; open letters. Such responses may seem ineffective when compared with the apparent successes of GLAAD but it is a question of resources and finance. BLGAMH has three organisational members who all have full-time jobs; GLAAD in New York is part of an established, properly funded and efficient US-wide alliance. The New York chapter of GLAAD has 10,000 members, though only four full-time staff (Sawyer 1993). While this highlights the huge differences between gay organisations in both countries it also illustrates the importance for effective resistance of alliances and coalitions across geographies and cultures.

In early March, the US television programme *The Tonight Show* withdrew its invitation for Shabba Ranks to appear because, as producer Bill Royce said, the controversy had become 'a human rights issue'. In Britain the Broadcasting Standards Authority in response to ten complaints against homophobic comments made on *The Word* stated that the original (Friday 4 December) programme was 'balanced in the context of a live programme, particularly in light of an intervention by presenter Mark Lamarr' (Castle 1993:3). However, they thought that the second programme should have edited out the 'crucifixion' comment because it could be 'perceived to be an incitement to violence'. In the same month the BBC1 programme *Top of the Pops* showed a video for Ranks' 'Mr Loverman'. OutRage! staged a protest at the BBC.[14]

During the first few weeks of March Shabba Ranks issued a statement via his agent in which he made an apology for what he said on *The Word*:

On 'The Word' I was asked to share my views regarding Buju Banton's controversial song ... I responded in support of Buju, who is a friend and colleague. My views were premised upon my support of Jamaican artists ... plus childhood religious training. In retrospect, I now realise that the comments were a mistake, because they advocated violence against gay men and lesbians. I regret having made such statements ... I do not approve of any act of violence against gay men or lesbians or any other human beings ... Gay-bashing is wrong. Everyone should live their own lives, and have their own beliefs without fear of being attacked or abused . . . I ask my fans to love, not hate.

quoted in Castle 1993:3

BLGAMH said in a statement to *Capital Gay* (Castle 1993:3):

International networking by lesbians and gays was crucial in obtaining the apologies from Shabba Ranks and Marky Mark . . . [But] Shabba's apology is incomplete; it was made only to benefit himself and other musicians, not because he admits his hate-filled remarks were harmful and wrong. [To Ranks:] Homophobia, racism and sexism are similar evils. It is both immature and immoral for you to use your platform, which is your gift for music, to support any of them. The issues are simple, why can't you understand?

BLGAMH, alongside GLAAD, have always targeted Banton, Ranks and Mark, but the media coverage, especially that in the UK's gay press (*Capital Gay*, *Gay Times* and *The Pink Paper*), have often singled out the two black performers. What triggered those comments from BLGAMH about Ranks' apology was that, in an exclusive interview with the London/Jamaican paper, *The Weekly Gleaner*, after his stated apology, Ranks said:

I am willing to appease those who might have been offended by my earlier statement. In carrying on the great work done by Bob Marley, I have opened the door for others to come through, and it is for this reason and none other why I am willing to come to this accord.

quoted in Castle 1993: 3

Therefore in a newspaper written by and for the Jamaican community in Britain and Jamaica, Ranks makes it clear that he only made the apology for the sake of Jamaican music and culture and for aspiring musicians. He makes no genuine attempt to encourage a change of heart and mind amongst his fans. As I will demonstrate below, the making of disingenuous apologies is not something only Ranks is guilty of; Marky Mark and Buju Banton have both demonstrated that they have not changed their homophobic views and have no real intention of doing so.

In April GLAAD threatened Calvin Klein with a consumer boycott because they employ Marky Mark as a model. Mark recanted and promised to make a public service announcement condemning anti-gay and racist violence. He argues that he was a product of his childhood environment and knew very little about the gay community:

Through my music I preach peace. Gays want the same freedoms I do. There's too much crazy garbage in the world to worry what's going on in people's bedrooms ... My own sexual preference is for females but I respect anybody's homosexuality.

quoted in Gay Times April 1993: 21

However in September Marky Mark allegedly assaulted a gay record executive and made 'disparaging remarks about homosexuals' (*Capital Gay* 1993e). GLAAD retracted from working with Mark for a PSA, arguing that he is a poor model for an anti-violence message.

In July Buju Banton was granted two awards at the British reggae awards in the Hackney Empire. In an interview with *i-D* magazine he said:

'I don't see myself as a homophobic artist so to speak. And furthermore where 'Boom, Bye, Bye' is concerned, I put that out in 1992 and I don't intend to live in the past and neither should you.'

quoted in Eshun 1993: 27

He reasserted that his religion does not accept homosexuality and so justifies the song. Although the song was released in 1992 he still performs it at his concerts.[15] Effectively, then, while all three homophobes have made some kind of public apologies, usually under pressure from their 'employers', all three have either reneged on those apologies or have demonstrated behaviour that indicates that they have not changed their minds.

In Britain, OutRage! maintained their campaign against the BBC which they argue tolerates homophobia because it allowed Ranks to perform on *Top of the Pops*. They organised sit-down demonstrations in the foyer of Broadcasting House and also jammed the switchboard for BBC's *Crimewatch* (a true-crime phone-in programme). In September the producer of BBC1's *Top of the Pops*, Stan Appel, admitted that the show is less likely to ban homophobic performers than performers with other prejudices, and that such a situation may have to be re-addressed in the future.[16]

What this section has shown is that there have been distinctive geographical locations of the discourse of resistance, namely London and New York. It has demonstrated the dynamism of resistance and the mechanisms of coalitions. The perceptible effects of the resistance are difficult to quantify. There were public apologies made but later these were shown to be insincere; the record was formally banned but is still widely played; concerts and television programmes still present Banton and Ranks; Ranks was dropped by several programmes and tours in the United States but remains a chart-topper in mainstream music ratings. What has been a tangible outcome, though, has been an airing of the debate about race and sexuality. There has been a greater visibility of the black gay community in both countries than ever before. Probably the greatest benefit which may come to fruition in the longer term has been the obvious need for the gay community to interrogate its own racism and attitudes towards those who are both black and gay. The following and final section investigates the articulation of the race/sexuality/masculinity vinculum.

THE RACE/SEXUALITY/MASCULINITY NEXUS

While the resistance against homophobic performance in Jamaican ragga continues, the focus of attention in the gay press (and in some limited way the mainstream press) throughout the latter months of 1993 was the reported growth in homophobic violence in South London in general, and Brixton in particular (see for example Burston 1993a, 1993b; *Capital Gay* 1993f, 1993g; *The Independent* 4 October 1993; *The Pink Paper* 22 October 1993).

In September 1993 *Capital Gay* carried the front page headline 'Attacks on the rise in South London'. Readers who reported the abuse and violence argue that it is in connection with Banton's song. 'Batty boy' is now a very common form of street abuse. However, once again the newspaper clearly positions itself and states: 'The problem is not a black–white one but a gay–straight one.' What emerges through the debate in these recent articles is the articulation of the complexities of the race/sexuality/masculinity nexus. In the *Capital Gay* report Ted Brown of BLGAMH argues that an explanation for an increase in the number of black youths being homophobic might be because attacks on Shabba Ranks and Buju Banton were seen as racist attacks on the black community.[17] He continues:

> One of the only tangible 'benefits' of racism is that black men are stereotyped as virile. So while a white racist might put black men down as stupid or violent, they don't accuse us of being weak. When you have people standing up saying 'Yes, we're black queens' it threatens that virility.
>
> *Capital Gay* 1993f: 1

In the same article, Harold Finley, who organised the arts festival *Black Queer and Fierce*, admits that throughout London, all his experiences of street homophobia have been from 'black kids'. He states:

> There's this cultural thing now that it's OK to be homophobic – everyone who is marginalised seems to be looking for someone else to marginalise. These kids have a very rough ride in society, and feel powerless. This gives people a false sense of power, with people like Banton telling kids that 'it's alright to take out your anger on fags because they're lower than you'. Instead of empowering themselves they're taking it out on an easy target.

In African-American academic debate there have been similar interlocutions around the subject of black masculinity and its linkage with a lack of power within racist/sexist societies. Patricia Hill Collins argues that a reconceptualisation of sexuality is essential for black empowerment (1991). She discusses the homophobia of black women in particular in the context of privilege; in a racist and sexist society the only privilege left to black women is that of being heterosexual. She also states that the African-American community has tried to ignore homophobia but argues that this detracts from the need to work at the transformation and empowerment of all in the black community. While such debates are located within the geography of African-American USA there are similar socio-cultural, sexual–political processes at work in British and Jamaican society. In his excellent essay 'Eloquence and epitaph' (1993), Phillip Harper discusses the silencing of the debate around sexuality in the African-American community, linking this to the continuing strength of influence of the Black Power and Black Nationalism movements of the 1960s. He identifies several cultural products which express homophobic violence. Such verbal violence, he argues, serves a dual purpose: it attests the performer's own aversion to homosexuality, as well as his own unquestionable masculinity. What Harper makes clear is that for African-American men the condemnation of homosexuality is proof positive of their masculinity. This, I would argue, is the same function songs like 'Boom, Bye, Bye' and calls for 'crucifixion' serve for many Jamaican men.

In *Black Looks*, bell hooks (1992) argues that in every black community there are men who represent all kinds of masculinity. She develops a critique of black nationalism and its followers' complete refusal to deconstruct phallocentric obsessions which establishes the 'ideal' masculinity through the possession of a penis and the use of that to assert (heterosexual) masculine status. Like Wallace (1990), hooks argues that the failure of black communities to debate sexual politics and phallocentric goals has been to the detriment of the community as a whole. Hooks concludes with cross-references to black gay male writers, arguing that there is a clear link between nationalist phallocentrism and homophobia:

Challenging black male phallocentrism would also make a space for critical discussion of homosexuality in black communities. Since so much of the quest for phallocentric manhood as it is expressed in black nationalist circles rests on the demand for compulsory heterosexuality, it has always promoted the persecution and hatred of homosexuals. This is yet another stance that has undermined black solidarity.

hooks 1992: 112

Returning to recent mainstream press articles, I want to critique the way the race/sexuality/masculinity nexus is presented there. In *The Independent* article 'Macho man music puts gays in fear for their lives' (Cusik 1993), the title alone establishes the binary division between machismo and gay masculinities. The opening paragraphs relate the story of a 'homosexual man being threatened by a black boy': white is set up against black. The fact that the boy is only about 12 or 13 accentuates the stereotypes of excessive black masculinity (hooks 1991; Majors and Billson 1992; Mercer and Julien 1988; Wallace 1990) and gay effeminacy. The journalist, James Cusick, talks of black homophobia being a new epidemic (echoes of AIDS) facing the 'large homosexual population living in south London'. Quoting a beat policeman near Brixton tube station, who comments that black homosexuals probably have a tougher time, Cusick states that the policeman is 'voicing the view that the growing problem is perhaps not exclusively racist'. However, the remaining incidents he outlines are all cases where young black men have publicly attacked homosexuals. His use of the word 'perhaps' is significant, as it makes some attempt to cast doubt on the contention that this is not a white–black issue. One has to question Cusick's motives behind writing in this way. I suggest that this article builds upon pre-existing stereotypes of macho black youth and effeminate gay adults in order to exacerbate tensions that already exist between the two communities.

In the *Guardian Weekend* colour supplement of 20 November 1993, Paul Burston (1993a) writes 'Batties bite back' . There are large colour photographs of black and white gay men and the article asks if gays have a legitimate complaint about black homophobia or whether their fear is fuelled by racism. The report begins in August 1992 when 'Boom, Bye, Bye' was released. It moves swiftly up to the present and focuses on accounts of increased homophobic attacks by 'straight black youths'. He leaves this phrase unqualified although material in the latter part of the report contradicts the idea that it is only black youths who are the attackers. He quotes Paul Sigel of Gay London Policing (GALOP) who says that GALOP receives more reports of attacks by white assailants. Burston also claims that 'attempts to organise around the problem have led to frictions which threaten to tear the black and white gay communities apart'. This is another inaccuracy considering the ways in which black and white gay action groups have worked together at several levels and especially at the level of the gay press coverage which has frequently included comment from activists within BLGAMH. On the second page of the article Burston quotes Andrew Loxton who said that 'the vast majority of the perpetrators are identified as straight black men, and stresses that the problem is one of straight on gay, not black on white – in some cases, victims are themselves black' (Burston 1993a: 36).

Surely 'and stresses' should read 'but stresses' to emphasise the rebuttal of the commonsense feeling that it is a black/white issue? Again Paul Sigel is quoted:

> While it's true to say that there does appear to be a specific homophobic problem among younger black men, it has to be contextualised. Homophobic violence

generally receives very little attention from the national press. The suspicion among some people is that the only reason these attacks are being so heavily publicised is because the attackers are black.

quoted in Burston 1993a: 36

Ted Brown of BLGAMH argues that the campaign against black homophobia runs the risk of becoming overtly racist, and that press coverage is liable to fall into that trap. He argues:

The press has given the impression that lesbians and gays were doing fine until these blacks came along and started beating them up. It's no wonder black lesbians and gays are refusing to come forward and protest about homophobia. They can see a racist element at work in the way the issue is being treated. If a choice is given to most black people between fighting for lesbian and gay rights or fighting for black rights, the black rights will win hands down ...

quoted in Burston 1993a: 36

The final section of the article quotes Mark Raddix, a black member of OutRage!, Oscar Watson, co-organiser of *Black, Queer and Fierce* and Aamir Ahmad, founder of the gay self-defence group, Queers Bash Back. The debate again foregrounds the fact that for black men the only sense of power they have is in the stereotype of being powerful and masculine and that for many the presence of a black gay man is a threat to that image and so is the focus for attack. As more men in the black community have the confidence to be open about their homosexuality then they force the black communities to come face to face with the question of sexuality and homophobia. These black commentators argue that to fight homophobia in the black community does not mean they lessen their fight against racism and fascism.

Burston's article for *The Guardian* presents various voices and contradictions within the debate but the first page and lead paragraphs establish the discourse of white gay against black youth. What this chapter has demonstrated is that the resistance has been much more complex and multi-faceted than that because of the complexities of identities for those involved and the need to perform and articulate those identities in ways which resist both homophobia and racism.

CONCLUSION

This chapter has mapped the resistance amongst the gay communities of London and New York, both black and white, against the homophobia of some Jamaican ragga music

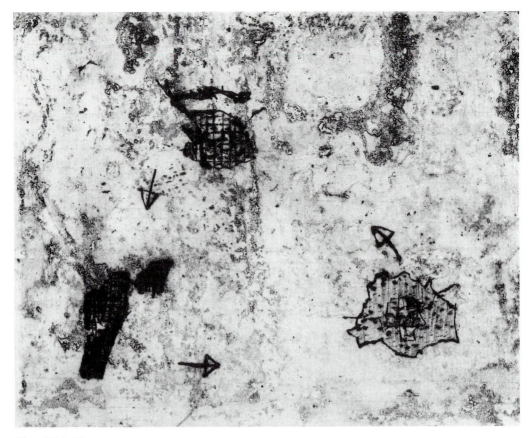

Plate 17.4 Cityscapes
Artist and ©: Michelle Keegan

and performers (Plate 17.4). Various strategies were employed to try and prevent the public performance of the homophobic song 'Boom, Bye, Bye' and restrict the access to public space for those performers who demonstrated homophobia. Within the United States the attempts to restrict access to performance spaces through the use of public space as a site of resistance were successful. In Britain there was not the same degree of success; the public space of cultural performance was deemed to be sacrosanct and protected.

On both sides of the Atlantic gay resistance groups have given voice to their sense of injustice and to their fears that homophobic cultural production (in this instance Jamaican) may provoke and encourage homophobic reactions and even lead to violence. In the case of Britain there has indeed been an escalation of violence, and more recent reports show that when gay men are attacked by black men the shout of 'batty boy' is a loud and clear slogan. However, the gay resistance continues. Homophobic attacks which appear to be linked with the song are reported in the gay press; BLGAMH continues to monitor and follow the statements and songs of the two Jamaican performers and those of Marky Mark;[18] Isaac Julien investigated the phenomena of the gun, misogyny and homophobia in Jamaican popular music and culture through his film 'The Dark Side of Black' for BBC2's *Arena*, broadcast on Saturday 12 February 1994.

In Britain the space of resistance has been predominantly the gay media. What makes the British situation complex is the way in which mainstream media lends a cultural space to the controversy and simultaneously places it within a discourse of racism and homophobia. The mainstream media establishes a binary based upon race, and another based on sexuality, rather than interrogating the race/sexuality/masculinity nexus. What this chapter has shown is that discourse, both within African-American academic space and black gay British activist arenas, foregrounds the need to examine this nexus, to re-define black masculinity and to engage with the complexities and contradictions of sexual politics within the black communities. What is beginning to emerge, but has a very long way to go, is the self-interrogation of white gay communities and groups who, while resisting Jamaican forms of homophobia, have to consider their own racism. Oppressive prejudices can be resisted through the use of various spaces; coalitions between groups who share those spaces are essential for success in such resistance.

ACKNOWLEDGEMENTS

I wish to thank the editors for their encouragement and comments on the earlier draft of this chapter. I am especially grateful to Jon Binnie at University College, London and Greg Woods at Nottingham Trent University, who have helped me enormously with the research for this chapter. Hardly a week went by without yet another article from the gay press appearing in my pigeonhole, so providing me with a wealth of material, in particular from London publications. Finally, warm thanks to Sue who gave me a peaceful space in which to work and write in Wales.

THE PRINTS AND ARTIST

The prints were made as specific illustrations of places that were important to the chapter – London, New York, Kingston. The boundaries of each urban place are delineated through texture, mark, compositional space and the print processes. Michelle Keegan is an artist, printmaker and lecturer. Her work is concerned with the manipulation of shapes gathered from specific locations and landscapes:

'The images emerge through allowing the print process, and the personality of the very metal the image is constructed on, to fertilise a visual dialogue only inherent through doing and making and responding to the drawn marks.' The artist retains the copyright of the prints.

NOTES

1 Homophobia is present in all Western societies and at all social levels, but while the author acknowledges that, this chapter is focusing on Jamaican homophobia as represented in ragga music.

2 Dave Hill in *The Guardian* (16 March 1993) defines ragga as 'the terse electro-burping reggae descendant fronted by a cast of rapid chatting, ripe-witted Jamaican knaves'. In *The Independent* (4 October 1993) James Cusick describes it as music which 'evolved out of Jamaican reggae and Afro-American rap', adding that much of it has 'lyrics that pay homage to macho virility'.

3 It is very important to stress that this chapter foregrounds a particular song and two ragga artists, Buju Banton and Shabba Ranks, and the various statements they have made. Ragga music, like all music, covers a spectrum of ideas and politics, not all defined as homophobic.

4 In Cuba I presented a paper entitled 'Domestic violence as patriarchal control: a Montserratian case study' (1991) and in Grenada the paper was 'Women, land and globalisation' (1992).

5 During September 1986 and August 1987 I was carrying out fieldwork for my PhD thesis, 'Women, men and power: Gender relations in Montserrat' and it was completed in September 1989.

6 I appreciate this is a value judgement but the comments made were indeed vitriolic and highly homophobic. However, it must be stated that other people in our group would not have defined them in this way.

7 When I discuss *homosexuality* in the context of Jamaican culture this is not because I prefer this clinical term rather than the more liberational terms *gay* or *queer* but because Jamaicans would very rarely use the term *gay*. The distance black homosexual people feel from the term *gay* is noted by Phillip Harper (1993).

8 Through my research for this chapter it appeared that publications foregrounding masculinity and sexuality very rarely addressed the question of black masculinities and the construction of racism. Exceptions to this trend were Craig (1992), Chapman and Rutherford (1988) and Rutherford (1992).

9 In January, February and March, *The Guardian* ran several articles on the subjects of WOMAD's near financial collapse, ragga and recording in Jamaica and the linkages between ragga/reggae and politics in its coverage of the run-up to the Jamaican general election. It did not mention the Buju Banton controversy with WOMAD, and it did not discuss the homophobia of ragga, although reference was made to ragga as 'a highly charged, often literally explosive, idiom that obsessively celebrates the joys of unsafe sex and the exploits

of the local gunmen' (O'Hagan 1993:4).

10 In relation to Banton's assertion that he was reflecting Caribbean beliefs and wanting to a be a voice for his community, the album he released soon after this controversy was entitled *Voice of Jamaica*, making clear his assumed role of spokesperson for Jamaica.

11 In fact WOMAD did intend to drop Buju Banton after Brighton City Council refused them a licence for the festival, but by then the organisation was already in financial difficulty (Myers 1993).

12 Noel (1993: 31) quotes from Natty B.:

Me an' Pimple ... sight two bwoy 'ug up inna
  dance hall
So one ah dem 'ead inna next one lap
So Pimple back he 'matie fi go lick two shot
Di bwoy dem say, 'Wait!' an' make a big
  splash ...

13 The ten signatories of the statement were: Linda Bellos (former Lambeth Council leader), Trevor Carter (author), Elayne (comic DJ), Justin Fashanu (footballer), Isaac Julien (film-maker), Kurshad Kharamanoglu (National and Local Government Officers' Association – NALGO), Martin Lindsay (National Union of Students), Bob Purkiss (Transport and General Workers' Union), Sanjiv Vedi (NALGO) and Mark Wadsworth (Anti-Racist Alliance) (Saxton 1993b).

14 Interestingly, while claiming to be opposed to any censorship of music, *Top of the Pops* was very careful with camera views of Jimmy Sommerville when he appeared on the programme in late March wearing a T-shirt which said 'Shabba Ranks is a Bigot'. There was one short flash when it was possible to read it, then every camera shot was either too far away to allow viewers to read it, or we were not shown his torso in close-up shots.

15 In October a NALGO coalition in Bristol reported how they tried to get Banton banned from a concert. The coalition was accused of racism and the concert went ahead; 'Boom, Bye, Bye' was the first song performed (*The Independent* 8 October 1993).

16 Appel said that they would not allow a band with Nazi or racist sympathies to perform, whether the particular song was prejudiced or not. Realising his inconsistencies Appel admitted that it might be something that he would have to think about in the future (*Capital Gay* 1993g).

17 Ted Brown of BLGAMH had appeared on *The Word* and condemned Buju Banton in December. Three days later he was attacked in his home by three men (two of them white) who asked him why he had to condemn Banton in public (Burston 1993a: Sawyer 1993).

18 Personal communication with Ted Brown, 9 February 1994.

THE DIVERSITY OF QUEER POLITICS
AND THE REDEFINITION OF SEXUAL IDENTITY
AND COMMUNITY IN URBAN SPACES

•

Tim Davis

Gay, lesbian and bisexual activists have used numerous strategies for social and cultural change. The list of strategies is long, but includes 'coming out', Gay Pride Parades, demonstrations, forming gay/lesbian religious organisations, and electing pro-gay or gay candidates. These strategies establish safe spaces, increase gay/lesbian/bisexual political power, and change culture and institutions (e.g. religious organisations and political parties) through visibility and education. American gay politics has historically depended upon the establishment and use of residential territories (known as gay territories, gay ghettos or liberated zones) as a survival tactic, as the centre for the creation of a common identity, as a base for electoral power and as a main focus of gay politics and gay/lesbian studies in geography and sociology. The gay/lesbian studies literature reveals that gay neighbourhoods were seen as spaces for the creation of distinct gay identity (D'Emilio 1981; Escoffier 1985). With the creation of a gay identity, these neighbourhoods could be used as a tool in establishing gay men and lesbians as a minority group that deserved a separate voice in local government.

Gay/lesbian/bisexual politics in America is at a strategic and philosophical crossroads, as the utopian idea implied by the term liberated zone has turned to a term of isolation and continued oppression – the gay ghetto. Gays and lesbians face a very different context than they did a decade ago, as society as a whole has changed, and the gay territories have themselves changed. Firstly, the power to create social and political change is no longer concentrated in government and a group of identifiable institutions, but has been dispersed in such a manner that progressive legislation can no longer be relied on to create wholesale change in society (Sassen 1988). Legislation is still needed to improve the position of gays, lesbians and bisexuals in society, but legislative victories are increasingly symbolic, when real acceptance can only be created in the cultural sphere.

Secondly, the gay and lesbian movement in America has also begun to examine more thoroughly internal differences and the impact of strategies and identity constructions upon various segments of the gay/lesbian/bisexual population. In many of America's largest cities, gays and lesbians have found a small niche in the local political structure, and the small measure of electoral and institutional power garnered through neighbour-hood control has largely benefited middle-class, gay white men. Like people of colour, gays in these cities have been able to use residential concentrations and voter turnout as a method of gay and lesbian empowerment. In cities like Boston, where no one minority group represents a large percentage of the population, minority groups have begun to realise the limitations of relying solely on minority neighbourhoods to elect people of colour. Sexuality cannot be equated with race, but gay activists have encouraged the notion that gays and lesbians constitute a minority that can elect leaders from gay neighbourhoods. Gay and lesbian activists of colour have been quick to point out that the essentialist evaluation of gay men and lesbians as a pseudo-ethnic minority tends to devalue the oppression and silences felt by people of colour (Fernandez 1991). As a part of the critique of the minority model of organising, Queer Nation, the Lesbian Avengers and ACT-UP (AIDS Coalition to Unleash Power), among others, are challenging the essentialist construction of community with new spatial tactics for social change, which reflect a shift towards an anti-essentialist understanding of identity and organising. Established gay/lesbian political institutions, as well as individual activists, continually rethink the meaning and types of strategies undertaken as old existing strategies are assessed, altered or abandoned. In Boston, this process of self-assessment has led to the creation of a wide range of groups, from the Log Cabin Club (gay Republicans) to Queer Nation to Latino, Asian and African-American gay and lesbian groups.

Thirdly, as can be seen in the work of Francis Fitzgerald (1986), attempts to create a safe space or 'liberated zone' in San Francisco and other cities have not come to fruition. Instead of safety, these areas, because of their visibility, have become the focus for many gay-bashings, and AIDS has had a profound impact upon the social structure of the gay scene and the life of these neighbourhoods. The dream of the liberated zone has been undermined, as other neighbourhoods and suburbs are considered viable destinations for many gay men and lesbians. In moving from a liberated zone to a gay ghetto, gay territories have lost their sheen. However, gay territories have played a profound role in increasing the power and visibility of gay and lesbian politics, and it is likely that the movement to a new form of 'Queer' politics could not have happened without the groundwork laid by the builders of these gay territories.

The Greater Boston Lesbian/Gay Political Alliance's activities around the 1993 Boston City Council redistricting; the Gay, Lesbian and Bisexual Irish Group of Boston, formed around the 1992/3 St Patrick's Day Parade controversy; and the oppositional tactics of the late Queer Nation/Boston reveal how gay, lesbian and bisexual politics in Boston are shifting. In each of these cases, there is a recognition of the need to move beyond the South End (Boston's 'gay ghetto') in order to better serve the needs of all gay men and lesbians. This reveals a movement away from a reliance on the gay ghetto as a base of strength, and each of these cases must be investigated for the way in which the concentration of power in the gay ghetto is undermined or reconstituted.

GEOGRAPHY MOVES BEYOND THE 'GHETTO'

The limited geographical and sociological literature on gay and lesbian communities from the late 1970s can be examined simultaneously as manifestations of varying notions of 'community' as well as windows upon essentialism in social science. 'Community' was conceptualised as a quasi-ethnic minority in which the politics of space are largely dependent upon the economic and social control of an individual neighbourhood. Sociologist Stephen Murray studied the way in which the institutional framework of gay communities is similar to that found in ethnic neighbourhoods (Murray 1979). Martin Levine's work closely relates to that of Murray, as he sought to describe the 'gay ghetto' and did not move beyond dot maps and explorations into the institutional completeness of specific gay communities (Levine 1979a). This work was politically motivated, as establishing the existence of gay territories worked to strengthen the notion that gays and lesbians functioned as an oppressed minority.

Castells and Murphy (1982) also mapped gay communities (though they did interrogate the methodological difficulty of this). Castells postulated that the late 1970s San Francisco gay male community attempted to build a self-sufficient community physically separated from neighbourhoods dominated by heterosexual, nuclear families (Castells 1983). Castells declared this attempt unsuccessful, not because there were no changes in society, but because his definition of success depended upon changes in the physical realm. Heterosexism and homophobia are located in the cultural, and are largely invisible in the physical landscape (Davis 1991).

In the mid-1980s, Larry Knopp's research began the process of moving beyond mere identification and description of gay ghettos with his work on the causes and impacts of gay gentrification in a New Orleans neighbourhood (Lauria and Knopp 1985; Knopp 1990b). Most recently, Knopp has taken another step by producing an examination of the intersection between gender, sexuality and capital (Knopp 1992). Knopp's writings point to the beginnings of a significant shift in the theoretical and philosophical underpinnings of geographical research on sexuality. The contents of this book are an example of how research now draws on and critiques the 'mapping' of a decade ago, and draw on a very different set of theories and literatures. Mapping gay spaces relied primarily on accepted sociological and geographical methodologies and understandings used to investigate ethnic and minority groups. These methodologies relied on inflexible notions of identity, did little to investigate culture, and could not take into account the way in which all spaces are sexed. Much of the current work interfaces with 'queer theory' and the growing field of gay and lesbian studies, and takes feminist theory and notions of the social construction of space and identity as starting points to study the relationship between sexuality and the creation of identity, community and citizenship.

Geographical research on sexuality has always been politically oriented, and geographers are responding to changes in gay/lesbian politics by moving beyond the ghetto and exploring the diversity of experiences and the multiplicity of sites and situations in which 'sexual dissidents' create spaces of safety and visibility. Gay ghettos, however, are still prominent, if not dominant, in American gay, lesbian and bisexual politics. As such, the power of the gay ghetto must be explored, not to celebrate it, but to problematise it and explore the ways in which gay territories remain effective tools for political action and limit the future and direction of the politics of sexuality.

MAPPING AN UNDERSTANDING OF GAY/ LESBIAN SOCIAL MOVEMENT STRATEGIES

My academic understanding of gay and lesbian experiences and politics is based on a range of literatures, each of which opened new doors of study and led to different paths of activism. My earliest research centred on the role of gentrification in building a gay territory in Minneapolis (Davis 1987). Taking a cue from Castells (1983) and the literature describing gay territories, I focused exclusively on bringing an area of new gay and lesbian gentrification to light and described its impact on the physical character of a neighbourhood. Like Castells, the physical, rather than the cultural, served as the centre of attention. A concurrent encounter with Queer Nation and the social constructionist and postmodernist debates exposed the limitations to research imposed by focusing exclusively on gay territories. Michel Foucault's *Discipline and Punish* (1979) opened a door on Queer Nation actions, as Foucault's description of the panopticon (a prison) seemed to be an appropriate metaphor for heterosexism. As Foucault states, the panopticon functions as 'a machine in which everyone is caught, those who exercise power just as much as those over whom it is exercised . . . [I]t becomes a machine no one owns' (Foucault 1977a: 156). Heterosexism, too, functions as a machine which has unreadily recognisable origins or control. Each one of us has been caught in its web, as the panoptic gaze exists in the form of heterosexism and internalised homophobia. Unlike Foucault's panopticon, heterosexism is not completely hegemonic, as its construction takes on diverse characteristics in different locations. Queer Nation actions appeared to be based upon an understanding of heterosexism as a spatially constituted discourse that can be interrupted and undermined. With Foucault serving as the theoretical basis of my work, Queer Nation became the centre of my research.

Researching the activities of Queer Nation and the Gay and Lesbian Irish Group of Boston revealed a need to examine these groups within the larger context of gay/lesbian social movements. In order to investigate the connection between group identity and the cultural production of spaces and places by social movements (e.g. the creation of neighbourhood character as 'ethnic' or as 'conservative'), I draw extensively on the work of new cultural geographers and historians, feminist geographers and social movement theorists. Feminist investigations of the relationship between definitions of masculinity/ femininity and gentrification (Rose 1984; Bondi 1991a 1992b; Warde 1991) and the cultural definition of African Caribbean (UK) neighbourhoods (Westwood 1990) have established that spaces are gendered, sexed and raced. Work by Kay Anderson (1987) has provided important contributions to this field by investigating *how* groups and neighbourhoods become racialised, and thus points to an understanding of how gay territories play a role in gay/lesbian/bisexual identity formation.

A number of writers (mostly historians) discuss the importance of ethnic parades as important moments in the creation and expression of group identity (Cohen 1982; Cottrell 1992; Davis 1986; Ellis 1993; Kertzer 1988; Marston 1989; Ryan 1989). Parades are gendered and raced (Davis 1986), and function as potent symbols of neighbourhood and group identity. As a result, any attempt to change the meaning of a parade is seen as a threat to the self-definition of a group. This work is very useful in theorising the role of symbolism in defining the cultural and physical character of neighbourhoods and urban spaces, and is especially useful in understanding the significance of gay/lesbian participation in the St Patrick's Day Parade, as well as Queer Nation tactics.

New theories of social movement activities, both from the resource mobilisation and new social movement perspectives, provide a context for understanding the role of culture and identity as strategies for social change. For example, Cohen and Arato (1992), as new social movement theorists, reveal how social movements should be judged by how discourses of domination and inclusion, in civil society and the state, shift as a result of social movement activity. This represents a move away from an exclusive focus on changes in culture. For the resource mobilisation perspective, social movements have been studied in relationship to government and institutions, to the neglect of culture. This has begun to change, as a number of writers have signified the importance of identity as a resource in itself, enhancing the field's ability to examine the resources necessary for particular types of strategies for social and cultural change (see Morris and McClary Mueller 1992). When examining gay, lesbian and bisexual social movement strategies, social movement approaches that combine an interest in the political and the cultural are needed if both institutional strategies such as voter mobilisation and the cultural tactics of Queer Nation are to be examined in tandem.

BOSTON LESBIAN/GAY/QUEER POLITICS:
THE LOCAL DIVERSITY OF STRATEGIES

My work to date has focused on the strategies and conceptualisations of community and identity expressed in three case studies: lobbying efforts around Boston City Council redistricting by the Greater Boston Lesbian/Gay Political Alliance (the 'Alliance'), Queer Nation/Boston, and lesbian, gay and bisexual participation in Boston's St Patrick's Day Parade ('GLIB'). Each case represents a different facet of Boston's gay/lesbian/bisexual political scene. The cases of the Alliance shed light on institutionalised gay/lesbian politics, and a focus on gay territories as a source of power. Queer Nation/Boston differs radically in its reliance on interrupting dominant cultural meanings in a multiplicity of sites. The recent St Patrick's Day controversy represents a strategy based in part on the politics of Queer Nation, but has had both institutional and cultural impacts.

Voting blocs, gay blocs

As a result of population shifts during the 1980s, the Boston City Council was forced to redistrict its nine city council districts (four others are elected at-large). The Redistricting Coalition, consisting of several minority organisations, seized this opportunity to increase representation of people of colour on the city council. Although 43 per cent of Boston residents are people of colour, they are 'packed' into two districts, creating two 'majority minority' districts, but diluting their power in other districts (Figure 18.1).

Following on the work around New York City council redistricting, completed by the Empire State Pride Agenda, I began my own research into the implications of redistricting for Boston's gay men and lesbians. In Boston's District 2, South Boston, a heavily conservative and Irish neighbourhood, is combined with the South End, a racially mixed and heavily gay neighbourhood. In 1991, Michael Cronin, an openly gay man from the South End, ran against James Kelly, the vocally homophobic and conservative councillor from South Boston (elected City Council President in January 1994). As the results reveal, councillor Kelly overwhelmingly won South Boston, but was unable to win much of the South End (Figure 18.2). This outcome revealed the inequities of the current council boundaries, and many South Enders expressed an interest in shifting the neighbourhood out of Kelly's district.

Using mailing lists with a combined total of over 7,000 names, Figure 18.3 was created, revealing a vivid picture of Boston's lesbian/gay concentrations. The Greater Boston Lesbian/Gay Political Alliance voted to support and lobby on behalf of the Redistricting Coalition's proposal, that would have created two additional racially *competitive* districts (a majority of residents would be people of colour, but no one racial

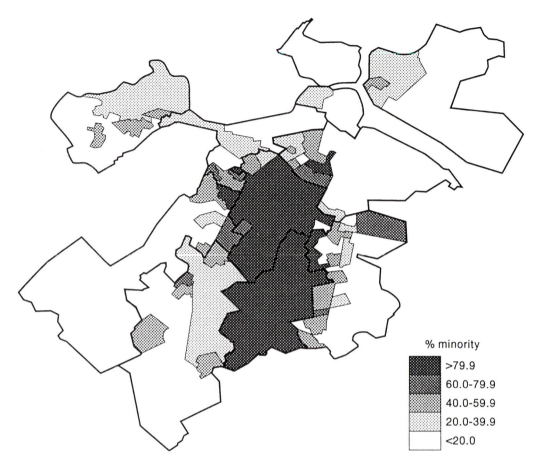

Figure 18.1 Boston: percentage minority by precinct
Source: Tim Davis

group would constitute a majority of the district), removed the South End from Councillor Kelly's District 2, and created the possibility of a gay/lesbian councillor from Jamaica Plain, Boston's other gay and lesbian residential concentration. Under James Kelly's control, the redistricting committee held one public hearing, and created and passed a status quo plan that shifted seven precincts of 252.

This case is applicable to this discussion as an example of an effort undertaken by an institutionally-oriented group such as the Alliance. Firstly, the very mapping of gay and lesbian concentrations essentialises sexuality and assumes that we identify gays and lesbians as a demographically distinct group, when in fact these maps only reveal

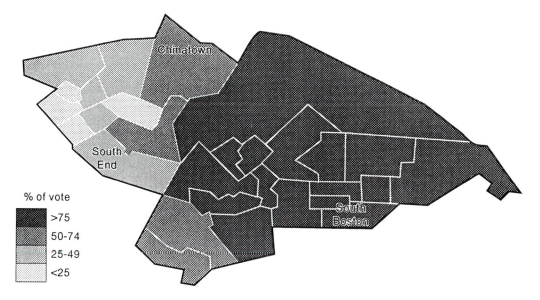

Figure 18.2 Boston City Council District 2, 1991 election results: percentage voting for James Kelly
Source: Tim Davis

concentrations of people who participate in or who are in some way part of the local gay/
lesbian 'scene'. More importantly, this strategy represents a continued reliance on gay/
lesbian territories to increase political power, and it largely benefits those identified
because of their activity in the gay scene. This strategy is also based on a notion that gays
and lesbians can gain political power as a 'minority' deserving a certain level of
representation. A strict dichotomy between gay/straight is maintained, allowing only for
the creation of a type of sexual pluralism.

The Alliance's work on redistricting did not simply re-establish the South End's gay
ghetto as a centre of power. Although Jamaica Plain has had a reputation as being a
lesbian/gay residential concentration (usually more associated with lesbians), few gay/
lesbian institutions are found in the neighbourhood. For this reason, revealing the size
and strength of this concentration has encouraged Alliance activists to work more to
include Jamaica Plain as a part of its grassroots organising. On the other hand, the
arguments made by the Alliance on behalf of the Redistricting Coalition's plan relied
solely on moving the South End into a new district. With conservative Councillor Kelly
as Redistricting Committee Chair, any mention of the prospects for a lesbian/gay
winnable seat in Jamaica Plain would have immediately killed the Redistricting
Coalition's proposal. Instead, the Alliance supported the Redistricting Coalition's
arguments that the Jamaica Plain district should be created because it could be Latino-

Figure 18.3 Boston: gay and lesbian residential concentrations
Source: Tim Davis

winnable. This had the effect of maintaining a status quo in which the South End remained the focus of gay and lesbian politics, to the detriment of creating visibility for Jamaica Plain.

At the same time that the South End served as the focus of the Alliance's redistricting arguments, the Alliance argued on behalf of the Redistricting Coalition's plan with the understanding that gay men, lesbians and bisexuals come from a variety of backgrounds, simply based on the societal demarcations of race, class and gender, as well as sexuality. What benefited people of colour is not always to the benefit of 'the gay agenda', but in this case, coalition building was an integral part of the Redistricting Coalition's and the

Alliance's understanding of how the city should move forward. In order to 'unpack' the minority districts and increase the power of people of colour, members of the Coalition argued that the new districts would be won through coalition building across the spectrum of identities, thus potentially empowering a number of groups (particularly Latinos and gay men and lesbians) that have been under-represented in the past. This new attention represents not only an understanding of the limitations to creating gay/lesbian winnable districts, but also a simultaneous movement among many African-American and Latino activists towards an understanding that limitations exist to the creation of 'minority–majority' districts.

THE LIFE AND DEATH OF A QUEER NATION

In Boston, a Queer Nation chapter burst forth with a great deal of energy in 1990, soon after the establishment of the New York chapter, and died a slow and quiet death in late 1992. Evidence of Queer Nation reveals a focus on a discourse intended to subvert the heterosexual assumption, and undermine the notion that we all must act 'straight' in public spaces. Queer Nation represents a distinct response to the changes in gay, bisexual and lesbian politics and circumstances, and depends upon a significantly different conceptualisation of community and strategies for social change. Underlying the 'kiss-ins', mock weddings (Plate 18.1), and queer shopping outings is the notion that *all* spaces are sexed (see Gill Valentine (1993a) for an excellent description), and that spaces are dominated by the heterosexist assumption. In this regard, Queer politics moves beyond the boundaries of physical gay spaces and a focus on the state, and challenges the heterosexist assumption in a diversity of locations, in so doing creating an idea of community based on 'We are everywhere'.

Here the words of Cindy Patton, AIDS activist and professor, are important:

> Gay people have been the perpetual victims of liberal notions of public and private: come out and be beaten up, stay in the closet and the government refuses to deal with the HIV epidemic. To cope with the spatial paradox, queer nationalism invades the mall, kisses in on the Supreme Court steps, unstraightens the Classics to signal that those territories have been co-occupied. Fundamentalists see queers and abortion-rights activists as the worst crimes against the Christian body.... Social space is leveled and zero-sum: the existence of any queer body anywhere reduces the space for the kingdom of God
>
> Patton 1993: 17

As an activist and academic, Patton draws on the notion that power is dispersed and

Plate 18.1 Queer Nation's August 1991 Queer Wedding, held adjacent to Boston's Cathedral of the Holy Cross
Photograph: Peter Erbland

variable, closely following the current understandings of Foucault's notion of power and discourse. Indeed, Patton's words reveal the interlocked character of Queer Nation activists and the academy. In this regard, the relationships between the social construction of identity and space are explicit in the writings and activities of Queer Nation. This is no more obvious than in the publication *Queers Read This*. An anonymous writer states: 'Let's make every space a Lesbian and Gay space. Every street a part of our sexual geography. A City of yearning and then total satisfaction. A City and a country where we can be safe and free and more' (Anonymous 1991a). Statements such as this have been asserted again and again within Queer action groups, revealing both the power and usage of space in challenging, altering or overcoming a hegemonic discourse such as heterosexism.

In Boston, Queer Nation actions closely followed the lead of the New York chapter, with kiss-ins at Faneuil Hall Marketplace (the earliest successful Rouse 'festival marketplace'), 'Queer Nights Out' at local sports bars, anti-violence demonstrations, a mock wedding with twenty couples next to Holy Cross Cathedral, and various 'non-traditional' activities at the Boston's 4th of July festivities on the Charles River Esplanade. The inherently temporary character of these activities made them most successful when the mass media took notice. Queer Nation/Boston could not point to legislative victories as a measure of effectiveness, so it is impossible to determine how Queer Nation/Boston altered the local character of homophobia and heterosexism.

Queer Nation/Boston may not have significantly undermined the heterosexist assumption, but it did make great strides towards undermining the South End gay ghetto as the centre of gay/lesbian politics and ideology. This is revealed in slogans such as 'Straight Acting/Appearing – NOT!', the continual pasting of posters in the South End attempting both to disturb and activate gay men, and the creation of the subgroup BRATS (Big-Honkin' Radical Anti-Assimilationist Terrorist Super-Queers), which wrote the following:

> We have declared war on heterosexism and homophobia. This of course means hets, but also the mainstream, nicey-nicey, assimilationist, 'We're just like you nice straight folks except for who we sleep with' gay and lesbian bowel movement. This movement needs to be flushed, and we are the tidy bowel queers! We will target many with our random, unpredictable, terrorist attacks who are deserving punishment. Hateful breeders who bash us and assimilationist scum that water down our angry voices and channel our efforts into a system that oppresses us are both cause to pull out the ammo and come out shootin'.
>
> BRATS, *Our Times* 1992

Although Queer Nation/Boston has disbanded, 'Queer activism' has continued, and thus

Plate 18.2 Boston St Patrick's Day Parade: GLIB
Photograph: Peter Erbland

the discourses of change that thrived within Queer Nation are being transplanted as former Queer Nation members find new niches in the political landscape. The case of the Gay and Lesbian Irish Group of Boston (GLIB) is one of these examples of continued activism by Queer Nationals, and perhaps the most profound.

LESBIANS/GAYS AND ST PATRICK'S DAY: A QUEER MOVEMENT?

In 1991, the St Patrick's Day Parade in New York City became a site of resistance, as the Irish Lesbian and Gay Organization (ILGO) attempted to gain entrance into the parade under their own banner. The group was denied the right to march as a distinct contingent (but did march with a sympathetic group) and insults and objects were hurled at the marchers. In 1992, ILGO New York again made an attempt to enter the New York parade, and as a sign of support, three Irish-American women involved in Queer Nation/ Boston formed the Irish-American Lesbian, Gay, and Bisexual Pride Committee, now known as GLIB (Plate 18.2).

These three initialised the strongest statement against homophobia and heterosexism in recent Boston politics. After a two-week battle in the media and the courts, gays, lesbians and bisexuals openly marched down the streets of socially conservative, Irish-American dominated South Boston. St Patrick's Day organisers, South Boston political leaders and parade spectators yelling 'Go home gays' saw this as an invasion of their neighbourhood, and thus an affront to their own identity, defined as Irish-American, Catholic and heterosexual (Plates 18.3 and 18.4). In fact, Irish and non-Irish gay men and lesbians have lived their entire lives in South Boston. This controversy represents an extremely important moment in the ongoing process of the definition and control of spaces.

Conflict over the meaning and image of a single parade may seem inconsequential in the grander scheme of neighbourhood and social movement politics, but for some neighbourhoods and identity-based social movements, the creation of 'communities' has been integral to their histories, and parades have served as a method of self-expression and self-definition, as well as a method of representation of the self to society as a whole. In the case of South Boston's St Patrick's Day Parade, gay men, lesbians and bisexuals inserted themselves into a heterosexist space, thus attempting to open up a space for the freedom to express their identity. Though a symbolic event, the St Patrick's Day Parade serves as a complex expression of ethnic and neighbourhood identity, and serves to redefine space, not just for the moment, but in general, by bringing new meaning and debates to the discourse of Irish and Irish-American identity.

It is in this case that I most actively draw on cultural historians' writings on the role of symbolism and parades in creating ethnic and group identity. Susan Davis (1986)

Plate 18.3 Boston St Patrick's Day Parade: homophobic protestors
Photograph: Peter Erbland

Plate 18.4 Boston St Patrick's Day Parade: spectator
Photograph: Peter Erbland

states, 'Parades were both tactics and the subject for new and ongoing controversies: Who should have the right to display themselves collectively in the streets?' In bringing out the more explicitly geographical aspects of ethnic parades, Susan Davis discusses the way in which parades were both gendered and raced, and therefore the use of public space in parades was important for social control. Davis states:

> Despite all this variety and possibility [of nineteenth-century ethnic parades], the right to street performances were significantly patterned by social differences as by space and time. By the early decades of the century, rights to ritual self-presentation roughly traced the definition of citizenship: elite manhood. This is not to say that only white males performed in the streets; rather, for all others, attempts at street performance could be physically or symbolically dangerous.
>
> Davis 1986: 45–6

St Patrick's Day Parades have served multiple functions, as sites of resistance, a form of celebration and sites for battles over the definition of identity. In this respect, the inclusion of gay men, lesbians and bisexuals represents another internal conflict over ethnic identity.

In the context of Boston's St Patrick's Day Parade, it is clear that it was physically dangerous for GLIB to march, and that the parade was both sexed and gendered in such a way as to make participation very significant. It is clear from the responses of the women who organised GLIB that they saw the homophobia and sexism of parade organisers, and that the organisers were as afraid of a particular kind of woman as they were of 'condom-throwing pedophiles' (a phrase used several times to link GLIB activists with ACT-UP activities). It is evident from my interviews that the women looked at the parade as it expressed both heterosexism and homophobia, as well as sexism. For example, one respondent said: 'I really grooved on the fact that we really penetrated in every sense of the word their little white-boy enclave.'

The lead 'spokesmodel', as she called herself, made it explicit that she organised GLIB with women and feminism in mind. She worked to ensure gender balance. Women served as the main spokespeople, the march contingent was thoughtfully balanced (not only by gender, but by age, ideology and other factors), and women were placed at the front of the contingent for better exposure. Adequate representation of women in the group was not the only goal, as undercutting stereotypes about women and femininity seemed to be as important as undercutting invisibility.

The media attention around this event was great enough that there is sufficient material to get an idea of South Boston residents' responses to GLIB's participation. South Boston residents saw this as *Southie's* parade, and any attempts to change it were seen as another attack on the neighbourhood – similar to past battles over housing and busing:

First the courts took our schools away with forced busing. Then the courts took our housing away with forced housing. Now they're taking our streets away with forced association. America has to wake up. Put down that TV clicker and take to the streets.

'Speak Out', *South Boston Tribune* 25 February 1993

This is geography. This is coming into South Boston for the sole purpose of disrupting us . . . This is not a sincere movement.

John 'Wacko' Hurley, parade organiser, *Boston Herald* 5 March 1992

GLIB, the gay, lesbian, and bisexual group of trouble makers who hate the Catholic Church and its teachings, are not welcome in South Boston's Evacuation Day Parade. If parading is so important to them, let them raise their own money, organize their own parade, and apply for a permit to march in downtown Boston to express their sexuality.

City Councillor James Kelly, in *South Boston Tribune* 25 February 1993

Interviews shed light on the understandings participants had of the importance of this parade and the radical impact of this event on its cultural meaning. Although founding members were also active in Queer Nation, this was not intended to be a Queer Nation action, and most of the gay/lesbian/bisexual participants had had little or no contact with Queer Nation. Queer Nation tactics have often included the use of drag, 'gender fuck', shouting and other theatrical activities. GLIB participants in the St Patrick's Day Parade did not resemble what many would consider 'Queer' activists. Instead, they wore 'normal' street clothes (including a lot of green), and a gay veteran wore his uniform. Despite this appearance of normality, the presence of GLIB was as threatening to many at the parade as Queer Nation could have ever been. Symbolically, the very existence of alternative sexualities was a threat to the locally prevailing notion of what it meant to be Irish. This is revealed in the following responses from GLIB marchers. The first is from a woman who had been in Queer Nation, and the second is from a man who states that he disagrees with the tactics of Queer Nation.

'What was really wild about South Boston was that we really didn't need to carry any signs. We didn't need to do shit. Just being there was being in their face.'

'Although this is a lot more controversial than anything ACT-UP or Queer Nation has ever done, I never thought it was going to get to that point. . . . All we did was march behind a banner and wave. As much as people think it was, I don't consider it particularly in your face.'

TIM DAVIS

Whether or not GLIB marching was 'in your face' is not agreed upon, but no matter how radical this event seemed to its participants, the sheer impact in the media and local neighbourhood discourse is astounding, and no Queer Nation/Boston action ever received this much attention. GLIB focused its energies on the creation of spaces for the free expression of an identity based in multiple communities – something that cut to the heart of heterosexism. At the same time, the Gay and Lesbian Irish presented a challenge to the institution of state and Constitution, as the group was allowed to march because, as a city-funded parade, the courts ruled the parade a public event. Over the last three years, in getting temporary injunctions against the parade organisers, and now working towards a permanent injunction, GLIB and parade organisers (the South Boston War Veterans' Council) have battled over the public/private character of this parade. The impact of GLIB on the legal position of gays, lesbians and bisexuals in Massachusetts cannot be determined at this time, but this group has succeeded in creating a stir in both politics and culture, bringing together the strategies of Queer Nation and the Alliance.

It is impossible to say to what extent South Boston has become a more friendly place for gay men, lesbians and bisexuals, but if the 1993 St Patrick's Day Parade is any indication, GLIB was greeted by fewer jeers, fewer projectiles and more cheers and waves from apartment windows and the crowds along the street. There may be a more immediate result of the St Patrick's Day controversies. Bisexual, lesbian and gay Irish and Irish-Americans are not only attempting to redefine what it means to be Irish in America, but have begun to re-examine the Irish past and are rediscovering a history of tolerance and inclusion. The work of Brendan Fay (1993) is important here. As the Irish-born founder of the Irish Lesbian and Gay Organization (New York), Fay has made great strides towards recovering and publicising this pre-colonial history of acceptance, and uses this to work as a political tool in combating the prevailing heterosexist definition of Irishness expressed in the New York and Boston controversies.

None of the GLIB (Boston) members interviewed had the same depth of historical knowledge as Brendan Fay, but there was a general sense that being Irish hinged on a history of oppression, and that the Irish should be able to identify with the oppression felt by gay men, lesbians and bisexuals. For Irish and Irish-American gay men, lesbians and bisexuals, there seems to be a new-found sense of pride in their ethnic identity, as an Irish history of inclusion and acceptance is reclaimed and new connections are created between those descended from different periods of immigrations, and a new, cross-national community is created that is based on a sense of belonging in many places.

CONCLUSION

Activists participating in each of the above cases sought to find ways to create political and cultural change in society as a whole while simultaneously altering the internal landscape of gay, lesbian and bisexual politics. While the degree to which these groups challenged a strategic reliance on gay territories varied widely, limitations to territorial-based strategies have been recognised, and the Boston lesbian, gay and bisexual political scene is definitely in transition. There is no doubt that the influence of the South End is on the wane, but the future is in doubt. A Queer politics based on visibility and the interruption of the dominant meaning of many spaces still exists in the hearts and minds of many, as well as in the politics of newer groups such as the Lesbian Avengers. In addition, a politics centred on networks that empower those physically and socially separate from the ghetto needs further investigation. The changing internal character of lesbian, bisexual and gay politics in American cities will have an impact not only on the character of gay/lesbian/queer identities and communities, but will also determine if geographically distinct communities exist at all. This will, in turn, impact on the structure and character of American cities. For this reason, further investigation into the shifting status and character of gay territories is necessary.

19

PERVERSE DYNAMICS, SEXUAL CITIZENSHIP AND THE TRANSFORMATION OF INTIMACY

•

David Bell

For a long time finding the right balance between the public and the private has been a major issue. The trouble is that the boundaries between the political, social and personal spheres of contemporary life are constantly shifting, or being shifted. The borderlines are extremely difficult to detect, let alone police, and where we should stake out the fences between public passion and private conscience, or between private needs and public indifference, are far from certain. In the increasingly complex cultural and moral universe characteristic of our age of uncertainty, boundary definition is difficult, and boundary conflict is pretty well inevitable.

Weeks 1992: 2

At the opening of this chapter I must make clear a point or two about both its title and its contents. The title is composed of three segments, each of which is borrowed – some might say stolen – from three significant (though not flawless) recent texts, each of which deals broadly with some of the aspects of sexuality (though not explicitly with their spatiality) that I want to engage with here. The phrase 'perverse dynamic' is lifted from Jonathan Dollimore's 1991 study, *Sexual Dissidence: Augustine to Wilde, Freud to Foucault*, where its conceptualisation and application takes up one-third of the book, sweeping through history, literature, psychology, sexology, sexual politics and post-modernity. *Sexual Citizenship* is the title of David Evans' ambitious and wide-ranging survey of what he calls in the book's subtitle 'the material construction of sexualities', and which takes up with homosexual, bisexual, transgender, children's and women's positions on the shifting map of sexual (and political) rights and obligations (Evans 1993). The final third of my chapter title comes from Anthony Giddens. In *The Transformation of Intimacy: Sexuality, love and eroticism in modern societies* (1992),

Giddens thinks through the social and personal implications of various 'sexual revolutions', using his concepts of confluent love, the pure relationship and plastic sexuality to work towards theorising the links between intimacy and democracy.

I also want to say at the outset that I don't intend to offer a close reading of these three texts. I don't have the audacity to tell you what the authors say, nor even what I *think* they're saying. The signalling in my title of these three particular works should instead be taken as a marker of certain broad theoretical agendas; readers who want or need to know more would be best served by the texts themselves. I want to use the figure of the 'pervert' – in this instance the practitioner of public (homo)sex[1] and the same-sex[2] sadomasochist – to examine rearticulations of and changing discourses about public spaces (spaces of citizenship) and private spaces (spaces of intimacy). The perverse dynamic, then, is a contradictory movement between the demands of and demands for these spaces: an uneasy oscillation, sometimes strategic, sometimes enforced.

The two particular bodies that I want to mobilise in thinking through my version of the perverse dynamic have been chosen for specific reasons. The practitioner of public (homo)sex – and here I want to think about cruising (but see David Woodhead in this volume) – is a figure whose presence in certain spaces and whose practising of certain (sex) acts embodies this oscillation very well: I want to theorise her/his[3] location in public (heterosexualised) space as being in tension with the desire for 'privacy', while at the same time trying to re-read this public/private tension and the position of the pervert within certain discourses of publicity and privacy. The same-sex sadomasochist is a currently-embattled figure here in the UK, thanks to a major police operation, the colloquially-named Operation Spanner, which resulted in the conviction in 1990 of sixteen men who had engaged in sadomasochist (SM) activities (Padfield 1992). More importantly, perhaps, is the fact that an appeal by some of the convicted before the Law Lords (*R. v. Brown, Laskey, Lucas, Jaggard, and Carter [1993] 2 All ER 75*; see Bix 1993; Hedley 1993; Mullender 1993) upheld their convictions, making the landmark judgement that 'consent' and 'privacy' have limits, and that these men had transgressed those limits (Bibbings and Alldridge 1993; Stanley 1993). Part of the (unintended) impact of this trial has been the eruption of a politicised SM and pro-SM protest body, named Countdown on Spanner, which has brought sadomasochism, and with it issues of sexual choices and freedoms more generally, out into the open (witness the number of supportive or sympathetic pieces written around Spanner in the quality British press). As Jo Eadie (1993a) has pointed out, the Countdown on Spanner campaign marks an important turning point in sexual politics and sexual citizenship, something it shares with certain AIDS activism agendas:

The productivity of such campaigns is precisely that while they require a form of consensus on one issue, there is no need for consensus on many others. They then become generative of debates which are possible because they do not undermine the basis of the collectivity, which lies elsewhere than the identity of the participants … Sexual citizenship, like sexual identity, is most powerful as a mobilising force, when it raises new questions rather than providing final answers.

Eadie 1993a: 167

In thinking through the very different historical geographies of my two pervert-figures, I hope to be able to examine how changes in private and public life (both those enforced by regulatory regimes and those borne out of resistance to that regulation) have impacted upon and been reconfigured by the practices of public (homo)sex and same-sex sadomasochism, suggesting that the perverse dynamic between citizenship and intimacy may be a way of opening up a new critique of both the exercise of sexual freedom and its containment.

PUBLIC (HOMO)SEX: PERVERSE PARADISE?

The term 'public (homo)sex' is already a contradictory one, for in some ways public (homo)sex can be very *private*. In terms of the location of the sex act, then, nominally it is taking place in *public space*: the park, the public toilet, the alley, the beach, the parking lot, the woods, the docks, the street. But in terms of the identities of the participants, their knowledge of each other, and the wider 'public' knowledge of the activities that go on in a particular setting, public (homo)sex can be very private, only attracting attention when the lives and loves of the rich and famous materialise there, or when the police or queerbashers target a particular site for their own kinds of nocturnal activities (usually dressed up as being 'in the public good'). Find an MP or a judge going down in the woods today, and there you have the created 'public' and media image of public (homo)sex: an image of shame, deceit, immorality – but more newsworthily, of scandal and of gossip. An image of a world inhabited by those sad creatures whose own internalised homophobia reduces them to self-loathing-filled anonymous encounters, and who would never 'come out', but who will live out their sex lives in this seedy underworld, a world far, far removed from notions of sexual identities, sexual communities and sexual politics. Public (homo)sex also runs against many societal constructs of intimacy, with the casual anonymous encounter being thought of as the very antipathy to the romantically charged (and heteronormative) model of sexual love:

'Impersonal', 'casual' or 'anonymous' sexual contacts had and still have a bad reputation among the majority of people. It is the kind of sex that violates notions of romantic love, steady relationships or longterm commitment, ideas which are widespread in our culture ... That this kind of sex is pursued and enjoyed as an end in itself seems shocking to them ... Public (homo)sexual encounters are contrary to conventional morality and (therefore) to legal rules.

Lieshout 1992: 3–4

Interestingly, Anthony Giddens characterises what he terms gay 'episodic sexuality' not as some kind of sad addiction, but as a

positive form of everyday experiment ... [which] expresses an equality which is absent from most heterosexual involvements, including transient ones. By its very nature, it permits power only in the form of sexual practice itself: sexual taste is the sole determinant. This is surely part of the pleasure and fulfilment that episodic sexuality can provide

Giddens 1992: 147

This emancipatory view of public (homo)sex is contrary even to some homosexual views: as Lieshout (1992: 4) points out, the Dutch gay movement once condemned cruising as an activity damaging to the 'desired image of the decent homosexual citizen' (see also Tucker (1991: 18–19) for the tale of Arizona State University's Lesbian and Gay Academic Union colluding with the cops to clean up 'anonymous men's room sex'). But public 'episodic' sex is often described by participants in idyllic tones; here's film-maker Derek Jarman on cruising Hampstead Heath:

The deep silence, the cool night air, the pools of moonlight and stars, the great oaks and beeches ... as always, once you are over the invisible border your heart beats faster and the world seems a better place.

Jarman 1992: 83

What is it that, in the cruising area, makes the world seem so much better? Perhaps the answer lies in Peter Keogh's (1992: 10) declaration that, on Hampstead Heath or in the cottage, he is 'experiencing pleasure which is not ... encoded in any way', an experience which means to him that 'On Hampstead Heath, I'm not myself'; or in the words of one Heath user, who said that what happens there 'goes to the very heart of what living in free Britain is all about' (quoted by Fielding 1992: 1). It certainly lies for many public (homo)sex participants in Scott Tucker's (1991: 17) observation that: 'Gay people often have no freedom to be gay in the privacy of their homes, due to family and neighborly [sic] pressures ... Lacking a secure privacy, they may find an insecure privacy and a selective publicity among similar seekers in [public] places.' In fact, as Jeffrey

Weeks (1985: 222) has said: 'Such places break with the conventional distinctions between private and public, making nonsense of our usual demarcations ... Most ostensibly public forms of sex actually involve a redefinition of privacy.' Such a redefinition, as Julia Cream (1993) has pointed out with reference to child sexual abuse, raises questions of where power lies and on what (and with whom) it rests. The freedom not to have to be 'out', not to have to subscribe to any identity or community, marks the cruising ground as the very site of this redefinition; the site of 'insecure privacy and selective publicity' for those who cruise it and use it, with 'private sex' taking place in 'public space'.

Running at times against this construction of public (homo)sex as anonymous and 'privatised', Maurice van Lieshout's (1992) paper 'Leather nights in the woods: homosexual encounters in a Dutch highway rest area' has pointed out that a liberalisation of attitudes in The Netherlands has brought cruisers out into the open, so to speak; they no longer need to be a 'silent community' and are afforded some state and legal protection (or at least a withdrawal of official harassment). The men who Lieshout talked to certainly countered the stereotype of public (homo)sex participants (embodied perhaps in the recent British tabloid scandalmongery about Tory MP Alan Amos, caught out on Hampstead Heath, who apologised for his 'childish' behaviour and for embarrassing his party and his family; see Fielding 1992). For many of the men in Lieshout's study visited het Mollebos, the cruising area, for reasons other than anonymous sex (although this did occur): for the pleasures of having sex outdoors, to have sex in front of an audience, for group sex, as an alternative to the bar and club scene, or just to meet up with friends and catch up on gossip. Their reasons for visiting the woodland echo Jarman's, as does their awareness (and eroticisation) of their surroundings, as this respondent shows:

Richard (27): 'Woodland in the dark has a promise of adventure. Perhaps it is the combination of fear and expectation. Behind each tree you may expect something or someone frightening or tempting or both. For me it is a very erotic scenery. When I have sex in the Mollebos I am always very aware of the surroundings. I also like other settings, but leathersex and a scenery like these woods belong to each other.'

Lieshout 1992: 16

Unfortunately, for every nation like The Netherlands where sexual expression is allowed some freedoms, there is one like Britain where it is severely limited. In terms of public (homo)sex between men, legal guidance is still given by the Sexual Offences Act 1967 (the Wolfenden Act), which decriminalised a very narrowly-defined homosexual sex act: the state and law could now tolerate sex occurring between two men, both aged over 21, in private (in February 1994 the House of Commons voted to reduce the age of consent for gay men to 18, after a bill proposing a reduction to 16 failed to get voted through[4]). As Angelia Wilson (1993: 174–5) says: 'Toleration has limits. When those limits are reached the tolerator has the power to criminalize and punish the tolerated. Since the 1967 Sexual Offences Act, homosexuals have been standing at those limits of toleration.' In Britain, sex between two men outside the confines of the bedroom is still often intolerable. Sex *in* the bedroom is intolerable if it involves anything but the most conventional, most 'straight', most 'vanilla' definition of what constitutes 'sex'. So is making a video, or taking photographs, of sexual activities (with the Obscene Publications Squad ever watchful). So is having sex in anything bigger than a twosome (by the Wolfenden Act's definition, sex with three or more men present, let alone participating, is 'public'). So is winking at a man in the street (this can be classified as procurement or soliciting). So is exhibiting a painting with a 'gay' theme, or even trying to publish an academic book on homosexuality (thanks to Section 28 of the Local Government Act; for a tale of publishers' squeamishness, see Shepherd and Wallis 1989). So is owning a vibrator or even a leather jacket (which can become, in the right setting, 'obscene material'). And so, of course, is being caught in the public toilets or in a cruising area. On top of this, the constant threat of homophobic violence which any and all of these activities also carries shows just how tightly the boundaries of tolerance are – how narrowly defined the private is – for British 'gay' men. I shall return to the issues I have hinted at here in the discussion below. But first I want to introduce my second pervert-figure, the same-sex sadomasochist, and begin to think about the location of sadomasochism within discourses of tolerance, privacy and citizenship.

SADOMASOCHISM: THE PARADOXICAL PERVERSE[5]

Sadomasochism (SM) is a troubled and troubling thing: a set of sexual practices – some might suggest an 'identity' or even a 'community' which can transcend sexual difference (Farshea 1993) – which has long been the centre of much debate about its 'political correctness'. This history of debate isn't something I want to take up time with; representative texts from the two sides include *Against Sadomasochism: A radical feminist analysis* (Linden *et al.* 1982) and Mark Thompson's 1991 collection *Leatherfolk: Radical sex, people, politics, and practice* (see also Binnie 1992b). Sadomasochism's trouble-causing status – by existing in a society which, as Stanley (1993: 214) says, 'allows the tickle but not the slap' – means that it is periodically assaulted by state and law via any means necessary, often with ancient statutes (like bawdy-house and slavery laws) being exhumed (see Califia 1993b; Rubin 1987). In my discussion of sadomasochism, I want to focus on a recent British legal case, referred to as Operation Spanner, in which sixteen men were charged, mainly under the Offences Against The Person Act 1861, for engaging in consensual same-sex SM (for detail see Bell 1994b; Bibbings and Alldridge 1993; Mullender 1993; Padfield 1992; Stanley 1993). Operation Spanner provides us with a useful example of how the tolerance conferred upon British 'gay' men by the Wolfenden Act has come to be eroded – especially around issues of privacy and consent. While an activity like sadomasochism had previously been projected into the 'privately legal yet immoral' domain (as Evans (1993: 52) calls it) along with many other dissident sexual activities and communities, the Spanner trial effectively redrew the boundary between 'crime' and 'sin', making it clear that 'the much vaunted assertion that there are immoral victimless consensual sexual behaviours which are not the law's business does not of course mean that they are simply consigned to absolute personal privacy' (Evans 1993: 63).

What exactly, then, does the figure of the pervert as constructed through the state and law's discourse on Operation Spanner tell us about the sexual citizenship, about the transformation of intimacy, and about the perverse dynamic? To begin with, it might be worth reflecting on this lengthy extract from an interview with the San Franciscan artist Nayland Blake:

> **Blake:** . . . The thing that is remarkable about 'Spanner' is that even if the people involved in the activity wanted to be involved, the State is given the final sanction on that activity.
>
> **Johnstone:** In this sense 'Spanner' might be an index of a particularly English relationship between individual behaviour and the gaze of the state. What is extraordinary about the Spanner prosecutions is the need to find someone who has been hurt.
>
> **Blake:** And posthumously, that is what is remarkable. Because, in general the State always attempts to find or formulate a victim, if you have a crime you have to have a victim. Victimless crimes do not exist very long because eventually somebody finds a victim for them, one way or another. Finally, this produces an extremely abstracted form of victim.
>
> **Johnstone:** Do you think the 'victim' in the 'Operation Spanner' case might be the public?
>
> **Blake:** Yes, which is an extremely abstracted notion because the State basically admits that the public does not reside in any of the people who participate in the activity or the people who know about the activity. So somewhere is this phantom public that is victimised.
>
> Gange and Johnstone 1993: 61

Seeing Spanner as 'an index of a particularly English relationship between individual behaviour and the gaze of the state' and especially law's gaze is perhaps a helpful starting point. Les Moran's (1993) critical historical interrogation of sex laws is useful here. Beginning with the Wolfenden Act, which, he writes, 'was used to define anew particular boundaries of the lawful male body and to inform and promote a renewed policing of that body of law' (109), Moran works back through the legal history of buggery, since the history of the homosexual male body in law 'has been dominated by ideas of sodomy, buggery, and more recently, (in)decency' (110). Tracking buggery in law back through time, Moran concludes, through a reading of seventeenth-century legal papers, that 'Englishness is that which is not buggery' (117), and vice versa. Perhaps, then, we could say that Operation Spanner uses law to suggest that Englishness is also that which is not male same-sex sadomasochism.

Spanner also reveals English law's ability to withhold freedoms and rights by deploying common law tangentially to prohibit and police them: by reinscribing the activities going on in SM scenes as crimes of interpersonal violence rather than sex acts, the 'idiosyncratic intelligibility and practices of the male body of law' (Moran 1993: 121) can survey, constrain and outlaw 'privately legal yet immoral' activities.[6] As Lois Bibbings and Peter Alldridge (1993: 364) put it, 'Sado-masochist [sic] activity seems to challenge and disrupt the system of differences between private/sexual and public/non-sexual activity': while the injuries incurred in sports, for example, were ruled as being inflicted 'for good reason' by the Law Lords, those occurring during SM scenes could not (and cannot) be condoned by law. Similarly, in a related case, body piercing for purely decorative purposes was deemed lawful, but the deriving of any sexual pleasure (from either piercer or client, or both) during the act of piercing commuted that act to an assault (Bibbings and Alldridge 1993).

Perhaps more central to decoding Spanner is Blake's statement that 'the public' is the 'victim' in the context of the case, and that this is problematic since the state 'admits that the public does not reside in any of the people who participate in the activity or the people who know about the activity'. It is this point of departure which I wish to discuss more fully next, reintroducing the 'public-(homo)sex pervert' to stand alongside the same-sex sadomasochist, so that both may articulate the perverse dynamic of public and private sex.

THE PUBLIC, THE PRIVATE, THE PERVERT

The notion of the public as being 'victimised' by Operation Spanner is one often deployed in state and legal discourses on 'sex crimes'. It is commonly articulated in the following terms: something is either for or against the common (public) good. If something is against the common good, albeit in a highly abstract way (presumably, in the case of both Spanner and public (homo)sex, against-common-good-because-immoral), then the public is victimised, and the perpetrators, who can no longer be tolerated, are thenceforth excluded from the public. Of course, this is a highly contradictory series of moves: the public must be made aware of something hitherto private (as Davina Cooper (1993: 269) puts it, 'the state and public generate a multiplicity of sexual discourse – the "unspeakable" they can never stop talking about'), sin then becomes crime, and the activity brought into the public is projected back into a reduced private (reduced because it has been partially and permanently ruptured: the once-private space of SM, in the case of Spanner, will forever now be not-private, since bringing it out into public only to expel it again robs it of its pre-public privacy, sending it into a not-public and not-private netherworld). SM, as Chris Stanley (1993: 219) puts it, is 'recognized, silenced and neutered as an activity that cannot come within orthodox reason and rationality'. Discussing the state's contradictory reactions to what Gayle Rubin (1989) calls 'scary sex', Cooper makes clear the selective deployment of discourses of public and private, and this need to make the private public (by what she refers to as 'sex talk') so that it can then be reprivatised:

> The [political] right emphasize the importance of keeping sexuality out of the public sphere and therefore the need for intervention – for a proliferation of sex-talk – to ensure this is achieved. However, when progressive forces attempt to deploy state power to achieve a more liberal sexual politics, the right focuses on the illegitimacy of sexuality as a topic of governmental concern, reinforcing this argument with claims that progressive forces intend to turn public what has previously functioned as a private terrain.
>
> Cooper 1993: 269

As Cooper notes, while there has rightly been much theoretical debate around the usefulness of a public/private divide, we must still recognise the 'ideological and normative power of the conceptual division' (269). The political tensions noted by Cooper situate the pervert exactly on the slash of this public/private split, irreducible to either domain. Jo Eadie's (1992) meditation of the position of bisexuality on the slash of the hetero/homo split is instructive here: echoing Barthes, he reads the separation around the slash as a 'panic function' to forbid transgression *across* the slash. In this way, then, just as bisexuality enacts a panic function across the hetero/homo dyad, so the pervert, inhabiting the space between the public and the private, threatens the collapse of both domains. In such a radical reading of the position of the pervert, however, the sacrifice of performing such a panic function, such a transgression, is not clearly articulated: the pleasures of perversion must be weighed against the dangers, as Spanner clearly shows. As Dollimore (1991: 230) writes: 'in creating a politics of the perverse we should never forget the cost: death, mutilation, and incarceration have been, and remain, the fate of those who are deemed to have perverted nature'.

The public/private dyad as it works around concepts of sexual citizenship has many more layers of complexity to it, and as such is the source of many more troubles. As I have already shown, law's eruptions into the private begin a process of reducing or even erasing the private as a site of pleasure, rendering pleasure a public – and by that a political – issue (thus transforming intimacy by removing it from an entirely private sphere). For sexual dissidents, there is an obvious tension between the desire for privacy and the need to be public, while state and law must draw things into the public only to thrust them back into the (reduced) private.

In much the same vein, Marilyn Frye (1983: 173) has shown how the lesbian is 'spat summarily out of reality, through the cognitive gap and into the negative semantic space', neatly articulating the tension between publicity and privacy: being *seen* as a lesbian means being instantly made *unseeable*, being projected into the 'stigma-impregnated space of refused recognition' (Sedgwick 1990: 63). And it is here that we find also find a difficulty for those wishing to revalorise both privacy and publicity as issues for sexual politics, since 'many of the same people who seek to politicise what has been too heavily privatised, also wish to affirm their "right to privacy" '(Weeks 1992: 3; see also Tucker 1991). This then is the pervert's paradox: a dynamic oscillation between realms, sometimes chosen and sometimes enforced, but always already moving.

MOMENTS IN LOVE: INTIMACY AND PERVERSION

Before concluding this essay, I want to return to the third element of my title; to the transformation of intimacy. As I mentioned earlier, Giddens highlights 'episodic sexuality' in 'gay' men as a model for a liberatory form of sexual intimacy freed from many of the power relationships which he sees as bogging down heterosexual love and sex. Now, while I have some serious reservations about Giddens' rather detached perspective on same-sex desire, I am glad that he doesn't follow the sociologists, psychologists and sexologists of old (and not so old), many of whom constructed episodic sex as compulsive, addictive, psychotic. But what about love? So far I've talked a lot about sex (and this is important because geographers haven't really begun to do this yet) and only a little about love. So I want here to briefly shift my emphasis, and to think about intimacy and perversion.

Love is a much more tricky thing to theorise than sex or sexuality. In a survey of sociological work on emotions, Stevi Jackson (1993) notes that, despite love's large role in public culture, it is often projected so far into the private (intimate) sphere as to be virtually untouchable, and also seen as so mysterious as to be untheorisable. Importantly, she notes how romantic love serves 'to validate sexual activity morally, aesthetically and emotionally' (210). And it might seem that 'love' is incompatible with the figure of the pervert described above. But texts written by advocates of public (homo)sex and SM do allude to romanticism and to fulfilments beyond the purely bodily (Jarman 1992; Thompson 1991; Tucker 1991). And with reference to cottaging, there's the statement from *Bisexual Lives* (OffPink Collective 1988: 59) that 'there's more love, more understanding, in many a men's toilet than in half the marriage beds'.[7] If we are to agree that the 'sex' involved in intimacy is constantly changing, that it has a history, then we must also see that the 'love' therein changes with it. But perhaps the spectacle of perverse love paraded in public would be more difficult for the 'moral majority' to deal with than perverse sex; for the image of two skinheads gently kissing might prove more destabilising than the image of them engaged in rough sex.[8]

Giddens talks of 'confluent love', love which is active and contingent and which, unlike romantic love, 'is not necessarily monogamous, in terms of sexual exclusiveness. What holds the pure relationship together is the acceptance on the part of each partner, "until further notice", that each gains sufficient benefit from the relationship to make its continuance worthwhile' (Giddens 1992: 63). Confluent love, Giddens states, is also far less founded on a heteronormative model than is romantic love. And it is a form of confluent love which we see acted out in deserted parking lots, in derelict East End warehouses and toilets at railway stations. For if Giddens is sceptical of the possibility of ever cleansing heterosexual sex of certain structural power imbalances (as he hints when saying that episodic 'straight' sex is still not as 'pure' as episodic 'gay' sex), then perhaps turning to the practices of public (homo)sex and same-sex sadomasochism might show him ways forward. Scott Tucker's (1991: 21) attempts to 'imagine a civilized erotic life, private and public' might take him in directions geometrically opposed to the ones Giddens takes,[9] but they are both out for the same thing: to theorise what Giddens (1992: chapter 10) calls 'intimacy as democracy'. His manifesto for a civilised erotic life could easily have been taken from Tucker's paper:

> The democratisation implied in the transformation of intimacy includes, but also transcends, 'radical pluralism'. No limits are set upon sexual activity, save for those entailed by the generalising of the principle of autonomy and by the negotiated norms of the pure relationship. Sexual emancipation consists of integrating plastic sexuality with the reflexive project of self. Thus, for example, no prohibition is necessarily placed on episodic sexuality so long as the principle of autonomy, and other associated democratic norms, are sustained on all sides.
>
> Giddens 1992: 194

Important for both Giddens and Tucker is thinking about power, which can all too often contaminate the pure relationship. But then there are certain sexed subjects who have always been aware of how power can be made performative or theatrical, and how reducing it to a game, to a set of roles, to a mask, can permit people to come together for the joy of sex which, by working on this performative power, can neutralise (and even eroticise) it:

> When I take a cock in my ass, I am actively taking power and pleasure, not simply reproducing a passive 'femininity;' and when I choose to give my partner the chief balance of power in sex, so that he strokes my cock with his asshole while I lie bound to a bed, then something is going on which is not reducible to the one word 'patriarchy.'
>
> Tucker 1991: 16

By eroticising anonymity and publicity, (some) participants in public (homo)sex are actively challenging hegemonic definitions of public and private,[10] while sadomasochists have reduced the power dimension of sex to the status of erotic play. As Wouter Geurtsen (1992: 4) says: 'Beneath the sexual game of power imbalance, the partners involved may very well have a strong sense of mutuality and equality. Most of the [SM] relationships do not involve feelings of inferiority or superiority outside the voluntarily adopted sexual or social setting'.[11] Finally, in the age of AIDS, the lessons of 'gay' men in 'how to have promiscuity in an epidemic' (Crimp 1987b) and the reconfiguring of sex away from what Linda Singer (1993: 122) calls 'ejaculatory teleology' and towards a more 'polymorphous decentred exchange', prominent in SM scenes, show that there are many other lessons in love that could be learnt from the pervert (Bell and Valentine 1995).

CONCLUSION

In this chapter I have sought to explore how the articulations of two particular sexed subjects – the cruiser and the sadomasochist – are useful for thinking about the relationship between perversity, citizenship and intimacy. For both cruisers and SMers *are* widely constructed as perverted by a heterosexist and profoundly vanilla society which can just about tolerate homosexuality provided it's out of sight and provided the participants conform with heteronormative constructions of love and sex; but the thought of people coming together in contexts beyond heteronormative romanticism and masculinist (hetero)sexuality, engaging in new forms of love and sex, is just too much to tolerate (Golding 1993a). And when the most heterosexualised bodies of them all, the state and law, catch sight of what goes on, then Davina Cooper's sex-talk begins, and suddenly the public is shown things so shocking that there's no alternative but to prohibit them and punish their practitioners. In the end, we come back to the debates around the Wolfenden Act: Lord Arran's especially-famous plea for homosexuals to 'comport themselves quietly and with dignity and to eschew any form of ostentatious behaviour or public flouting' (quoted by A.R. Wilson 1993: 175) remains a powerful marker of public, legal and governmental toleration of perversion. As Jeffrey Weeks (1992: 5) says: 'The freedoms of everyday life are constantly governed by a host of assumptions embedded in the practices of public life about what constitutes proper behaviour, and these shape what should be properly regarded as appropriately private'. In tension with state and legal policing of the boundaries of the public and private, the perverse dynamic redefines, reworks and reconfigures the public and the private, making new claims on citizenship and creating new patterns of intimacy.

ACKNOWLEDGEMENTS

Segments of this chapter, and a lot of the ideas contained within it, have been performed in front of live audiences at Royal Holloway and Bedford New College (January 1993), Staffordshire University (December 1993) and Sheffield University (March 1994). I am grateful to those who offered constructive comments, and to the Sexuality and Space Network posse for their inspiration. This essay is dedicated to Daisy Holliday, in the hope that she may grow up to see the boundaries of tolerance stretching outwards.

NOTES

1 I am using the slightly clumsy term public (homo)sex to differentiate it from public sex between opposite-sex participants; while this may have some things in common with public sex between parties of the same sex, the pressures of homophobia make them very different activities. I also want to point out that by using the term (homo) I only mean 'same'; I don't want to suggest that those taking part would necessarily want to class themselves as 'homosexual'. Hence I also use the word 'gay' in inverted commas, to mark it as an identity which need not bear relation to the sex act between those of the same sex.

2 For the same reasons as stated in note 1, I prefer the term 'same-sex' to the more commonly-used 'gay': the personal details of the Spanner men reveal that they too might not have subscribed to an identity such as 'gay' (see Bibbings and Alldridge 1993). At the same time, the implications of the Spanner case have impacted upon all SM practitioners, whatever their sexualities (although the fact of the Spanner men's 'homosexuality' was a central point in their trial, feeding off the AIDS panic;

see Farshea 1993).

3 While cruising is conventionally seen as a ('gay') male activity, I want to point out that 'lesbians' also cruise, and point you towards the story in Sue Golding's paper 'Sexual manners' (1993c). However, as Scott Tucker (1991: 19) points out, the threat of sexual violence against all women makes 'lesbian' cruising less common.

4 The bill was introduced by Conservative MP Edwina Currie, and voting took place on the night of 21 February. The proposed reduction of the age of consent to 16, in line with heterosexual sex, was outvoted by 307 to 280 votes, with the 'compromise' of a reduction to 18 being supported by 427 MPs, with only 162 voting against. The case for full equality is now being taken before the European Court.

5 Parts of this section also appear in modified form in Bell (1995).

6 In a frightening series of moves, the legal system managed to find charges for all the Spanner men, including the preposterous charge of aiding and abetting assault on oneself: since the 'bottoms' (masochists) freely consented to the 'injuries' inflicted upon them, they were all charged with conspiracy, too, whereas if they hadn't have consented, they would have been classed as the 'victims' of assault (Bibbings and Alldridge 1993; Farshea 1993).

7 This quotation was used by me in a minor skirmish on the pages of geography journal *Area* concerning geographers and love (see Bell 1992, 1994; Hay 1991, 1992; Robinson 1994 for the whole sorry story).

8 This point, suggested by Jon Binnie, is made in a discussion of gay skinheads and the performativity of 'transgressive' sexual identities (Bell *et al.* 1994).

9 A significant difference is that Giddens employs the work of John Stoltenberg in an uncritical way (Giddens 1992: 199–200), whereas a sustained attack on Stoltenberg's *Refusing To Be A Man* is at the centre of Tucker's essay.

10 I say '(some) participants' here because while there are some who are free to choose public places as erotic space (like Tucker himself, or those enjoying leather nights in the woods in Lieshout's paper), for many 'gay' men – those who live in rural areas, for example – there are just no other options.

11 The issue of power in SM is a hotly-debated one, with supporters arguing that parodying power destabilises it and detractors arguing that it reinforces it (see Binnie 1992b).

LIST OF CONTRIBUTORS

•

David Bell is a Research Fellow in the Department of Geography at the University of Sheffield, UK, where he is working on two research projects with Gill Valentine. His work focuses on identity, citizenship and sexual geographies. He is a founder member of the Sexuality and Space Network. He also teaches geography part-time at Staffordshire University.

Jon Binnie is a Lecturer in Human Geography at Liverpool John Moores University, UK. His current research interests include sexuality, consumption and the state. He is currently engaged in a comparative study examining theories of nationalism and European sexual citizenship.

Michael Brown completed his doctorate at the University of British Columbia, Canada. His research investigates new forms and locations of city politics through the responses to AIDS in Vancouver. He has recently been appointed as a Lecturer in Geography at the University of Canterbury, New Zealand.

Julia Cream is at the Department of Geography, University College London, UK. She is interested in constructions of the body and her most recent work has focused on women on the contraceptive pill in the 1950s/1960s.

Tim Davis, on leave from the PhD programme at the Graduate School of Geography, Clark University, Worcester, MA, USA, is administrator for South End Co-operative Housing in Boston, Massachusetts. Tim founded the Lesbian, Gay and Bisexual Caucus of the Association of American Geographers and serves as a board member for the Greater Boston Lesbian/Gay Political Alliance.

Glen Elder is completing a PhD dissertation on gender identities and access to housing in post-apartheid South Africa at the Graduate School of Geography, Clark University, Worcester, MA, USA. He received a Bachelor of Arts with Honours (1989) from the University of Witwatersrand, Johannesburg, South Africa. His current research interests include gender studies, links between the regulation of sexuality through space and urban social theory.

Angie Hart is a Lecturer in Sociology and Anthropology at LSU University College, Southampton, UK. She is currently completing a book entitled *Buying Power: Prostitution in Spain*, which is based on her DPhil. research. She is also working on a comparative study of lesbian and gay families in Britain and Denmark.

Clare Hemmings is a research student with the Centre for Women's Studies, University of York, UK. Her thesis is concerned with genealogies of bisexuality, the relationships between acts and identities, and feminist theories of bodies and subjects. She co-ordinates *Bi-Academic Intervention* – a network of people researching bisexuality – with Ann Kaloski. Her most recent publication is 'Resituating the bisexual body: from identity to difference', in J. Bristow and A. Wilson (1993) (eds) *Activating Theory: Lesbian, gay and bisexual politics*, London: Lawrence and Wishart. She is currently on a Fullbright Research Scholarship in the Five Colleges Women's Research Project, at Mount Holy Oak College, Massachusetts, USA.

Lynda Johnston is a graduate student in Geography at the University of Waikato, New Zealand. She is currently writing a Masters thesis on women body-builders and their training environments. Her research interests include the intersections between feminist and postmodernist thought, in particular focusing on the areas of corporeality and lesbian identity.

Lawrence Knopp is Associate Professor of Geography at the University of Minnesota-Duluth and Adjunct Associate Professor of Geography at the University of Minnesota-Twin Cities, USA. He has written articles on social movements, urbanisation, gentrification and sexuality and is currently working on a book-length manuscript concerning geography and gay cultural identities in the US, UK and Australia.

Jerry Lee Kramer hails from rural North Dakota, and is presently pursuing his doctorate in geography from the University of Minnesota in Minneapolis, USA. His interests focus on gay and lesbian demography, specifically behavioural aspects of inter- and intra-regional migration, settlement patterns, and the impact of gay and lesbian communities on individual queer identity development. His other major interests involve examining the global diffusion of Western gay culture and its impact on non-Western peoples.

Linda McDowell is a Fellow of Newnham College and a Lecturer in the Department of Geography at the University of Cambridge, UK, where she teaches urban and social geography. Her interests are in feminist theory and labour market restructuring and urban change in the UK. She is the author of numerous papers in geographical journals and co-author or co-editor of several books. Her most recent book (with Rosemary Pringle) is *Defining Women* (1992, Cambridge: Polity Press). She is the British editor of *Antipode: A radical journal of geography*.

Sally Munt is Postdoctoral Fellow in American Studies at the University of East Anglia, UK. She is author of *Murder by the Book? Feminism and the crime novel* (1994, London: Routledge) and the editor of *New Lesbian Criticism* (1992, Harvester Wheatsheaf/ Columbia University Press). She is presently researching a new book on urban dyke identity.

Alison Murray studied geography at Oxford and the Australian National University. Her PhD was published by Oxford University Press (1991) as *No Money, No Honey: A study of street traders and prostitutes in Jakarta*. She is currently a fellow at the Australian National University working on sexuality and HIV/AIDS in South-East Asia. She is also researching a book on traditional tattoo.

Tamar Rothenberg, who grew up in Park Slope, Brooklyn, USA, is a doctoral candidate in geography at Rutgers University, USA. She got her BA degree from Wesleyan University. Her work follows interests in urban geography, gender issues, race issues, popular culture and intellectual history.

Tracey Skelton is a Lecturer in the Department of International Studies, the Nottingham Trent University, UK, and teaches geography and cultural studies. Much of her research is Caribbean in focus and she has lived in and travelled widely in the region. She is currently co-editing a book on aspects of culture and development within the South and is a contributing author for a reader in feminist geography.

Gill Valentine is a Lecturer in the Department of Geography, University of Sheffield, UK, where she teaches a course on society and space. She has written on women's fear of male violence, and on lesbian geographies. She is presently working on three research projects (two with David Bell) about children's safety, food, and rural lesbian and gay communes.

David Woodhead is a research student in the School of Education and Health Studies, South Bank University, London, UK. His research interests include gay men's health issues, cultural politics, space and identity.

Gregory Woods teaches in the Department of English and Media Studies at the Nottingham Trent University, UK. He is author of *Articulate Flesh: Male homo-eroticism in modern poetry* (1987, Yale University Press) and of an acclaimed collection of poems, *We Have the Melon* (1992, Carcanet Press). His essays and reviews on gay culture and on AIDS have been published in Britain for over a decade. He has also written for the *New Statesman and Society*, the *Times Higher Education Supplement* and the *Times Literary Supplement*.

A GUIDE TO FURTHER READING

•

Trying to compile a list of useful books and articles is a very difficult task; but, as any of the contributors to this volume would tell you, getting started on geographical research into sexualities is even more difficult. We hope that our rather skimpy literature review in the introductory chapter provides a useful opening, by surveying the somewhat limited work explicitly looking at sexual matters from a geographer's viewpoint. We do not intend to duplicate that listing here, but rather to broaden our sweep somewhat, to indicate material which may be useful or inspiring, and which may (but may not) have some geographical or spatial perspective to it.

As ever with reading lists and bibliographies, a number of disclaimers apply. This listing is intended to be indicative rather than definitive, and a beginning rather than an end. It inevitably reflects the interests and knowledge of its compilers, and is frozen in time: assembled in July 1994, it will be out of date by the time it gets printed. Nevertheless, we hope it will prove to be of some use, especially to anyone about to move to uncharted terrain – and perhaps even for those familiar with the lay of the land, it might provide the occasional new departure point.

SEXUALITY AND SPACE

As we say above, a good place to start might be our introductory chapter, which surveys the geographical literature on sexualities. The most widely available and resoundingly geographical material is contained in papers such as those by Sy Adler and Johanna Brenner, David Bell, Larry Knopp, Gill Valentine – although all of this tends to focus on lesbian and gay experiences in the urban realm. Work by feminist geographers also contains valuable insights, especially in terms of a critique of geography's masculine logic: Gillian Rose's *Feminism and Geography* (1993) and Doreen Massey's *Space, Place and Gender* (1994) are recent contributions to this ongoing critique, while the journal *Gender Place and Culture*, which first appeared in January 1994, should establish itself as a forum for discussing sexual geographies. Collections such as Michael Keith and Steve Pile (eds) *Place and the Politics of Identity* (1993), *Mapping the Futures* (edited by Jon Bird, Barry Curtis, Tim Putnam, George Robertson and Lisa Tickner, 1993) and Erica Carter, James Donald and Judith Squires (eds) *Space and Place* (1993), while dealing with a lot of issues besides sexualities, are worth consulting for their theoretical takes on the links between identity and geography, as is Paul Rodaway's mapping of our senses, *Sensuous Geographies* (1994). Further, the collection

Nationalisms and Sexualities (Andrew Parker, Mary Russo, Doris Sommer and Patricia Yaeger, 1992) has some exemplary essays on the intersections of these twin powerful discourses. And finally, of course, there is Beatriz Colomina's collection *Sexuality and Space* (1992) – an intense (and difficult) engagement with architectural theory, social theory and sexuality.

LESBIAN AND GAY STUDIES

The lesbian and gay studies literature is huge, and ever-expanding. A very handy further reading guide, slightly biased Stateside but nevertheless comprehensive and user-friendly, is contained in *The Lesbian and Gay Studies Reader* (edited by Henry Abelove, Michele Aina Barale and David Halperin, 1993); there's no point in our retreading their footsteps. The contents pages of periodicals such as the *Journal of Homosexuality*, *GLQ*, *Perversions* and the *Journal of the History of Sexuality*, not to mention feminist theory and cultural theory publications like *New Formations*, *Differences*, *Hypatia* and *Social Text* are always worth checking out. And from the ever-expanding world of queer theory, authors from the United States, such as Judith Butler, Diana Fuss and Eve Kosofksy Sedgwick continue to write dazzling and captivating texts (most recently, Butler's *Bodies that Matter* (1993) and Sedgwick's *Tendencies* (1994), not to mention, say, Lee Edelman's *Homographesis* (1994) or Michael Warner's (ed.) *Fear of a Queer Planet* (1993), have shown that the queer contains much that geographers could deploy in their own readings of the world). Recent UK collections such as *Activating Theory* (edited by Jo Bristow and Angelia Wilson, 1993) and *Pleasure Principles* (edited by Victoria Harwood, David Oswell, Kay Parkinson and Anna Ward, 1993) show that here, too, lively debate and wild theorising are *de rigeur*. Of course, we might trace an intimate link between queer theory and work from feminist theorists, and so it is worth mentioning a few recent feminist texts: among the many wonderful books and papers recently emerging, we might point out as especially relevant Vikki Bell's *Interrogating Incest* (1993), Elspeth Probyn's *Sexing the Self* (1993), the collection *Beyond Equality and Difference* (edited by Gisela Bock and Susan James, 1992), Elizabeth Reba Weise (ed.) *Closer to Home* (1992), Judith Butler and Joan Scott's edited volume *Feminists Theorize the Political* (1992), Linda Alcoff and Elizabeth Potter's collection *Feminist Epistemologies* (1993), and Arthur and Marilouise Kroker's (eds) *The Hysterical Male* (1991). Away from the queer lexicon, important work engaging with issues of sexual rights and democracy includes Anthony Giddens' *The Transformation of Intimacy* (1992), David Evans' *Sexual Citizenship* (1993), John D'Emilio's *Making Trouble* (1992), Richard Mohr's *Gay Ideas* (1992) and Davina Cooper's *Sexing the City* (1994). And from a hybrid anthropology-sociology-history canon come works such as *American Sexual*

Politics (ed. John Fout and Maura Shaw Tantillo, 1993), *Modern Homosexualities* (ed. Ken Plummer, 1992), Gilbert Herdt's collection *Gay Culture in America* (1992) and *Dislocating Masculinity* (ed. Andrea Cornwall and Nancy Lindisfarne, 1994).

Work concentrating on literary and cultural readings of the sexual continues to flourish, with texts such as *New Lesbian Criticism* (ed. Sally Munt, 1992), *Sexual Dissidence* (Jonathan Dollimore, 1991), *The Book of Sodom* (Paul Hallam, 1993) and Mark Lilly's *Gay Men's Literature in the Twentieth Century* (1993) focusing predominantly on the written word; *Queer Looks* (edited by Martha Gever, John Greyson and Pratibha Parmar, 1993), Richard Dyer's *The Matter of Images* (1993), and Bad Object Choices' *How Do I Look?* (1991) on film and video; and Susie Bright's *Sexual Reality* (1992), Dyer's *Only Entertainment* (1992) and the collection *Madonnarama* (Lisa Frank and Paul Smith, 1993) on other popular cultural productions. Meanwhile, reading the body as cultural production remains in vogue: texts such as Arthur and Marilouise Kroker (eds) *The Last Sex* (1993), Michael Ryan and Avery Gordon (eds) *Body Politics* (1994), Marjorie Garber *Vested Interests* (1992), Julia Epstein and Kristina Straub (eds) *Body Guards* (1991) and Anthony Synnott *The Body Social* (1993) all produce readings of particular bodies, often in particular places.

Life stories, oral histories and travelogues similarly continue to map the experiences of particular sexed bodies in particular places and times: from Edmund White's pioneering *States of Desire* (new edition, 1986) and Lillian Faderman's tremendously important *Odd Girls and Twilight Lovers* (1992) to the oral histories of *Growing Up Before Stonewall* (Peter Nardi, David Sanders and Judd Marmor, 1994) and *Between the Acts* (Kevin Porter and Jeffrey Weeks, 1991); from the life stories collected by the UK's National Lesbian and Gay Survey, and published in two volumes, *What a Lesbian Looks Like* (1993b) and *Proust, Cole Porter, Michelangelo, Marc Almond and Me* (1993a) to the community histories contained in *Boots of Leather, Slippers of Gold* (Elizabeth Lapovsky Kennedy and Madeline Davis, 1993) and *Cherry Grove, Fire Island* (Esther Newton, 1993b), this work contains incredibly rich material on individual and collective lives. And, of course, we can also learn a lot about present-day sexual geographies by looking at gay travel guides and the like – a book like Betty and Pansy's *Severe Queer Review of San Francisco* (1993), for example, is at once an indispensable guide to the city's erotic topographies and a unique source of situated knowledge.

AIDS

As our introduction suggests, geographers who have turned their attention to AIDS have largely done so from the perspective of medical mapping: they have used diffusion patterns and transmission models, but have rarely gone beyond clinical exercise to consider what we might call the cultural geography of HIV and AIDS. So, rather than point to a literature which seems 'objective' and apolitical to the point of erasing the lives of those affected by the epidemic, we would like instead to signal a small number of works which might help us to think more critically about the ways in which geographers can – indeed should – be researching AIDS. A better starting point than Cindy Patton's *Inventing AIDS* (1990) would be difficult to find. Edward King's *Safety in Numbers* (1994) is a similarly vital, and more recent, look at the construction of AIDS, and Linda Singer's *Erotic Welfare* (1993) should also be highlighted as a collection of essays which deal with the themes of sexuality, illness and politics in creative and insightful ways. Texts focusing on activist and cultural responses to the crisis could also usefully inform geographers: Philip Kayal's *Bearing Witness* (1993), Douglas Crimp and Adam Rolston's *AIDS Demo Graphics* (1990), and Tessa Boffin and Sunil Gupta's *Ecstatic Antibodies* (1990) spring to mind.

BIBLIOGRAPHY

•

Abelove, H., Barale, M. and Halperin, D. (eds) (1993) *The Lesbian and Gay Studies Reader*, New York: Routledge.

Achilles, N. (1967) 'The development of the homosexual bar as an institution', in J. Gagnon and W. Simon (eds) *Sexual Deviance*, New York: Harper and Row.

Acker, J. (1990) 'Hierarchies, jobs, bodies: A theory of gendered organisations', *Gender and Society* 4, 139–58.

Adesina, Z. (1993) 'The oppressed homophobe', *Capital Gay* 15 October: 13.

Adler, S. and Brenner, J. (1992) 'Gender and space: lesbians and gay men in the city', *International Journal of Urban and Regional Research* 16: 24–34.

AIDS Vancouver (1992) *Annual Report*, Vancouver: AIDS Vancouver.

Aitkin, S. and Zonn, L. (1993) 'Weir(d) sex: representations of gender-environment relations in Peter Weir's *Picnic at Hanging Rock* and *Gallipoli*', *Environment and Planning D: Society and Space* 11: 191–212.

Albro, J.C. and Tully, C. (1979) 'A study of lesbian lifestyles in the homosexual micro-culture and the heterosexual macro-culture', *Journal of Homosexuality* 4, 4: 331–44.

Album Amsterdam (1992) *Album Amsterdam – Een kijk op je voorkeur in de stad van voorkeur – Adviesnota voor de vestiging in Amsterdam van een internationaal expositie- en informatiecentrum*, Amsterdam: Stichting Album Amsterdam.

Alcoff, L. and Potter, E. (eds) (1993) *Feminist Epistemologies*, New York: Routledge.

Alcorn, K. (1991) 'Boyz are in the pink', *The Guardian* 8 July: 25.

—— (1992) 'Queer and now', *Gay Times* May: 20–4.

Aldrich, R. (1993) *The Seduction of the Mediterranean: Writing, art and homosexual fantasy*, London: Routledge.

Alger, D. (1990) 'The world relation of cities: closing the gap between social science paradigms and everyday human experience', *International Studies Quarterly* 34: 493–518.

Allan, G. (1979) *A Sociology of Friendship and Kinship*, London: Allen and Unwin.

—— (1989) 'Insiders and outsiders: boundaries around the home', in G. Allan and G. Crow (eds) *Home and Family: Creating the domestic sphere*, Basingstoke: Macmillan.

Allan, G. and Crow, G. (eds) (1989) *Home and Family: Creating the domestic sphere*, Basingstoke: Macmillan.

Altman, D. (1982) *The Homosexualization of America, the Americanization of the Homosexual*, Boston: Beacon Press.

Anderson, B. (1983) *Imagined Communities: Reflections on the origin and spread of nationalism*, London: Verso.

—— (1991) *Imagined Communities: Reflections on the origin and spread of nationalism*, revised edition, London: Verso.

Anderson, K. (1987) 'The idea of Chinatown: the power of place and institutional practice in the making of a racial category', *Annals of the Association of American Geographers* 77: 580–98.

Anlin, S. (1989) *Out But Not Down! The Housing Needs of Lesbians*, London: Homeless Action.

[Anonymous] (1991a) 'An army of lovers cannot lose', in *Queers Read This*, leaflet 'published anonymously by queers'.

[Anonymous] (1991b) 'Return to the Blue Lagoon', *My Guy* 8 August: 16.

[Anonymous] (1991c) 'Return to the Blue Lagoon', *Young Americans*, 15: 22–5.

Anzaldua, G. (1987) *Borderlands/La Frontera: The new mestiza*, San Francisco: Spinsters/Aunt Lute.

Ardill, S. and O'Sullivan, S. (1986) 'Upsetting the applecart: difference, desire and lesbian sadomaso-chism', *Feminist Review* 23: 31–59.

Arrabal, F. (1967) *The Architect and the Emperor of Assyria*, London: Calder and Boyars.

Ashworth, G.J., White, P.E. and Winchester, H.P.M. (1988) 'The red-light district in the West European city: a neglected aspect of the urban landscape', *Geoforum* 19, 2: 201–12.

Atkinson, D. (1993) 'Working at Asda', *The Guardian*, 11 November: 14.

BRATS (1992) *Our Times*, flyer published by group involved in Queer Nation/Boston.

Bacchi, C. (1990) *Same Difference: Feminism and sexual difference*, Sydney and London: Allen and Unwin.

Bad Object Choices (1991) *How Do I Look?*, Seattle: Bay Press.

Bailey, M.E. (1993) 'Foucauldian feminism: contesting bodies, sexuality and identity', in C. Ramazano-glu (ed.) *Up Against Foucault: Explorations of some tensions between Foucault and feminism*, London: Routledge.

Ballantyne, R.M. (1979) *The Coral Island*, London and Glasgow: Blackie.

Barnes, T., Edgington, D., Denike, K. and McGee, T. (1992) 'Vancouver, the province, and the Pacific Rim', in G. Wynn and T. Oke (eds) *Vancouver and its Region*, Vancouver: University of British Columbia Press.

Barrett, M. and McIntosh, M. (1984) 'Ethnocentrism and socialist–feminist theory', *Feminist Review* 20: 23–49.

Barrie, J.M. (1945) *The Admirable Crichton*, London: Hodder and Stoughton.

Bashford, K., Laybutt, J., Munster, A. and O'Sullivan, K. (1993) *Kink*, Sydney: Wicked Women Publications.

Baudrillard, J. (1988) *America*, London: Verso.

Bauman, Z. (1992) *Intimations of Postmodernity*, London: Routledge.

Beam, J. (ed.) (1986) *In the Life: A black gay anthology*, Boston: Allyson.

Beauregard, R.A. (1986) 'The chaos and complexity of gentrification', in N. Smith and P. Williams (eds) *Gentrification of the City*, Boston: Allen and Unwin.

Bech, H. (1992) 'The disappearance of the modern homosexual, or: homo-genizing difference', paper presented at Forum on Sexuality Conference, Sexual Cultures in Europe, Amsterdam.

—— (1993) 'Citysex: representing lust in public', presented at Geographies of Desire Conference, Netherlands' Universities Institute for Co-ordination of Research in Social Sciences, Amsterdam.

Bell, D. (1991) 'Insignificant others: lesbian and gay geographies', *Area* 23: 323–9.

—— (1992) 'What we talk about when we talk about love: a comment on Hay (1991)', *Area* 24: 409–10.

—— (1993a) 'Back to basic instincts: bisexual tropes and troubles', paper presented at the 11th National Bisexual Conference, Nottingham, September.

—— (1993b) 'The politics of sex: queer as fuck?', paper presented at New Theoretical Directions in Political Geography, University of Birmingham, UK, September.

—— (1994) 'In bed with M.E. Robinson', *Area* 26: 86–9.

—— (1995) 'Pleasure and danger: the paradoxical spaces of sexual citizenship', *Political Geography*, forthcoming.

Bell, D. and Valentine, G. (1994) 'Queer country: rural lesbian and gay lives', presented at the Rural Economy and Society Study Group Conference, Cheltenham and Gloucester College of Further Education, September.

—— (1995) 'The sexed self: strategies of performance, sites of resistance', in S. Pile and N. Thrift (eds) *Mapping the Subject: Geographies of cultural transformation*, London: Routledge.

Bell, D., Binnie, J., Cream, J. and Valentine, G. (1994) 'All hyped up and no place to go', *Gender, Place and Culture* 1: 31–47.

Bell, S. (1993) *Reading, Writing and the Prostitute Body*, Bloomington: Indiana University Press.

Bell, V. (1993) *Interrogating Incest: Feminism, Foucault and the law*, London: Routledge.

Benchley, P. (1979) *The Island*, London: André Deutsch.

Benjamin, J. (1980) 'The bonds of love: rational violence and erotic domination', in H. Eisenstein and A. Jardine (eds) *The Future of Difference*, New Brunswick and London: Rutger's University Press.

—— (1986) 'A desire of one's own: psychoanalytic feminism and intersubjective space', in T. de Lauretis (ed.) *Feminist Studies/Critical Studies*, Bloomington: Indiana University Press.

Bentley, P. (1993) 'Banton convicted', *The Pink Paper* 18 April: 4.

Bernardin de Saint-Pierre, J.-H. (1989) *Paul et Virginie*, Harmondsworth: Penguin.

Berrill, K. (1992) 'Anti-gay violence and victimisation in the United States: an overview', in G.M. Herek and K.T. Berrill (eds) *Hate Crimes: Confronting violence against lesbians and gay men*, London: Sage.

Betty and Pansy (1993) *Betty and Pansy's Severe Queer Review of San Francisco*, San Francisco: Bedpan Productions.

Beyer, J. (1992) 'Sexual minorities and geography', paper presented at 27th International Geographical Congress, Washington, DC.

Bhabha, H.K. (1990) 'The third space', in J. Rutherford (ed.) *Identity: Community, culture, difference*, London: Lawrence and Wishart.

Bianchini, F. and Parkinson, M. (eds) (1993) *Cultural Policy and Urban Regeneration: The West European experience*, Manchester: Manchester University Press.

Bibbings, L. and Alldridge, P. (1993) 'Sexual expression, body alteration, and the defence of consent', *Journal of Law and Society* 20: 356–70.

Billings, D. and Urban, T. (1982) 'The socio-medical construction of transsexualism: an interpretation and critique', *Social Problems* 29: 266–82.

Binnie, J. (1992a) 'Fucking among the ruins: postmodern sex in postindustrial places', paper presented at Sexuality and Space Network Conference on Lesbian and Gay Geographies?, University College London.

—— (1992b) 'An international cock-tail party: the spatiality of fetishism', paper presented at Sexuality and Space Network Conference on Lesbian and Gay Geographies?, University College London.

—— (1993a) 'Invisible cities/hidden geographies: sexuality and the city', paper presented at Social Policy and the City Conference, University of Liverpool, July.

—— (1993b) 'Everybody in the house of love', paper presented at the Place of Music Conference, University College London, September.

—— (1993c) 'Notes towards a political geography of sexuality', paper presented at the New Theoretical Directions in Political Geography Conference, University of Birmingham, UK, September.

—— (1994) 'Coming out of geography: notes towards a queer epistemology', unpublished paper.

Bird, T., Curtis, B., Putnam, T., Robertson, G. and Tickner, L. (eds) (1993) *Mapping the Futures: Local cultures, global change*, London: Routledge.

Birke, L. (1991) ' "Life" as we have known it: feminism and the biology of gender', in M. Benjamin (ed.) *Science and Sensibility: Gender and scientific enquiry, 1780–1945*, Oxford: Blackwell.

Bix, B. (1993) 'Assault, sado-masochism and consent', *Law Quarterly Review* 109: 540–4.

Blankley, E. (1984) 'Return to Mytilene: Renée Vivien and the city of women', in S.M. Squier (ed.) *Women Writers and the City*, Knoxville: University of Tennessee Press.

Boal, F.W. (1976) 'Ethnic residential segregation', in D.T. Herbert and R.J. Johnston (eds) *Spatial Processes and Form*, London: Wiley.

Bock, G. and James, S. (eds) (1992) *Beyond Equality and Difference: Citizenship, feminist politics and female subjectivity*, London: Routledge.

Boffin, T. and Gupta, S. (eds) (1990) *Ecstatic Antibodies: Resisting the AIDS mythology*, London: Rivers Oram Press.

Bolin, A. (1988) *In Search of Eve: Transsexual rites of passage*, Massachusetts: Bergin and Garvey.

Bondi, L. (1991a) 'Gender divisions and gentrification: a critique', *Transactions of the Institute of British Geographers* NS 16: 190–8.

—— (1991b) 'Women, gender relations and the "inner city" ', in M. Keith and A. Rogers (eds) *Hollow Promises: Rhetoric and reality in the inner city*, New York: Mansell.

—— (1992a) 'Gender and dichotomy', *Progress in Human Geography* 16, 1: 98–104.

—— (1992b) 'Gender symbols and the urban landscape', *Progress in Human Geography* 16: 157–70.

—— (1992c) 'Sexing the city', presented at annual meetings of the Association of American Geographers, San Diego, USA.

—— (1993) 'Gender and geography: crossing boundaries', *Progress in Human Geography* 17, 2: 241–6.

Borchert, J.R. (1987) *America's Northern Heartland: An economic and historical geography of the Upper Midwest*, Minneapolis: University of Minnesota Press.

Bordo, S. (1991) ' "Material Girl": the effacements of postmodern culture', in L. Goldstein (ed.) *The Female Body: Figures, styles, speculations*, Ann Arbor: University of Michigan Press.

—— (1992) 'Postmodern subjects, postmodern bodies', *Feminist Studies* 18: 159–75.

—— (1993) *Unbearable Weight: Feminism, western culture and the body*, London: University of California Press.

Bourdieu, P. (1989) *In Other Worlds: Essays in reflexive sociology*, Cambridge: Polity Press.

—— (1991) *Language and Symbolic Power*, Cambridge: Polity Press.

Boyz (1993a) 'Euroboyz', 19 June, special issue.

—— (1993b) 'Boyz Soho Special', 22 May, special issue.

Bozzoli, B. (1983) 'Marxism, feminism and South African studies', *Journal of Southern African Studies* 9: 139–71.

Bradby, B. (1993) 'Lesbians and popular music: does it matter who is singing?', in G. Griffen (ed.) *Outwrite: Lesbianism and popular culture*, London: Pluto Press.

Bradshaw, P. (1992) 'Maria always knew she was a woman . . .', *Evening Standard* 10 February: 10.

Bravmann, S. (1994) 'The lesbian and gay past: it's all Greek to whom?', *Gender, Place and Culture*, 1:149–67.

Bright, S. (1992) *Susie Bright's Sexual Reality: A virtual sex world reader*, Pittsburgh: Cleis Press.

Bristow, J. (1991) *Empire Boys*, London: HarperCollins.

—— (1992) *Sexual Sameness: Textual differences in lesbian and gay writing*, London: Routledge.

Bristow, A. and Pearn, P. (1984) 'Comment on Krieger's "Lesbian identity and community: recent social science literature" ', *Signs* 9: 729–32.

Bristow, J. and Wilson, A. (eds) (1993) *Activating Theory: Lesbian, gay, bisexual politics*, London: Lawrence and Wishart.

Brown, M. (1992) ' "Queer politics" reinforces ghettoisation', letter to *Gay Times* October: 30.

Brown, M.P. (1994) 'The work of city politics: citizenship through employment in the local response to AIDS', *Environment and Planning A* 26, 6: 873–94.

—— (1995) 'Ironies of distance: an ongoing critique of the geographies of AIDS', *Environment and Planning D: Society and Space* 13 (forthcoming).

Brown, R. (1977) *Rubyfruit Jungle*, New York: Bantam Books.

Budd, L. and Whimster, S. (1992) *Global Finance and Urban Living: A study of metropolitan change*, London: Routledge.

Bunton, R. and MacDonald, G. (eds) (1992) *Health Promotion: Disciplines and diversity*, London: Routledge.

Burston, P. (1993a) 'Batties bite back', *The Guardian Weekend* 20 November: 34–9.

—— (1993b) 'Homophobia grips Brixton', *Time Out* 15 September: 13.

Burton, Sir R. (1970) 'Terminal essay [to *The Arabian Nights*]', in B. Reade (ed.) *Sexual Heretics: Male homosexuality in English literature from 1850 to 1900*, London: Routledge and Kegan Paul.

Butler, J. (1989) 'Gendering the body: Beauvoir's philosophical contribution', in A. Garry and M. Pearsell (eds) *Women, Knowledge and Reality*, London: Unwin Hyman.

—— (1990) *Gender Trouble: Feminism and the subversion of identity*, New York: Routledge.

—— (1991) 'Imitation and gender insubordination', in D. Fuss (ed.) *Inside/Out: Gay theories, lesbian theories*, New York: Routledge.

—— (1993) *Bodies That Matter: On the discursive limits of 'sex'*, New York: Routledge.

Butler, J. and Scott, J. (eds) (1992) *Feminists Theorize the Political*, New York: Routledge.

Califia, P. (1988) *Macho Sluts*, Boston: Allyson.

—— (1993a) *Melting Point*, Boston: Allyson.

—— (1993b) '*Sex* and Madonna, or, what do you expect from a girl who doesn't put out on the first five dates?', in L. Frank and P. Smith (eds) *Madonnarama: Essays on sex and popular culture*, Pittsburgh: Cleis Press.

Cameron, E. (1993) 'Sexual orientation and the constitution: a test case for human rights', *Lawyers for Human Rights* 1: 32–42.

Campbell, D. (1993a) 'Gay myths, criminal realities' (part 1), *Gay Scotland* issue 71: 9–10, 20.

—— (1993b) 'Gay myths, criminal realities' (part 2), *Gay Scotland* issue 72: 9–10, 12, 25.

Cantarella, E. (1992) *Bisexuality in the Ancient World*, New Haven and London: Yale University Press.

Capital Gay (1993a) 'Shabba Ranks: "Gay bashing is wrong" ', 26 March: 4.

—— (1993b) 'OutRage! to quiz Beeb chiefs', 23 April: 3.

—— (1993c) 'Schoolkids throw rocks at OutRage!', 18 June: 9.

—— (1993d) Letters page, 16 July: 2.

—— (1993e) 'Ad campaign drops Marky Mark', 3 September: 11.

—— (1993f) 'Attacks "on the rise" in South London', 10 September: 1.

—— (1993g) 'Readers report spate of vicious attacks', 17 September: 1.

Carr, D. (1993) *Feminine Genders*, Sydney: unpublished film script.

Carter, E., Donald, J. and Squires, J. (eds) (1993) *Space and Place: Theories of identity and location*, London: Lawrence and Wishart.

Case, S.E. (1989) 'Toward a butch-femme aesthetic', *Discourse* 11, 55–73.

Castel, P. (1994) 'A day in the life of a mistress', *Wicked Women* 2, 9: 9–11.

Castells, M. (1983) *The City and the Grassroots*, Berkeley: University of California Press.

Castells, M. and Murphy, K. (1982) 'Cultural identity and urban structure: the spatial organization of San Francisco's gay community', in N. Fainstein and S. Fainstein (eds), *Urban Policy under Capitalism*, Beverly Hills: Sage.

Castle, A. (1993) 'Word is out on Ranks' "crucify" jibe', *The Pink Paper* 26 March: 3.

Cather, W. (1983a) *My Antonia*, London: Virago.

—— (1983b) *O Pioneers!* London: Virago.

Chapman, R. and Rutherford, J. (eds) (1988) *Male Order: Unwrapping masculinity*, London: Lawrence and Wishart.

(charles), H. (1993) 'A homogenous habit? Heterosexual display in the English holiday camp', in S. Wilkinson and C. Kitzinger (eds) *Heterosexuality*, London: Sage.

Cheney, J. (ed.) (1985) *Lesbian Land*, Minnesota: Word Weavers.

Child, R. (1993) 'The economics of homosexuality', in K. Waaldijk and A. Clapham (eds) *Homosexuality: A European Community issue*, Dordrecht: Martinus Nijhoff.

City of Vancouver (1988) *Downtown South: Towards a new neighborhood*, Planning Department, City of Vancouver.

Cixous, H. (1975) 'Sorties: out and out: attacks/ways out/forays', in H. Cixous and C. Clement (eds) *The Newly Born Woman*, trans. B. Wing, introduction by S. Gilbert, Manchester: Manchester University Press.

Clarke, S. and Kirby, A. (1989) 'In search of the corpse: the mysterious case of local politics', *Urban Affairs Quarterly* 25: 389–412.

Clegg, S.R. (1989) *Frameworks of Power*, London: Sage.

Cockburn, C. (1983) *Brothers: Male dominance and technological change*, London: Pluto Press.

Cohen, A. (1982) 'A polyethnic London carnival as a contested cultural performance', *Ethnic and Racial Studies* 5: 23–41.

Cohen, A.P. (1985) *The Symbolic Construction of Community*, London: Routledge.

Cohen, J. and Arato, A. (1992) *Civil Society and Political Theory*, Cambridge, MA: MIT Press.

Coleman, E. (1987) 'Developmental stages of the coming-out process', in J. Gonsiorek (ed.) *Homosexuality and Psychotherapy: A practitioner's handbook of affirmative models*, New York: Haworth.

Collins, P. (1991) *Black Feminist Thought*, London: Routledge.

Colomina, B. (ed.) (1992) *Sexuality and Space*, Princeton: Princeton Architectural Press.

Comstock, G.D. (1989) 'Victims of anti-gay/lesbian violence', *Journal of Interpersonal Violence* 4: 101–6.

Connell, R. (1987) *Gender and Power*, Cambridge, MA: Polity Press.

Connell, R. and Dowsett, G. (1993) ' "The Unclean Motion of the Generative Parts": frameworks in Western thought on sexuality', in R. Connell and G. Dowsett (eds) *Rethinking Sex: Social theory and sexuality research*, Philadelphia: Temple University Press.

Cooper, C. (1992) *Noises in the Blood: Orality and the 'vulgar' body of Jamaican popular culture*, Basingstoke: Macmillan Caribbean.

Cooper, D. (1993) 'An engaged state: sexuality, governance, and the potential for change', *Journal of Law and Society* 20: 257–75.

—— (1994) *Sexing the City: Lesbian and gay politics within the activist state*, London: Rivers Oram.

Cornwall, A. and Lindisfarne, N. (eds) (1994) *Dislocating Masculinity: Comparative ethnographies*, London: Routledge.

Corzine, J. and Kirby, R. (1977) 'Cruising the truckers: sexual encounters in a highway rest area', *Urban Life* 6, 2: 171–92.

Cottrell, M. (1992) 'St Patrick's Day Parades in nineteenth century Toronto: a study of immigrant adjustment and elite control', *Histoire Sociale – Social History* 25: 57–73.

Coward, R. (1984) *Female Desire: Women's sexuality today*, London: Granada.

Cox, K. (1991) 'Questions of abstraction in studies in the new urban politics', *Journal of Urban Affairs* 13: 267–80.

—— (1993) 'The global and the local in the new urban politics', *Environment and Planning D: Society and Space* 11: 433–48.

Cox, K. and Mair, A. (1988) 'Locality and community in the politics of local economic development', *Annals of the Association of American Geographers* 78: 307–25.

Craig, S. (ed.) (1992) *Men, Masculinity and the Media*, London: Sage.

Cream, J. (1993) 'Child sexual abuse and the symbolic geographies of Cleveland', *Environment and Planning D: Society and Space* 11: 231–46.

Creet, J. (1991) 'Daughter of the movement: the psychodynamics of lesbian SM fantasy', *Differences: A Journal of Feminist Cultural Studies*, 3: 135–59.

Cresswell, T. (1994) 'The geography of transgressions: its limits and uses', paper presented at the New Theoretical Directions in Social and Cultural Geography Conference, Manchester, February.

Crimp, D. (1987a) 'AIDS: cultural analysis/cultural activism', *October* 43: 3–16.

—— (1987b) 'How to have promiscuity in an epidemic', *October* 43: 237–71.

Crimp, D. and Rolston, A. (1990) *AIDS Demo Graphics*, Seattle: Bay Press.

Cruikshank, M. (1992) *The Gay and Lesbian Liberation Movement*, London: Routledge.

Cusick, J. (1993) 'Macho man music puts gays in fear for their lives', *The Independent* 4 October: 3.

Czyselska, J. (1993) 'The shabbiness of Shabba', *All Points North* June: 34–5.

D'Augelli, A.R. (1989) 'Lesbians' and gay men's experiences of discrimination and harassment in a university community', *American Journal of Community Psychology*, 17: 317–21.

D'Augelli, A.R. and Hart, M. (1987) 'Gay women, men and families in rural settings: towards the development of helping communities', *American Journal of Community Psychology* 15: 70–93.

D'Augelli, A.R., Collins, C. and Hart, M. (1987) 'Social support patterns of lesbian women in a rural helping network', *Journal of Rural Community Psychology*, 15, 1: 79–93.

D'Emilio, J. (1981) 'Gay politics, gay communities: the San Francisco experience', *Socialist Review* 55: 77–104.

—— (1983) *Sexual Politics, Sexual Communities: The making of a homosexual minority in the United States, 1940–1970*, Chicago: University of Chicago Press.

—— (1989) 'Not a simple matter: gay history and gay historians', *Journal of American History* 76, 2: 435–42.

—— (1992) *Making Trouble: Essays on gay history, politics, and the university*, New York: Routledge.

Dann, G. (1987) *The Barbadian Male: Sexual attitudes and practice*, Basingstoke: Macmillan Caribbean.

Daumer, E. (1992) 'Queer ethics, or the challenge of bisexuality to lesbian ethics', *Hypatia* 7: 91–105.

Davis, S. (1986) *Parades and Power: Street theater in 19th century Philadelphia*, Berkeley: University of California Press.

Davis, T. (1987) 'Gay gentrification in the Central [Minneapolis] neighbourhood', unpublished research paper, Macalester College.

—— (1991) ' "Success" and the gay community: reconceptualizations of space and urban social movements', paper presented at the First Annual National Graduate Student Conference on Lesbian and Gay Studies, Milwaukee.

—— (1992) 'Where should we go from here? Towards an understanding of gay and lesbian communities', paper presented at the 27th International Geographical Congress, Washington DC, August.

Defoe, D. (1719) *Robinson Crusoe*, London: Thomas Nelson.

De Jongh, J.L. (1990) 'The image of black Harlem in literature', in C. Mulvey and J. Simons (eds) *New York: City as text*, Basingstoke: Macmillan.

De Lauretis, T. (1991) 'Queer theory: lesbian and gay sexualities: an introduction', *Differences: A Journal of Feminist Cultural Studies* 3: 1–10.

DeLeon, R. (1992) *Left Coast City: Progressive politics in San Francisco 1975–1991*, Lawrence: University of Kansas Press.

De Lynn, J. (1990) *Don Juan in the Village*, New York: Pantheon.

De Monteflores, C. and Schultz, S.J. (1978) 'Coming out: similarities and differences for lesbian and gay men', *Journal of Social Issues* 34, 3: 59–72.

Denzin, N. (1990) 'Reading "Wall Street": postmodern contradictions in the American social structure', in B. Turner (ed.) *Theories of Modernity and Postmodernity*, London: Sage, pp. 31–44.

De Vito, J.A. (1981) 'Educational responsibilities to gay male and lesbian students', in J.W. Chesebro (ed.) *Gayspeak: Gay male and lesbian communication*, New York: Pilgrim Press.

Doane, M. (1990) 'Technophilia: technology, representation, and the feminine', in M. Jacobus, E. Fox Keller and S. Shuttleworth (eds) *Body/Politics: Women and the discourse of science*, London: Routledge.

Doll, L.S. (1992) 'Homosexually and nonhomosexually identified men who have sex with men: a behavioural comparison', *Journal of Sex Research* 29, 1: 1–14.

Dollimore, J. (1991) *Sexual Dissidence: Augustine to Wilde, Freud to Foucault*, Oxford: Clarendon Press.

Dooley, L. (1985) 'A study of the bedroom as an area of personal space', Occasional Paper No. 1, University of Waikato, New Zealand.

Dorn, M. and Laws, G. (1994) 'Social theory, body politics, and medical geography: extending Kearns's invitation', *Professional Geographer* 46: 106–10.

Douglas, M. (1973) *Natural Symbols*, London: Barrie and Jenkins.

—— (1978) *Purity and Danger*, London: Routledge.

Dreuilhe, E. (1987) *Mortal Embrace: Living with AIDS*, London: Faber and Faber.

Dromgoole, W.A. (1913) *The Island of Beautiful Things*, London: Sir Isaac Pitman.

Dube, G. and Smailes, A. (1992) *AIDS and Human Rights: Sharing the challenge*, Vancouver: Vancouver World AIDS Group.

Duberman, M., Vicinus, M. and Chauncey, G. (1989) *Hidden from History: Reclaiming the gay and lesbian past*, New York: NAL Books.

Duden, B. (trans. T. Dunlap) (1991) *The Woman Beneath the Skin: A doctor's patients in eighteenth century Germany*, Cambridge, MA: Harvard University Press.

Duncan, J.S. (1981) 'Introduction', in J.S. Duncan (ed.) *Housing and Identity*, London: Croom Helm.

Duyves, M. (1992a) 'The inner-city of Amsterdam: gay show-place of Europe?', paper presented at Forum on Sexuality Conference, Sexual Cultures in Europe, Amsterdam, June.

—— (1992b) 'In de ban van de bak: openbaar ruimtegebruik naar homoseksuele voorkeur in Amsterdam', in J. Burgers (ed.) *De Uitstaad: Over stedelijk vermaak*, Utrecht: Uitgeverij Jan van Arkel.

—— (1993) 'The sexual and the spatial – framing preferences', paper presented at Geographies of Desire Conference, Amsterdam, June.

Dyer, R. (1992) *Only Entertainment*, London: Routledge.

—— (1993) *The Matter of Images: Essays on representation*, London: Routledge.

Eadie, J. (1992) 'The motley crew: what's at stake in the production of bisexual identity', paper presented at the Sexuality and Space Network Conference, Lesbian and Gay Geographies?, September.

—— (1993a) 'Activating bisexuality: towards a bi/sexual politics', in J. Bristow and A.R. Wilson (eds) *Activating Theory: Lesbian, gay, bisexual politics*, London: Lawrence and Wishart.

—— (1993b) 'We should be there bi now', *Rouge* 12: 26–7.

Echols, A. (1985) 'The taming of the id: feminist sexual politics, 1968–83', in C. Vance (ed.) *Pleasure and Danger: Exploring female sexuality*, Boston: Routledge and Kegan Paul.

Edelman, L. (1994) *Homographesis: Essays in gay literary and cultural theory*, New York: Routledge.

Egerton, J. (1990) 'Out but not down: lesbians' experiences of housing', *Feminist Review*, 35: 75–88.

Ehrenreich, B. and English, D. (1990) *For Her Own Good: 150 years of expert advice for women*, London: Pluto Press.

Eisenstein, H. (1980) 'Preface', in H. Eisenstein and A. Jardine (eds) *The Future of Difference*, New Brunswick and London: Rutger's University Press.

—— (1984) 'Sexual politics revisited: pornography and sadomasochism', *Contemporary Feminist Thought*, London: Allen and Unwin.

Eisenstein, Z. (1988) *The Female Body and the Law*, Berkeley and London: University of California Press.

Elkin, S. (1987) *City and Regime in the American Republic*, Chicago: University of Chicago Press.

Ellis, J. (1993) 'Where you could walk . . . The Orange Parade as popular culture', paper presented at the Seventh Annual Graduate Irish Studies Conference, Harvard University and Boston College, March.

Ellison, R. (1982 [1952]) *Invisible Man*, Harmondsworth: Penguin.

Emberley, J. (1988) 'The fashion apparatus and the deconstruction of postmodern subjectivity', in A. Kroker and M. Kroker (eds) *Body Invaders*, Basingstoke: Macmillan.

England, K. (1991) 'Gender relations and the spatial structure of the city', *Geoforum* 22: 135–47.

—— (1994) 'Getting personal: reflexivity, positionality and feminist research', *Professional Geographer* 46, 1: 80–9.

Epstein, J. (1990) 'Either/or – neither/both: sexual ambiguity and the ideology of gender', *Genders* 7: 99–142.

Epstein, J. and Straub, K. (eds) (1991) *Body Guards: The cultural politics of gender ambiguity*, New York: Routledge.

Epstein, S. (1987) 'Gay politics, ethnic identity: the limits of social constructionism', *Socialist Review* 17: 9–54.

Escoffier, J. (1985) 'Sexual revolution and the politics of gay identity', *Socialist Review* 15: 119–53.

Eshun, K. (1993) 'Ragga on trial', *i-D* August: 26–7.

Ettorre, E.M. (1978) 'Women, urban social movements and the lesbian ghetto', *International Journal of Urban and Regional Research* 2: 499–520.

Evans, D.T. (1993) *Sexual Citizenship: The material construction of sexualities*, London: Routledge.

Ewles, L. and Simmett, I. (1992) *Promoting Health: A practical guide*, London: Scutari Press.

Eyles, J. (1984) *Senses of Place*, Warrington: Silverbrook Press.

Faderman, L. (1992) *Odd Girls and Twilight Lovers: A history of lesbian life in twentieth century America*, Harmondsworth: Penguin.

Fallon, M. (1989) *Working Hot*, Melbourne: Sybella.

Farshea, K. (1993) 'Operation Spanner', paper presented at the 11th National Bisexual Conference, Nottingham, October.

Farwell, M. (1990) 'Heterosexual plots and lesbian subtexts: toward a theory of lesbian narrative space', in K. Jay and J. Glasgow (eds) *Lesbian Texts and Contexts: Radical revisions*, New York: New York University Press.

Fay, B. (1993) 'From hidden Ireland – Irish lesbians and gay men from the Celtic period to modern times – a celebration and lamentation', paper presented at the Seventh Annual Graduate Irish Studies Conference, Harvard University and Boston College, March.

Feinberg, L. (1993) *Stone Butch Blues*, Ithaca, NY: Firebrand Books.

Ferguson, A. *et al.* (1984) 'Forum: the feminist sexuality debates', in E.B. Freedman and B. Thorne (eds) 'Introduction to the feminist sexuality debates', *Signs* 10: 106–35.

Fernandez, C. (1991) 'Undocumented aliens in the Queer Nation', *Out/Look*. 12: 20–3.

Fiedler, L.A. (1960) *Love and Death in the American Novel*, New York: Criterion.

Fielding, H. (1992) 'If you go down on the Heath today . . .', *Sunday Times* Style and Travel section, 15 March: 1–2.

Finn, K. and O'Connor, D. (1994) 'It's simple justice', *Boston Globe* 19 March: 18.

Fiske, J. (1993) *Power Plays Power Works*, Verso: London.

Fitzgerald, F. (1986) *Cities on a Hill*, London: Picador.

Ford, R. (1993) 'Soho woos the pink pound', *The Times* 18 June: 7.

Forrest, D. (1994) '"We're here, we're queer, and we're not going shopping": changing gay male identities in contemporary Britain,' in A. Cornwall and N. Lindisfarne (eds) *Dislocating Masculinity: Comparative ethnographies*, London: Routledge.

Foster, S. (1992) 'Dancing bodies', in J. Crary and S. Kwinter (eds) *Incorporations*, New York: Zone Books.

Foucault, M. (1977a) (orig. published 1975) *Discipline and Punish*, London: Allen Lane.

—— (1977b) 'The eye of power', in C. Gordon (ed.) *Power/Knowledge*, New York: Pantheon.

—— (1977c) 'Questions on geography', in C. Gordon (ed.) *Power/Knowledge*, Brighton: Harvester Press.

—— (trans. A. Sheridan) (1979) *Discipline and Punish*, New York: Vintage.

—— (1980) *Herculine Barbin*, Brighton: Harvester Press.

—— (1982) 'Afterword', in H. Dreyfuss and P. Rabinow (eds) *Michel Foucault: Beyond structuralism and hermeneutics*, Chicago: University of Chicago Press.

—— (1984) *The History of Sexuality. Vol. 1: An Introduction*, Harmondsworth: Penguin.

Fout, J. and Tantillo, M.S. (eds) (1993) *American Sexual Politics: Sex, gender, and race since the Civil War*, Chicago: University of Chicago Press.

Fox, N.J. (1993) *Postmodernism, Sociology and Health*, Milton Keynes: Open University Press.

Franzen, T. (1993) 'Differences and identities: feminism and the Albuquerque lesbian community', *Signs* 18: 891–906.

Frank, L. and Smith, P. (eds) (1993) *Madonnarama*, Pittsburgh: Clesis Press.

Fraser, N. and Boffin, T. (1991) *Stolen Glances: Lesbians take photographs*, London: Gay Men's Press.

Frye, M. (1983) *The Politics of Reality: Essays in feminist theory*, Trumansburg, NY: Crossing Press.

Fuss, D. (1989) *Essentially Speaking: Feminism, nature and difference*, New York: Routledge.

—— (ed.) (1991) *Inside/Out: Lesbian theories, gay theories*, London: Routledge.

Gadsby, P. (1988) 'Mapping the epidemic: geography as destiny', *Discovery* April: 28–31.

Gagnon, J. and Simon, W. (eds) (1967) *Sexual Deviance*, New York: Harper and Row.

Gallagher, C. and Laqueur, T. (1987) *The Making of the Modern Body*, Berkeley and London: University of California Press.

Gallop, J. (1992) *Around 1981: Feminist literary theory*, New York: Routledge.

Gange, J. and Johnstone, S. (1993) 'Believe me, everybody has something pierced in California: an interview with Nayland Blake', *New Formations* 19: 51–68.

Garber, M. (1992) *Vested Interests: Cross dressing and cultural anxiety*, London: Routledge.

Gay Times (1993) 'Threatened with boycott, Marky Mark regrets homophobia', April: 21.

Geltmaker, T. (1992) 'The queer nation acts up: health care, politics, and sexual diversity in the county of angels', *Environment and Planning D: Society and Space* 10: 609–50.

George, S. (1993) *Women and Bisexuality*, London: Scarlet Press.

Geurtsen, W. (1992) 'The pleasure of pain; sado-masochism as leisure', paper presented at the Leisure Studies Association Conference, Bodily Matters, Sheffield, September.

Gever, M. (1990) 'The names we give ourselves', in R. Ferguson, M. Gever, T.M. Trinh and C. West (eds) *Out There: Marginalization and contemporary cultures*, New York: New Museum of Contemporary Art.

Gever, M., Greyson, J. and Parmar, P. (eds) (1993) *Queer Looks: Perspectives on lesbian and gay film and video*, New York: Routledge.

Giddens, A. (1979) *Central Problems in Social Theory*, Berkeley: University of California Press.

——— (1984) *The Constitution of Society*, Cambridge: Polity Press.

——— (1992) *The Transformation of Intimacy: Sexuality, love and eroticism in modern societies*, Cambridge: Polity Press.

Golding, S. (1993a) 'The excess: an added remark on sex, rubber, ethics, and other impurities', *New Formations* 19: 23–8.

——— (1993b) 'Quantum philosophy, impossible geographies and a few small points about life, liberty and the pursuit of sex (all in the name of democracy)', in M. Keith and S. Pile (eds) *Place and the Politics of Identity*, London: Routledge.

——— (1993c) 'Sexual manners', in V. Harwood, D. Oswell, K. Parkinson and A. Ward (eds) *Pleasure Principles: Politics, sexuality and ethics*, London: Lawrence and Wishart.

Golding, W. (1960) *Lord of the Flies*, Harmondsworth: Penguin.

Goldsby, J. (1993) 'Queen for 307 days: looking b(l)ack at Vanessa Williams and the sex wars', in A. Stein (ed.) *Sisters, Sexperts, Queers: Beyond the lesbian nation*, New York: Plume.

Goldsmith, M. and Wolman, H. (1992) *Urban Politics and Policy: A comparative approach*, Oxford: Blackwell.

Goodman, G., Lakey, G., Lashof, J. and Thorne, E. (1983) *No Turning Back: Lesbian and gay liberation for the '80s*, Philadelphia: New Society Publishers.

Gottdiener, M. (1985) *The Social Production of Urban Space*, Austin: University of Texas Press.

——— (1987) *The Decline of Urban Politics*, Newbury Park, CA: Sage.

Gould, P. (1989) 'Geographic dimensions of the AIDS epidemic', *Professional Geographer* 41, 1: 71–7.

——— (1993) *The Slow Plague: A geography of the AIDS pandemic*, Oxford: Blackwell.

Grebinoski, J. (1993) 'Out north: gays and lesbians in the Duluth, Minnesota – Superior, Wisconsin area', presented at annual meeting of the Association of American Geographers, April (Atlanta).

Griffin, G. (ed.) (1993) *Outwrite: Lesbianism and popular culture*, London: Pluto Press.

Gross, H.E. and Merritt, S. (1981) 'Effect of social/organisation context on gatekeeping in lifestyle pages', *Journalism Quarterly* 58, 3: 420–7.

Grosz, E. (1990) 'Inscriptions and bodymaps: representations and the corporeal', in T. Threadgold and A. Grany-Francis (eds) *Feminine, Masculine and Representation*, Sydney: Allen and Unwin.

——— (1991) 'Introduction', *Hypatia* 6: 1–3.

—— (1992) 'Bodies-cities', in B. Colomina (ed.) *Sexuality and Space*, New York: Princeton Architectural Press.

—— (1993) 'Bodies and knowledges: feminism and the crisis of reason', in L. Alcoff and E. Potter (eds) *Feminist Epistemologies*, New York: Routledge.

Grover, J. (1992) 'AIDS, keywords and cultural work', in L. Grossberg, C. Nelson and P. Treichler (eds) *Cultural Studies*, New York: Routledge.

Gutstein, D. (1990) *The New Landlords: Asian investment in Canadian real estate*, Victoria, BC: Porcepic Books.

Hall, A. (1992) 'Abuse in lesbian relationships', *Trouble and Strife* 23: 12–15.

Hallam, P. (1993) *The Book of Sodom*, London: Verso.

Hamblim, A. (1983) 'Is a feminist heterosexuality possible?', in S. Cartledge and J. Ryan (eds) *Sex and Love: New thoughts on old contradictions*, London: Women's Press.

Hamer, D. (1990) ' "I am a woman": Ann Bannon and the writing of lesbian identity in the 1950s', in M. Lilly (ed.) *Lesbian and Gay Writing*, Basingstoke: Macmillan.

Hanke, R. (1992) 'Redesigning men: hegemonic masculinity in transition', in S. Craig (ed.) *Men, Masculinity and the Media*, London: Sage.

Haraway, D. (1991) *Simians, Cyborgs, and Women: The reinvention of nature*, London: Free Association Books.

Harper, P. (1993) 'Eloquence and epitaph: black nationalism and the homophobic impulse in reponses to the death of Max Robinson', in T. Murphey and S. Poirier (eds) *Writing AIDS: Gay literature, language and analysis*, New York: Columbia University Press.

Harry, J. (1974) 'Urbanization and the gay life', *Journal of Sex Research* 10, 3: 238–47.

—— (1992) 'Conceptualising anti-gay violence', in G.M. Herek and K.T. Berrill (eds) *Hate Crimes: Confronting violence against lesbians and gay men*, London: Sage.

Harvey, D. (1985) *The Urbanization of Capital*, Baltimore: Johns Hopkins University Press.

—— (1989) *The Condition of Postmodernity*, Baltimore: Johns Hopkins University Press.

—— (1992) 'Social justice, postmodernism and the city', *International Journal of Urban and Regional Research* 16: 588–601.

—— (1993) 'From space to place and back again: reflections on the condition of postmodernity', in J. Bird, B. Curtis, T. Putnam, G. Robertson and L. Tickner (eds) *Mapping the Futures: Local cultures, global change*, London: Routledge.

Harwood, V., Oswell, D., Parkinson, K. and Ward, A. (eds) (1993) *Pleasure Principles: Politics, sexuality and ethics*, London: Lawrence and Wishart.

Hausman, B. (1992) 'Demanding subjectivity: transsexualism, medicine and the technologies of gender', *Journal of the History of Sexuality* 3: 270–302.

Hay, R. (1991) 'Parallels between love relations and our relations with place', *Area* 23: 256–9.

—— (1992) 'Being politically correct or enquiring: a reply to Bell', *Area* 24: 411–12.

Haywood, L. *et al.* (1989) *Understanding Leisure*, Cheltenham: Thornes.

Hedley, S. (1993) 'Sado-masochism, human rights and the House of Lords', *Cambridge Law Journal* 52: 194–6.

Heidegger, M. (1972) *On Time and Being*, New York: Harper and Row.

Heim, M. (1992) 'The erotic ontology of cyberspace', in M. Benedikt (ed.) *Cyberspace: First steps,*

Cambridge, MA: MIT Press.

Hekma, G. (1992) 'The Amsterdam bar culture and changing gay/lesbian identities', paper presented at Forum on Sexuality Conference, Sexual Cultures in Europe, Amsterdam, June.

Hekma, G. *et al.* (1992) *De Roze Rand Van Donker Amsterdam: De opkomst van een homoseksuele kroegcultuur 1930–1970*, Amsterdam: Van Gennep.

Hemmings, C. (1993) 'Resituating the bisexual body: from identity to difference', in J. Bristow and A. Wilson (eds) *Activating Theory: Lesbian, gay, bisexual politics*, London: Lawrence and Wishart.

Herdt, G. (ed.) (1992) *Gay Culture in America*, Boston: Beacon Press.

Herzlich, C. and Pierret, J. (trans. E. Fraser) (1987) *Illness and Self in Society*, Baltimore and London: Johns Hopkins University Press.

Hill, D. (1993) 'Notes from the underground', *The Guardian* 16 March: 5.

Hinvest, C. (1993) 'Theories, practices, possibilities: safer-sex information for women', unpublished dissertation, University of Westminster, London.

Hochschild, A. (1983) *The Managed Heart: Commercialization of human feeling*, Berkeley: University of California Press.

Holcomb, B. (1986) 'Geography and urban women', *Urban Geography* 7: 448–56.

Hollibaugh, A. and Moran, C. (1992) 'What we're rollin' round in bed with: sexual silences in feminism: a conversation towards ending them', in J. Nestle (ed.) *A Persistent Desire*, Boston: Allyson.

Hollway, W. (1983) 'Heterosexual sex: power and desire for the Other', in S. Cartledge and J. Ryan (eds) *Sex and Love: New thoughts on old contradictions*, London: Women's Press.

hooks, b. (1984) *Feminist Theory: From margin to center*, Boston: Allyson.

—— (1989) *Talking Back: Thinking feminist – thinking black*, London: Sheba.

—— (1991) *Yearning*, London: Turnaround.

—— (1992) *Black Looks: Race and representation*, London: Turnaround.

Hope, S.J. (1992) 'From foreground to margin: female configurations and masculine representation in Black Nationalist fiction', in A. Parker, M. Russo, D. Sommer and P. Yaeger (eds) *Nationalisms and Sexualities*, New York: Routledge.

Hulme, P. (1986) *Colonial Encounters: Europe and the native Caribbean, 1492–1797*, London: Routledge.

Humphrey, R. (1992) 'Three Caribbean islands' interest in popularity of Caribbean music', *Caribbean Studies* 25: 123–32.

Humphries, L. (1970) *Tearoom Trade: Impersonal sex in public places*, Chicago: Aldine.

Hunt, L. (1992) 'Three sisters, one dilemma', *The Independent* 15 December: 12.

Hunt, P. and Frankenberg, R. (1981) 'Home: castle or cage?', in *An Introduction to Sociology*, Milton Keynes: Open University Press.

Hunt, P. and Milton, J. (1982) *South African Criminal Law and Procedure*, volume two, Cape Town: Juta and Co.

Huxley, A. (1976) *Island*, London: Granada.

Ingram, G. (1993) 'Queers in space: towards a theory of landscape, gender and sexual orientation', presented at Queer Sites Conference, University of Toronto.

Irigaray, L. (1981) 'And the one does not stir without the other', *Signs* 7: 60–7.

—— (1985a) *Speculum of the Other Woman*, Ithaca, NY: Cornell University Press.

—— (1985b) *This Sex Which is Not One*, Ithaca, NY: Cornell University Press.

Isaacs, G. and McKendrick, B. (1992) *Male Homosexuality in South Africa: Identity formation, culture, and crisis*, Cape Town: Oxford University Press.

Jackson, P. (ed.) (1987) *Race and Racism: Essays in social geography*, London: Allen and Unwin.

—— (1989) *Maps of Meaning*, London: Unwin Hyman.

—— (1991) 'The cultural politics of masculinity: towards a social geography', *Transactions of the Institute of British Geographers* NS 16: 199–213.

—— (1993) 'Changing ourselves: a geography of position', in R.J. Johnston (ed.) *The Challenge for Geography: A changing world, a changing discipline*, London: Blackwell.

—— (1994) 'Black male: advertising and the cultural politics of masculinity', *Gender, Place and Culture* 1, 1: 49–60.

Jackson, S. (1993) 'Even sociologists fall in love: an exploration in the sociology of emotions', *Sociology* 27: 201–20.

Jameson, F. (1984) 'Post-modernism, or the cultural logic of late capitalism', *New Left Review* 146: 53–92.

Jardine, A. (1980) 'Prelude: the future of difference', in H. Eisenstein and A. Jardine (eds) *The Future of Difference*, New Brunswick and London: Rutger's University Press.

Jarman, D. (1992) *Modern Nature: The journals of Derek Jarman*, London: Vintage.

Johnson, J. (1994) 'Politics and passion or pilsner and porn?', *Rouge* 16: 14.

Johnson, L. (1989) 'Embodying geography – some implicatons of considering the sexed body in space', *New Zealand Geographical Society Proceedings of the 15th New Zealand Geography Conference*, Dunedin: NZGS.

—— (1990) 'New patriarchal economies in the Australian textile industry', *Antipode* 22: 1–32.

—— (1992) 'Housing desire: a feminist geography of suburban housing', *Refractory Girl: A feminist journal* 42: 40–7.

—— (1993) 'Managing sexed bodies in an industrial landscape', unpublished paper.

—— (1994) 'What future for feminist geography?', *Gender, Place and Culture* 1, 1: 103–13.

Johnston, R.J., Gregory, D. and Smith, D.M. (eds) (1986) *The Dictionary of Human Geography*, Oxford: Blackwell.

Jonas, A. (1994) 'The scale politics of spatiality', *Environment and Planning D: Society and Space* 12: 257–64.

Jones, G. (1986) *Workers at Play*, London: Routledge and Kegan Paul.

Jones, H. (1993) 'Rewards for companies that go straight to the gay market', *The Independent* 26 September: 27.

Judd, D. and Swanstrom, T. (1994) *City Politics: Private power and public policy*, Toronto: HarperCollins.

Julien, I. (1992) 'Black is . . . black ain't: notes on de-essentializing black identities', in M. Wallace (ed.) *Black Popular Culture*, Seattle: Bay Press.

Juno, A. (1991) *Angry Women*, San Francisco: Re/Search.

Kaahumanu, L. and Hutchins, L. (1991) *Bi Any Other Name: Bisexual people speak out*, Boston: Allyson.

Kader, C. and Piontek, T. (1992) 'Introduction', *Discourse* 15: 5–10.

Kaloski, A. (1994) 'Following my thumbs: towards a methodology for a bisexual reading', unpublished University of York DPhil. thesis.

Karp, D.A., Stone, G.P. and Yoels, W.C. (1991) *Being Urban: A sociology of city life*, second edition, New York: Praeger.

Katz, C. (1992) 'All the world is staged: intellectuals and the project of ethnography', *Environment and Planning D: Society and Space* 10: 495–510.

Katz, J. (1976) *Gay American History*, New York: Thomas Crowell.

Kayal, P. (1993) *Bearing Witness: Gay Men's Health Crisis and the politics of AIDS*, Oxford: Westview.

Kearns, G. and Withers, C. (1991) *Urbanising Britain*, Cambridge: Cambridge University Press.

Keene, N. (1964) *Twice as Gay*, New York: After Hours Books.

Keith, M. (1992) 'Angry writing: (re)presenting the unethical world of the ethnographer', *Environment and Planning D: Society and Space* 10: 551–68.

Keith, M. and Pile, S. (eds) (1993) *Place and the Politics of Identity*, London: Routledge.

Kellogg, E. (ed.) (1983) 'Literary visions of homosexuality', *Journal of Homosexuality* 8, whole issue.

Kelly, J. (1993) 'City Hall column', *South Boston Tribune* 25 February: 8.

Kennedy, H. (1992) 'Gay group fails to reach deal on Parade', *Boston Herald* 3 May: 1, 28.

Kennedy, E.L. and Davis, M.D. (1992) ' "They was no-one to mess with": the construction of the butch role in the lesbian community of the 1940s and 1950s', in J. Nestle (ed.) *A Persistent Desire*, Boston: Allyson.

—— (1993) *Boots of Leather, Slippers of Gold: The history of a lesbian community*, New York: Routledge.

Keogh, P. (1992) 'Public sex: spaces, acts, identities', presented at Lesbian and Gay Geographies Conference, University College London.

Kerouac, J. (1972) *On the Road*, Harmondsworth: Penguin.

Kertzer, D. (1988) *Ritual, Politics, and Power*, New Haven: Yale University Press.

Kessler, S. (1990) 'The medical construction of gender: case management of intersexed infants', *Signs* 16: 3–26.

Ketteringham, W. (1979) 'Gay public space and the urban landscape: a preliminary assessment', presented at annual meeting of the Association of American Geographers.

—— (1983) 'The broadway corridor: gay businesses as agents of revitalization in Long Beach, California', presented at annual meeting of the Association of American Geographers, Denver.

King, D. (1992) *The Transvestite and the Transsexual: Public categories and private identities*, Aldershot: Avebury.

King, E. (1993) *Safety in Numbers*, London: Cassell.

King, K. (1992) 'Audre Lorde's lacquered layerings: the lesbian bar as a site of literary production', in S.R. Munt (ed.) *New Lesbian Criticism: Literary and cultural readings*, Hemel Hempstead: Harvester Wheatsheaf.

Kirby, A. (1990) 'And the bandwagon drove on: geographers and the AIDS pandemic', *Focus* 40, 10–14.

Kitzinger, C. and Wilkinson, S. (eds) (1993) *Heterosexuality: A Feminism and Psychology reader*, London: Sage.

Klein, M. (1988) 'Envy and gratitude', *Envy and Gratitude and Other Works 1946–1963*, London: Virago.

Knopp, L. (1987) 'Social theory, social movements and public policy: recent accomplishments of the gay and lesbian movements in Minneapolis, Minnesota', *International Journal of Urban and Regional Research* 11: 243–61.

—— (1990a) 'Some theoretical implications of gay involvement in an urban land market', *Political Geography Quarterly* 9: 337–52.

—— (1990b) 'Exploiting the rent-gap: the theoretical significance of using illegal appraisal schemes to encourage gentrification in New Orleans', *Urban Geography* 11: 48–64.

—— (1992) 'Sexuality and the spatial dynamics of capitalism', *Environment and Planning D: Society and Space*, 10: 651–69.

—— (1994) 'Rings, circles and perverted justice: gay judges and moral panic in contemporary Scotland', presented at annual meeting of the Association of American Geographers, San Francisco, April.

—— (forthcoming) 'Social justice, *sexuality*, and the city', *Urban Geography*.

Koenders, M. (1987) *Het Homomonument*, Amsterdam: Stichting Homomonument.

Kotz, L. (1992) ' "The body you want": an interview with Judith Butler', *Artforum* November: 82–9.

Krieger, S. (1982) 'Lesbian identity and community: recent social science literature', *Signs* 8: 91–108.

—— (1983) *The Mirror Dance: Identity in a women's community*, Philadelphia: Temple University Press.

Kroker, A. and Kroker, M. (eds) (1991) *The Hysterical Male: New feminist theory*, Basingstoke: Macmillan.

—— (eds) (1993) *The Last Sex: Feminism and outlaw bodies*, Basingstoke: Macmillan.

Kuhn, A. (1988) 'The body and cinema: some problems for feminism', in S. Sheridan (ed.) *Grafts*, London: Verso.

Lake, M. (1994) 'Between old world "barbarism" and stone age "primitivism": the double difference of the white Australian feminist', in N. Greive and A. Burns (eds) *Feminist Questions for the Nineties*, Oxford: Oxford University Press.

Laqueur, T. (1990) *Making Sex: Body and gender from the Greeks to Freud*, Cambridge, MA, and London: Harvard University Press.

Larsen, N.E. (1992) 'The politics of prostitution control: interest group politics in four Canadian cities', *International Journal of Urban and Regional Research* 16, 2: 169–89.

Lauria, M. and Knopp, L. (1985) 'Toward an analysis of the role of gay communities in the urban renaissance', *Urban Geography* 6: 152–69.

Lee, A. (1990) 'For the love of separatism', in J. Allen (ed.) *Lesbian Philosophies and Cultures*, New York: SUNY Press.

Lee, J.A. (1978) 'Meeting males by mail', in L. Crew (ed.) *The Gay Academic*, Palm Springs: ETC Publications.

Lee, W. (1993) 'Prostitution and tourism in South-East Asia', in N. Redclift and M.T. Sinclair (eds) *Working Women: International perspectives on labour and gender ideology*, London: Routledge.

Leech, K. (1993) 'Let us play', letter to *Time Out* 30 June–7 July: 170.

Lefebvre, H. (1991) *The Production of Space*, Oxford: Blackwell.

Leidner, R. (1991) 'Serving hamburgers and selling insurance', *Gender and Society* 5, 154–77.

—— (1993) *Fast Food, Fast Talk: The routinisation of everyday life*, Berkeley and London: University of California Press.

Levi, D. (1992) 'Fragmented nationalism: Jamaica since 1938', *History of European Ideas* 15: 413–17.

Levine, M. (1979a) 'Gay ghetto', *Journal of Homosexuality*, 4: 363–77.

—— (ed.) (1979b) *Gay Men: The sociology of male homosexuality*, New York: Harper and Row.

Lewis, M. (1989) *Liar's Poker: Two cities, true greed*, London: Hodder and Stoughton.

Lewis, M. (1992) 'A brief ethnographic study of the gay scene at Newcastle-upon-Tyne', paper presented at the Sexuality and Space Network Conference, Lesbian and Gay Geographies?, London, September.

Ley, D. (1994) 'Gentrification and the politics of the new middle class', *Environment and Planning D: Society and Space* 12: 53–74.

Ley, D. and Olds, K. (1988) 'Landscape as spectacle: world's fairs and the culture of heroic consumption', *Environment and Planning D: Society and Space* 6: 191–212.

Ley, D., Hiebert, D. and Pratt, G. (1992) 'Time to grow up? From urban village to world city, 1966–1991', in G. Wynn and T. Oke (eds) *Vancouver and its Region*, Vancouver: University of British Columbia Press.

Lieshout, M. van (1992) 'Leather nights in the woods: homosexual encounters in a Dutch highway rest area', paper presented at the Forum on Sexuality Conference, Sexual Cultures in Europe, Amsterdam, June.

Lilly, M. (1993) *Gay Men's Literature in the Twentieth Century*, Basingstoke: Macmillan.

Lindemann, G. (1992) 'The differences in the process of gender-change and the differences of gender', presented at the Forum on Sexuality Conference, Sexual Cultures in Europe, Amsterdam, June.

Linden, R., Pagano, O., Russell, D. and Star, S. (1982) *Against Sadomasochism: A radical feminist analysis*, East Palo Alto, CA: From Frog in the Wall.

Locke, A. (ed.) (1975) *The New Negro*, New York: Atheneum.

Lofland, L. (1973) *A World of Strangers*, New York: Basic Books.

Logan, J. and Molotch, H. (1987) *Urban Fortunes: The political economy of place*, Berkeley: University of California Press.

Longhurst, R. (1994) 'The strange case of the missing body in geography', paper presented at annual meeting of the Association of American Geographers, San Francisco, April.

Lorber, J. and Farrell, S. (eds) (1991) *The Social Construction of Gender*, Newbury Park, CA: Sage.

Lorde, A. (1984) *Sister Outsider: Essays and speeches*, California: Crossing Press.

Lugones, M.C. and Spelman, E. (1984) 'Have we got a theory for you! Feminist theory, cultural imperialism and the demand for "the woman's voice" ', *Women's Studies International Forum* 6: 573–81.

Lynch, F. (1987) 'Non-ghetto gays: a sociological study of suburban homosexuals', *Journal of Homosexuality* 13, 4: 13–42.

Lynch, L. (1990) 'The swashbuckler', in J. Nestle and N. Holoch (eds) *Women on Women: An anthology of America lesbian short fiction*, New York: Plume Books.

Lyod, B. and Rowntree, L. (1978) 'Radical feminists and gay men in San Francisco: social space in dispersed communities', in D. Lanegran and R. Palm (eds) *An Invitation to Geography*, New York: McGraw-Hill.

MacCowan, L. (1992) 'Re-collecting history, renaming lives: femme stigma and the feminist seventies and eighties', in J. Nestle (ed.) *The Persistent Desire: A femme-butch reader*, Boston: Allyson.

McDonald, G.J. (1982) 'Individual differences in the coming out process for gay men: implications for theoretical models', *Journal of Homosexuality* 8, 1: 47–60.

McDowell, L. (1983) 'City and home: urban housing and the sexual division of space', in M. Evans and C. Ungerson (eds) *Sexual Divisions: Patterns and processes*, London: Tavistock.

McDowell, L. (1990) 'Sex and power in academia', *Area* 22, 4: 323–32.

McDowell, L. and Court, G. (1994) 'Missing subjects: gender, power and sexuality in merchant banking', *Economic Geography* 70: 229–51.

McGrath, R. (1990) 'Dangerous liaisons: health, disease and representation', in T. Boffin and S. Gupta (eds) *Ecstatic Antibodies: Resisting the AIDS mythology*, London: Rivers Oram.

McNee, B. (1984) 'If you are squeamish . . .', *East Lakes Geographer* 19: 16–27.

—— (1985) 'It takes one to know one', *Transition* 14: 2–15.

Madigan, R., Munro, M. and Smith, S.J. (1990) 'Gender and the meaning of home', *International Journal of Urban and Regional Research* 14: 625–47.

Magnusson, W. (1992) 'The constitution of movements versus the constitution of the state: rediscovering the local as a site for global politics', in H. Lustiger-Thaler (ed.) *Political Arrangements: Power in the city*, Montreal: Black Rose.

Majors, R. and Billson, J. (1992) *Cool Prose: Dilemmas of black manhood in America*, New York: Lexington.

Malamud, B. (1983) *God's Grace*, Harmondsworth: Penguin.

Mann, L. (1993) 'Domestic violence within lesbian relationships', paper presented at the British Sociological Association, University of Essex.

Marcuse, P. (1986) 'Abandonment, gentrification, and displacement: the linkages in New York City', in N. Smith and P. Williams (eds) *Gentrification of the City*, Boston: Allen and Unwin.

Mark, R. (1992) 'Why am I whispering? Including sexuality in the multi-cultural dialogue', *Women's Studies* 20: 230–4.

Marston, S. (1989) 'Public rituals and community power: St Patrick's Day Parades in Lowell, Massachusetts 1841–1874', *Political Geography Quarterly* 8: 255–69.

Martin, B. (1992) 'Sexual practice and changing lesbian identities', in M. Barrett and A. Phillips (eds) *Destabilizing Theory: Contemporary feminist debates*, Cambridge, MA: Polity Press.

Martin, E. (1987) *The Woman in the Body: A cultural analysis of reproduction*, Milton Keynes: Open University Press.

—— (1992) 'Body narratives, body boundaries', in L. Grossberg, C. Nelson and P. Treichler (eds) *Cultural Studies*, New York: Routledge.

Mason, J. (1989) 'Reconstructing the public and the private: the home and marriage in later life', in G. Allan and G. Crow (eds) *Home and Family: Creating the domestic sphere*, Basingstoke: Macmillan.

Massey, D. (1991) 'Flexible sexism', *Environment and Planning D: Society and Space* 9: 31–57.

—— (1992) 'A place called home?' *New Formations* 17: 3–15.

—— (1993) 'Power-geometry and a progressive sense of place', in J. Bird, B. Curtis, T. Putnam, G. Robertson and L. Tickner (eds) *Mapping the Futures: Local cultures, global change*, London: Routledge.

—— (1994) *Space, Place and Gender*, Cambridge: Polity Press.

Mathew, P. (1988) 'On being a prostitute', *Journal of Homosexuality* 15: 119–35.

Mercer, K. (1990) 'Welcome to the jungle: identity and diversity in postmodern politics', in J. Rutherford (ed.) *Identity: Community, culture, difference*, London: Lawrence and Wishart.

—— (1992) 'Skin head sex thing: racial difference and the homoerotic imaginary', *New Formations* 16: 1–23.

Mercer, K. and Julien, I. (1988) 'Race, sexual politics and black masculinity: a dossier', in R. Chapman and J. Rutherford (eds) *Male Order: Unwrapping masculinity*, London: Lawrence and Wishart.

Metcalf, A. and Humphries, M. (eds) (1988) *The Sexuality of Men*, London: Pluto Press.

Meyerowitz, J. (1990) 'Sexual geography and gender economy: the furnished room districts of Chicago, 1890–1930', *Gender and History* 2: 274–96.

Middleton, P. (1992) *The Inward Gaze: Masculinity, subjectivity in modern culture*, London: Routledge.

Miles, P. (1992) 'Giving a platform to intolerance and hate' (letter), *Gay Times* November: 31.

Miller, D.A. (1992) *Bringing Out Roland Barthes*, Berkeley: University of California Press.

Miller, J. (1993) *The Passion of Michel Foucault*, London: HarperCollins.

Miller, N. (1989) *In Search of Gay America: Women and men in a time of change*, New York: Harper and Row.

Moers, E. (1977) *Literary Women: The great writers*, New York: Doubleday Anchor Press.

Mohlabe, (1970) 'Moral effects of the system of migratory labour on the labourer and his family', paper presented to the Consultation on Migrant Labour and Church Involvement, Natal, August.

Mohr, R. (1992) *Gay Ideas: Outing and other controversies*, Boston: Beacon Press.

Molnar, J.J. and Lawson, W.D. (1984) 'Perceptions of barriers to black political and economic progress in rural areas', *Rural Sociology* 49, 2: 261–83.

Moodie, T.D. (1983) 'Mine cultures and miners' identity on the South African gold mines', in B. Bozzoli (ed.) *Town and Countryside in the Transvaal*, Johannesburg: Raven Press.

—— (1986) 'The moral economy of the black miner's strike of 1946', *Journal of Southern African Studies* 13: 1–35.

—— (1988) 'Migrancy and male sexuality on the South Africa gold mines', *Journal of Southern African Studies* 14: 228–56.

Moran, L. (1993) 'Buggery and the tradition of law', *New Formations* 19: 110–24.

Morris, A. and McClary Mueller, C. (eds) (1992) *Frontiers in Social Movement Theory*, New Haven: Yale University Press.

Mort, F. (1987) *Dangerous Sexualities: medico-moral politics in England since 1830*, London: Routledge and Kegan Paul.

Moses, A.E. and Buckner, J.A. (1980) 'The special problems of rural gay clients', *Human Services in the Rural Environment* 5: 22–7.

Mosse, G. (1985) *Nationalism and Sexuality*, New York: Fertig.

Mullender, R. (1993) 'Sado-masochism, criminal law and adjudicative method: *R. v Brown* in the House of Lords', *Northern Ireland Legal Quarterly* 44: 380–7.

Mulvey, C. (1990) 'The black capital of the world', in C. Mulvey and J. Simons (eds) *New York: City as text*, Basingstoke: Macmillan.

Mulvey, C. and Simon, J. (eds) *New York: City as text*, Basingstoke: Macmillan.

Munt, S. (1992) 'Introduction', in S. Munt (ed.) *New Lesbian Criticism: Literary and cultural readings*, Hemel Hempstead: Harvester Wheatsheaf.

Murray, A. (1991) 'Kampung culture and radical chic in Jakarta', *Review of Indonesian and Malaysian Affairs* 25: 1–17.

——— (1993) 'City, subculture and sexuality: alternative spaces in Jakarta', *Development Bulletin* 27: 35–8.

Murray, S. (1979) 'The institutional elaboration of a quasi-ethnic community', *International Review of Modern Sociology* 9: 165–77.

Myers, P. (1993) 'New world order', *The Guardian* 12 January: 20.

Namaste, K. (1992) 'If you're in clothes, you're in drag', *Fuse* Fall: 7–9.

Nardi, P., Saunders, D.A. and Marmor, J. (1994) *Growing Up Before Stonewall*, New York: Routledge.

National Lesbian and Gay Survey (1993a) *Proust, Cole Porter, Michelangelo, Marc Almond and Me: Writings by gay men on their lives and lifestyles*, London: Routledge.

——— (1993b) *What a Lesbian Looks Like: Writings by lesbians on their lives and lifestyles*, London: Routledge.

Nelson, E. (1991) 'Critical deviance: homophobia and the reception of James Baldwin's fiction', *Journal of American Culture* 14: 91–6.

Nestle, J. (1987a) *A Restricted Country*, Ithaca: Firebrand.

——— (1987b) 'Lesbians and prostitutes: a historical sisterhood', in *Good Girls/Bad Girls*, Toronto: Women's Press.

——— (1988) *A Restricted Country*, London: Sheba Feminist Press.

——— (ed.) (1992) *The Persistent Desire: A femme-butch reader*, Boston: Allyson.

Newton, E. (1993a) 'My best informant's dress: the erotic equation in fieldwork', *Cultural Anthropology* 8: 2–23.

——— (1993b) *Cherry Grove, Fire Island*, Seattle: Bay Press.

New York Department of City Planning (1984) *Private Reinvestment and Neighborhood Change*.

NiCarthy, G. (1982) *Getting Free: A handbook for women in abusive situations*, London: Journeyman Press.

Noel, P. (1993) 'Batty Boys in Babylon: West Indian gay culture comes out in Brooklyn, and so does violence', *The Village Voice* 12 January: 29–36.

Norris, K. (1993) *Dakota: A spiritual geography*, New York: Ticknor and Fields.

Nye, S. (1994) 'Slow hand, easy touch', *The Pink Paper* 10 June: 28.

O'Flaherty, M.C. (1994) 'Wallpapering the watering hole', *The Pink Paper* 10 June: 17.

O'Hagan, S. (1993) 'On location', *The Guardian* 15 February: 4.

O'Sullivan, K. (1994) 'A mardi-gras omission', *Sydney Star Observer* 11 March: 40.

Oakely, A. (1976) *Housewife*, Harmondsworth: Penguin.

——— (1979) *Becoming a Mother*, London: Martin Robertson.

OffPink Collective (1988) *Bisexual Lives*, London: OffPink.

Ogborn, M. (1992) 'Love-state-ego: "centres" and "margins" in 19th century Britain', *Environment and Planning D: Society and Space* 10: 287–305.

Otey, A.M. (1991) 'Call a stop to violence: slope march against anti-gay acts', *The Phoenix* 14 March: 1, 5.

Padfield, N. (1992) 'Consent and the public interest', *New Law Journal* 142: 430–2.

Paglia, C. (1990) *Sexual Personae: Art and decadence from Nefertiti to Emily Dickinson*, New Haven and London: Yale University Press.

Paris, C. (1983) 'The myth of urban politics', *Environment and Planning D: Society and Space* 1: 89–108.

Park, R. (1928) 'Foreword', in L. Wirth (ed.) *The Ghetto*, Chicago: Chicago University Press.

Parker, A., Russo, M., Sommer, D. and Yaeger, P. (eds) (1992) *Nationalisms and Sexualities*, New York: Routledge.

Patton, C. (1985) *Sex and Germs*, Boston: South End Press.

—— (1990) *Inventing AIDS*, New York: Routledge.

—— (1993) 'Public enemy: fundamentalists in your face', *Voice Literary Supplement* February: 17–18.

Peacock, I. (1993) 'Comptomania!', *Boyz* 14 August: 7.

Peake, L. (1993) 'Race and sexuality: challenging the patriarchal structuring of urban social space', *Environment and Planning D: Society and Space* 11: 415–32.

Perkins, R. and Bennett, G. (1985) *Being a Prostitute*, Sydney: Allen and Unwin.

Perks, W. and Jamieson, W. (1991) 'Planning and development in Canadian cities', in T. Bunting and P. Filion (eds) *Canadian Cities in Transition*, Toronto: Oxford University Press.

Peterson, P. (1981) *City Limits*, Chicago: University of Chicago Press.

Pfohl, S. (1993) 'Venus in microsoft: male mas(s)ochism and cybernetics', in A. Kroker and M. Kroker (eds) *The Last Sex: Feminism and outlaw bodies*, Basingstoke: Macmillan.

Pheterson, G. (ed.) (1989) *A Vindication of the Rights of Whores*, Seattle: Seal Press.

Philo, C. (1992) 'Neglected rural geographies', *Journal of Rural Studies* 8, 2: 193–207.

Philo, C. and Kearns, G. (1994) *Selling Places*, London: Paul Chapman.

Pile, S. (1993) 'Human agency and geography revisited. A critique of "new models" of the self', *Transactions of the Institute of British Geographers* NS 18: 122–39.

Plummer, K. (ed.) (1992) *Modern Homosexualities: Fragments of lesbian and gay experience*, London: Routledge.

Polak, F. (1973) *The Image of the Future*, New York: Elsevier.

Ponse, B. (1976) 'Secrecy in the lesbian world', *Urban Life* 5, 3: 313–38.

Ponte, M.R. (1974) 'Life in the parking lot: an ethnography of a homosexual drive-in', in J. Jacobs (ed.) *Deviance: Field studies and self-disclosures*, Palo Alto: National.

Porter, K. and Weeks, J. (eds) (1991) *Between the Acts: Lives of homosexual men, 1885–1967*, London: Routledge.

Porter, R. and Tomaselli, S. (1989) *The Dialectics of Friendship*, London: Routledge.

Power, L. (1992) 'Diversity and denial', plenary address at Activating Theory Conference, York, October.

Pratt, M.B. (1984) 'Identity: skin blood heart', in E. Bulkin, M.B. Pratt and B. Smith (eds) *Yours in Struggle: Three feminist perspectives on anti-Semitism and racism*, New York: Long Haul Press.

Pringle, R. (1993) 'Femininity and performance in the medical profession', paper presented at

Newnham College Geographical Society, Cambridge, UK, February.

Probyn, E. (1992) 'Technologizing the self: a future anterior for cultural studies', in L. Grossberg, C. Nelson and P. Treichler (eds) *Cultural Studies*, New York: Routledge.

—— (1993) *Sexing the Self: Gendered positions in cultural studies*, London: Routledge.

—— (1995) 'Lesbians in space: gender, sex and the structure of missing', *Gender, Place and Culture* 2, forthcoming.

Puar, J. (1993) 'Challenging the white gaze: the deconstruction of relational discourses', paper presented at Development of a European Curriculum in Women's Studies from a Multi-Cultural Perspective Conference, Utrecht, June.

Ramazanoglu, C. (1986) 'Ethnocentrism and socialist-feminist theory: a response to Barrett and McIntosh', *Feminist Review* 22: 83–91.

—— (1989) *Feminism and the Contradictions of Oppression*, London: Routledge.

Ramazanoglu, C. (ed.) (1993) *Up against Foucault: Explorations of some tensions between Foucault and feminism*, London: Routledge.

Ramazanoglu, C. and Holland, J. (1993) 'Women's sexuality and men's appropriation of desire', in C. Ramazanoglu (ed.) *Up Against Foucault: Explorations of some tensions between Foucault and feminism*, London: Routledge.

Ransby, B. and Matthews, T. (1993) 'Black popular culture and the transcendence of patriarchal illusions', *Race and Class* 35: 57–68.

Rapoport, A. (1981) 'Identity and environment: a cross-cultural perspective', in J.S. Duncan (ed.) *Housing and Identity*, London: Croom Helm.

Reagan, S. (1994) 'No fun in Soho', letter to *Capital Gay*, 3 June: 2.

Reich, J. (1992) 'Genderfuck: the law of the dildo', *Discourse* 15: 112–27.

Rich, A. (1979) *Of Woman Born: Motherhood as experience and institution*, London: Virago Press.

—— (1980) 'Compulsory heterosexuality and lesbian existence', in C.R. Stimpson and E.S. Person (eds) *Women: Sex and sexuality*, Chicago and London: University of Chicago Press.

Riggs, M. (1991) 'Black macho revisited: reflections of a snap! queen', *Black American Literature Forum* 25: 389–94.

Robinson, M.E. (1994) 'Ask not for whom: a comment on Bell', *Area* 26: 84–6.

Robson, E. (1991) 'Space, place and sexuality in Hausaland, Northern Nigeria', paper presented at ERASMUS Geography and Gender Course, University of Durham, UK, April.

Rodaway, P. (1994) *Sensuous Geographies*, London: Routledge.

Rodgers, S. (1981) 'Women's space in a men's house: the British House of Commons', in S. Ardener (ed.) *Women and Space: Ground rules and social maps*, London: Croom Helm, pp. 50–71.

Rojek, C. (1985) *Capitalism and Leisure Theory*, London: Tavistock.

Rose, D. (1984) 'Rethinking gentrification: beyond the uneven development of Marxist urban theory', *Environment and Planning D: Society and Space* 1: 47–74.

Rose, G. (1993a) 'Some notes towards thinking about spaces of the future', in J. Bird, B. Curtis, T. Puttnam, G. Robertson and L. Tickner (eds) *Mapping the Futures: Local cultures, global change*, London: Routledge.

—— (1993b) *Feminism and Geography: The limits to geographical knowledge*, Cambridge: Polity Press.

Rothenberg, T. and Almgren, H. (1992) 'Social politics of space and place in New York City's lesbian and gay communities', presented at 27th International Geographical Congress, Washington, DC.

Rounds, K. (1988) 'AIDS in rural areas: challenges to providing care', *Social Work* May–June: 257–61.

Rubin, G. (1987) 'The leather menace: comments on politics and S/M', in Samois (eds) *Coming to Power: Writings and graphics on lesbian S/M*, Boston: Allyson.

—— (1989) 'Thinking sex: notes for a radical theory of the politics of sexuality', in C. Vance (ed.) *Pleasure and Danger: Exploring female sexuality*, second edition, London: Pandora.

Russell, M. (1992) 'It doesn't matter if you're black or white', *The Pink Paper* 20 December: 4.

Rust, P. (1993) ' "Coming out" in the age of social constructionism: sexual identity formation among lesbian and bisexual women', *Gender and Society* 7: 50–77.

Rutherford, J. (1992) *Men's Silences: Predicaments in masculinity*, London: Routledge.

Ryan, M. (1989) 'The American parade: representations of the nineteenth century social order', in L. Hunt (ed.) *The New Cultural History*, Berkeley: University of California Press.

Ryan, M. and Gordon, A. (eds) (1994) *Body Politics: Disease, desire, and the family*, Oxford: Westview.

Sage, V. (1984) 'Crusoe', in *Dividing Lines*, London: Chatto and Windus/Hogarth.

Said, E.W. (1978) *Orientalism*, London: Routledge and Kegan Paul.

Sassen, S. (1988) *The Mobility of Capital and Labor*, New York: Cambridge University Press.

Saunders, P. (1986) *Social Theory and the Urban Question*, London: Hutchinson.

—— (1989) 'The meaning of "home" in contemporary English culture', *Housing Studies* 4: 177–92.

Sawyer, M. (1993) 'Shhhh! Why Shabba had to shut up', *Select* June: 40–3.

Saxton, A. (1993a) 'Knocked about a bit', *The Pink Paper* 8 October: 15.

—— (1993b) 'These songs can kill', *Capital Gay* 5 March: 4.

—— (1993c) 'Financial query', *The Pink Paper* 3 September: 3.

Sayer, A. (1989) 'The "new" regional geography and problems of narrative', *Environment and Planning D: Society and Space* 7: 253–76.

Schulman, S. (1986) *Girls, Visions and Everything*, London: Sheba Press.

Seabrook, J. (1988) *The Leisure Society*, Oxford: Blackwell.

Seamon, D. (1979) *The Geography of the Lifeworld*, New York: St Martin's Press.

Sedgwick, E.K. (1985) *Between Men: English literature and male homosocial desire*, New York: Columbia University Press.

—— (1990) *Epistemology of the Closet*, Berkeley: University of California Press.

—— (1993) 'Queer performativity: Henry James's *The Art of the Novel*', *GLQ* 1: 1–16.

—— (1994) *Tendencies*, New York: Routledge.

Seidler, V.J. (1994) *Unreasonable Men: Masculinity and social theory*, London: Routledge.

Seidman, S. (1992) *Embattled Eros: Sexual politics and ethics in contemporary America*, New York: Routledge.

Sennett, R. (1991) *The Conscience of the Eye: The social life of cities*, London: Faber and Faber.

Shapiro, J. (1991) 'Transsexualism: reflections on the persistence of gender and the mutability of sex', in J. Epstein and K. Straub (eds) *Body Guards: The cultural politics of gender ambiguity*, New York: Routledge.

Sharrock, D. (1993a) 'Jamaica set for snap election', *The Guardian* 10 March: 14.

—— (1993b) 'Gun culture rules ghettoes as Jamaica goes to ballot', *The Guardian* 29 March: 15.

Shepherd, S. and Wallis, M. (eds) (1989) *Coming on Strong: Gay politics and culture*, London: Unwin Hyman.

Shields, R. (1991) *Places on the Margin: Alternative geographies of modernity*, London: Routledge.

Shilling, C. (1993) *The Body and Social Theory*, London: Sage.

Short, B. (1993) 'Up Queer Street', in *Lesbian and Gay Pride – Official Souvenir Programme*: 16–19.

Silvera, M. (1992) 'Man royals and sodomites: some thoughts on the invisibility of Afro-Caribbean lesbians', *Feminist Studies* 18: 521–32.

Silverstone, R. (1994) *Television and Everyday Life*, London: Routledge.

Simpson, M. (1992a) 'Gay marketing comes out of the closet', *The Independent* 22 September: 15.

—— (1992b) 'Purchasing power of pink pound makes gays an attractive market', *The Guardian*: 21 November: 24.

Simpson, M. (1994) *Male Impersonators*, London: Cassell.

Singer, L. (1993) *Erotic Welfare: Sexual theory and politics in the age of epidemic*, New York: Routledge.

Smith, B. (1993) 'Homophobia: why bring it up?' in H. Abelove, M. Barale and D. Halperin (eds) *The Lesbian and Gay Studies Reader*, New York: Routledge.

Smith, N. (1983) 'Toward a theory of gentrification: a back to the city movement of capital, not people', in R.W. Lake (ed.) *Readings in Urban Analysis: Perspectives on urban form and structure*, New Brunswick: Center for Urban Policy Research.

—— (1993) 'Homeless/global: scaling places', in J. Bird, B. Curtis, T. Putnam, G. Robertson and L. Tickner (eds) *Mapping the Futures: Local cultures, global change*, London: Routledge.

Smith, N. and Katz, C. (1993) 'Grounding metaphor: towards a spatialized politics', in M. Keith and S. Pile (eds) *Place and the Politics of Identity*, New York: Routledge.

Smyth, C. (1992) *Lesbians Talk Queer Notions*, London: Scarlet Press.

Snitow, A. (1990) 'A gender diary', in M. Hirsch and E.F. Keller (eds) *Conflicts in Feminism*, New York and London: Routledge.

Solomon, R. (1987) 'Heterosex', in E. Shelp (ed.) *Sexuality and Medicine*, volume 1, Dordrecht: Reidel.

Somerville, P. (1992) 'Homelessness and the meaning of home: rooflessness or rootlessness?' *International Journal of Urban and Regional Research* 16: 528–39.

Sontag, S. (1979) 'Introduction', in W. Benjamin (ed.) *One Way Street and Other Writings*, London: New Left Books.

South Boston Tribune (1993) 'Speak Out', 25 February: 16.

Spark, M. (1964) *Robinson*, Harmondsworth: Penguin.

Spelman, E. (1990) *Inessential Woman: Problems of exclusion in feminist thought*, London: Women's Press.

Spivak, G.C. (1988) *In Other Worlds: Essays in cultural politics*, New York: Routledge.

Stacpoole, H. de V. (1980) *The Blue Lagoon*, London: Futura.

Stafford, J. (1986) *Bizarro in Love*, San Francisco: Cheap Shots.

Stanley, C. (1993) 'Sins and passions', *Law and Critique* 4: 207–26.

Steakley, J. (1975) *The Homosexual Emancipation Movement in Germany*, New York: Arno.

Stein, E. (1993) 'Evidence for queer genes: an interview with Richard Pillard', *GLQ* 1: 93–110.

Stelter, G. and Artibise, A. (1986) *Power and Place: Canadian urban development in the North*

American perspective, Vancouver: University of British Columbia Press.

Stinchcombe, A.L. (1990) 'Work institutions and the sociology of everyday life', in K. Erikson and S.P. Vallas (eds) *The Nature of Work: Sociological perspectives*, New Haven: Yale University Press.

Stoddard, C.W. (1987) *Cruising the South Seas*, San Francisco: Gay Sunshine.

Stone, C. and Sanders, H. (1987) *The Politics of Urban Development*, Lawrence: University of Kansas Press.

Stone, S. (1991) 'The empire strikes back: a posttranssexual manifesto', in J. Epstein and K. Straub (eds) *Body Guards: The cultural politics of gender ambiguity*, New York: Routledge.

Strathern, M. (1989) 'Between a Melanesianist and a deconstructive feminist', *Australian Feminist Studies* 10, 49–69.

Street, B.V. (1975) *The Savage in Literature: Representations of 'primitive' society in English fiction, 1858–1920*, London: Routledge and Kegan Paul.

Styles, J. (1979) 'Insider/outsider: researching gay baths', *Urban Life* 8: 135–52.

Sunindyo, S. and Sabaroedin, S. (1989) 'Notes on prostitution in Indonesia', in G. Pheterson (ed.) *A Vindication of the Rights of Whores*, Seattle: Seal Press.

Symanski, R. (1974) 'Prostitution in Nevada', *Annals of the Association of American Geographers* 64: 357–77.

—— (1981) *The Immoral Landscape: Female prostitution in western societies*, Toronto: Butterworth.

Synnott, A. (1992) 'Tomb, temple, machine and self: the social construction of the body', *British Journal of Sociology* 43: 79–110.

—— (1993) *The Body Social*, London: Routledge.

Tatchell, P. (1992) *Europe in the Pink: Lesbian and gay equality in the New Europe*. London: Gay Men's Press.

The Best Guide to Amsterdam and Benelux (1992) Fourth edition, Amsterdam: Bookscene.

The Independent (1993) 'Ragga, "racism" and attacks on gays', 8 October: 25.

The Pink Paper (1993a) 'Banton wins reggae awards', 30 July: 3.

—— (1993b) 'Trouble in take-away', 22 October: 3.

The Voice (1993) 'Talking turkey', 5 January: 28–9.

Thomas, B. (1992) 'Caribbean Black Power – from slogan to practical politics', *Journal of Black Studies* 22: 392–410.

Thompson, M. (ed.) (1991) *Leatherfolk: Radical sex, people, politics, and practice*, Boston: Allyson.

Thrift, N. (1983) 'On the determination of social action in space and time', *Environment and Planning D: Society and Space* 1: 23–57.

—— (1993) 'For a new regional geography 3', *Progress in Human Geography* 17: 92–100.

Thrift, N. and Johnston, R.J. (1993) 'The futures of *Environment and Planning A*', *Environment and Planning A, Anniversary Issue* 83–102.

Thrift, N., Leyshon, A. and Daniels, P. (1987) 'Sexy greedy: the new international financial system, the City of London and the South East of England', Working Papers in Producer Services No. 7, University of Bristol, UK.

Tong, R. (1989) *Feminist Thought: A comprehensive introduction*, Boulder: Westview Press.

Tournier, M. (1974) *Friday or the Other Island*, Harmondsworth: Penguin.

Tranter, J.E. (1994) 'Soho – an emerging gay territory?', unpublished West London Institute of Higher Education dissertation.

Troiden, R.R. and Goode, E. (1980) 'Variables related to the acquisition of a gay identity', *Journal of Homosexuality* 5, 4: 383–92.

Tucker, S. (1991) 'Gender, fucking, and utopia: an essay in response to John Stoltenberg's *Refusing to be a Man*', *Social Text* 29: 3–34.

Turner, B. (1984) *The Body and Society*, Oxford: Blackwell.

—— (1992) *Regulating Bodies: Essays in medical sociology*, London: Routledge.

Tyler Bennett, D. (1993) *Djuna Barnes*, unpublished doctoral dissertation, University of Loughborough.

Valentine, G. (1989) 'The geography of women's fear', *Area* 21, 4: 385–90.

—— (1993a) '(Hetero)sexing space: lesbian perceptions and experiences of everyday spaces', *Environment and Planning D: Society and Space* 11: 395–413.

—— (1993b) 'Negotiating and managing multiple sexual identities: lesbian time–space strategies', *Transactions of the Institute of British Geographers*, NS 18: 237–48.

—— (1993c) 'Desperately seeking Susan: a geography of lesbian friendships', *Area* 25: 109–16.

—— (1995a) 'Out and about: geographies of lesbian landscapes', *International Journal of Urban and Regional Research*, forthcoming.

—— (1995b) 'Creating transgressive space: the music of Lang', paper available from the author.

Valverde, M. (1985) 'Heterosexuality: contested ground', in M. Valverde (ed.) *Sex, Power and Pleasure*, Toronto: Women's Press.

Van Every, J. (1992) 'Who is the "family"? The assumptions of British social policy', *Critical Social Policy* 33: 62–75.

Van Gelder, L. and Brandt, P.R. (1992) *Are You Two . . . Together? A gay and lesbian travel guide to Europe*, London: Virago.

Vicqua, J.H. (1993) 'Lesbians light up love industry/personal preference as pathology', *Aktion Surreal* women's issue: 32–3.

Vincent, S. (1993) 'Lost boys', *Guardian Weekend* 16 October: 4–8 and 37.

Waaldijk, K. (1993) 'The economic situation in member states', in K. Waaldijk and A. Clapham (eds) *Homosexuality: A European Community issue*, Dordrecht: Martinus Nijhoff.

Wallace, M. (1990) *Black Macho and the Myth of Superwoman*, London: Verso.

Wallis, L. (1993) 'Pink pounding', *The Guardian* 25 August: 14–15.

Warde, A. (1991) 'Gentrification as consumption: issues of class and gender', *Environment and Planning D: Society and Space* 9: 223–32.

Ware, V. (1993) *Beyond the Pale: White women, racism and history*, London: Verso.

Warner, M. (1993) 'Introduction', in M. Warner (ed.) *Fear of a Queer Planet: Queer politics and social theory*, Minneapolis: University of Minnesota Press.

Warner, S.T. (1978) *Mr Fortune's Maggot*, London: Virago.

Wa Sibuyi, M. (1993) 'Tinkoncana Etimayinini: the wives of the mines', in M. Krause and K. Berman (eds) *Invisible Ghetto: Lesbian and Gay Writing from South Africa*, Johannesburg: COSAW.

Watson, S. (1993) 'Is Sir Humphrey dead? The changing culture of the Civil Service', *Working Paper 103*, School of Advanced Urban Studies, University of Bristol, UK.

Watson, S. and Austerberry, H. (1986) *Housing and Homelessness: A feminist perspective*, London: Routledge and Kegan Paul.

Weeks, J. (1977) *Coming Out: Homosexual politics in Britain from the nineteenth century to the present*, London: Quartet.

—— (1985) *Sexuality and its Discontents: Myths, meanings and modern sexualities*, London: Routledge and Kegan Paul.

—— (1992) 'Changing sexual and personal values in the age of AIDS', paper presented at the Forum on Sexuality Conference, Sexual Cultures in Europe, Amsterdam, June.

Weightman, B. (1980) 'Gay bars as private places', *Landscape* 23: 9–16.

—— (1981) 'Commentary: towards a geography of the gay community', *Journal of Cultural Geography* 1: 106–12.

Weinberg, T.S. (1978) 'On "doing" and "being" gay: sexual behaviour and homosexual male self-identity', *Journal of Homosexuality* 4, 2: 143–56.

Weir, A. and Wilson, E. (1992) 'The Greyhound bus station in the evolution of lesbian popular culture', in S.R. Munt (ed.) *New Lesbian Criticism: Literary and cultural readings*, Hemel Hempstead: Harvester Wheatsheaf.

Weise, E.R. (ed.) (1992) *Closer to Home: Bisexuality and feminism*, Seattle: Seal Press.

Wekker, G. (1993) 'Mati-ism and black lesbianism – two ideal-typical expressions of female homosexuality in black communities of the Diaspora', *Journal of Homosexuality* 24: 145–58.

Wells, H.G. (1933) *Mr Blettsworthy on Rampole Island*, London: Benn.

—— (1993 [1896]) *The Island of Dr Moreau*, London: Everyman.

West, C. and Zimmerman, D.H. (1987) 'Doing gender', *Gender and Society* 1, 125–51.

West, C. and Zimmerman, D. (1991) 'Doing gender', in J. Lorber and S. Farrell (eds) *The Social Construction of Gender*, Newbury Park, CA: Sage.

Westwood, S. (1990) 'Racism, black masculinity, and the politics of space', in J. Hearn and D. Morgan (eds) *Men, Masculinities and Social Theory*, London: Allen and Unwin.

White, E. (1980) *States of Desire: Travels in gay America*, London: André Deutsch.

—— (1986) *States of Desire: Travels in Gay America*, London: Picador.

Wiggins, M. (1989) *John Dollar*, Harmondsworth: Penguin.

Wigley, M. (1992) 'Untitled: the housing of gender', in B. Colomina (ed.) *Sexuality and Space*, Princeton: Princeton Architectural Press.

Williams, P. (1986) 'Class constitution through spatial reconstruction? A re-evaluation of gentrification in Australia, Britain and the United States', in N. Smith and P. Williams (eds) *Gentrification of the City*, Boston: Allen and Unwin.

Williams, W. (1993) 'Being gay and doing research on homosexuality in non-western cultures', *Journal of Sex Research* 30: 115–20.

Williamson, J. (1992) 'Images of "Woman": the photography of Cindy Sherman', in H. Crowley and S. Himmelweit (eds) *Knowing Women: Feminism and knowledge*, Cambridge: Polity Press.

Wilson, A. (1992) 'Just add water: searching for a bisexual politics', *Out/Look* 16: 22–8.

Wilson, A.R. (1993) 'Which equality? Toleration, indifference or respect', in J. Bristow and A.R. Wilson (eds) *Activating Theory: Lesbian, gay, bisexual politics*, London: Lawrence and Wishart.

Wilson, E. (1991) *The Sphinx in the City*, London: Virago.

———— (1992) 'The invisible flâneur', *New Left Review* 195: 90–110.

———— (1993) 'Is transgression transgressive?', in J. Bristow and A.R. Wilson (eds) *Activating Theory: Lesbian, gay, bisexual politics*, London: Lawrence and Wishart.

Wilson, F. (1972) *Migrant Worlds*, Johannesburg: SPRO-CAS.

Winchester, H. and White, P. (1988) 'The location of marginalised groups in the inner city', *Environment and Planning D: Society and Space*: 37–54.

Winters, C. (1979) 'The social identity of evolving neighborhoods', *Landscape* 23: 8–14.

Wirth, L. (1928) *The Ghetto*, Chicago: University of Chicago Press.

Woolaston, G. (1991) 'The good life', *Gay Times* May: 39–40.

Wolf, D.G. (1979) *The Lesbian Community*, Berkeley: University of California Press.

Wolf, N. (1990) *The Beauty Myth*, London: Chatto and Windus.

Wolfe, M. (1992) 'Invisible women in invisible places: lesbians, lesbian bars, and the social production of people/environment relationships', *Architecture and Behaviour* 8: 137–58.

Woods, G. (1987) *Articulate Flesh: Male homo-eroticism and modern poetry*, New Haven and London: Yale University Press.

Wotherspoon, G. (1991) *City of the Plain*, Sydney: Hale and Iremonger.

Wyss, J.R. (1910) *The Swiss Family Robinson*, London and New York: Everyman.

Yanagisako, S. and Collier, J. (1990) 'The mode of reproduction in anthropology', in D. Rhode (ed.) *Theoretical Approaches on Sexual Difference*, New Haven and London: Yale University Press.

Young, I.M. (1990a) *Justice and the Politics of Difference*, Princeton: Princeton University Press.

———— (1990b) 'The ideal of community and the politics of difference', in L. Nicholson (ed.) *Feminism/Postmodernism*, New York: Routledge.

Zaretsky, E. (1976) *Capitalism, the Family and Personal Life*, New York: Harper Colophon.

Zita, J. (1992) 'The male lesbian and the postmodernist body', *Hypatia* 7: 106–27.

INDEX

•